Balancing and Sequencing of Assembly Lines

Contributions to Management Science

Harald Dyckhoff/Ute Finke
Cutting and Packing in Production and Distribution
1992. ISBN 3-7908-0630-7

Richard Flavell (Ed.)
Modelling Reality and Personal Modelling
1993. ISBN 3-7908-0682-X

Michael Hofmann/Monika List (Eds.)
Psychoanalysis and Management
1994. ISBN 3-7908-0795-8

Rita L. D'Ecclesia/Stavros A. Zenios (Eds.)
Operations Research Models in Quantitative Finance
1994. ISBN 3-7908-0803-2

Mario S. Catalani/Giuseppe F. Clerico
Decision Making Structures
1996. ISBN 3-7908-0895-4

M. Bertocchi/E. Cavalli/S. Komlósi (Eds.)
Modelling Techniques for Financial Markets and Bank Management
1996. ISBN 3-7908-0928-4

Holger Herbst
Business Rule-Oriented Conceptual Modeling
1997. ISBN 3-7908-1004-5

Constantin Zopounidis
New Operational Approaches for Financial Modelling
1997. ISBN 3-7908-1043-6

Klaus Zwerina
Discrete Choice Experiments in Marketing
1997. ISBN 3-7908-1045-2

Giovanni Marseguerra
Corporate Financial Decisions and Market Value
1998. ISBN 3-7908-1047-9

WHU Koblenz – Otto Beisheim Graduate School of Management (Ed.)
Structure and Dynamics of the German Mittelstand
1999. ISBN 3-7908-1165-3

Armin Scholl

Balancing and Sequencing of Assembly Lines

Second revised edition

With 71 Figures
and 75 Tables

Springer-Verlag Berlin Heidelberg GmbH

Series Editors
Werner A. Müller
Martina Bihn

Author
Dr. Armin Scholl
Institut für Betriebswirtschaftslehre
Fachgebiet Operations Research
Technische Universität Darmstadt
Hochschulstraße 1
D-64289 Darmstadt, Germany

The first edition was published in
the series "Production and Logistics".

ISBN 978-3-7908-1180-3

Cataloging-in-Publication Data applied for
Die Deutsche Bibliothek – CIP-Einheitsaufnahme
Scholl, Armin: Balancing and sequencing of assembly lines: with 75 tables / Armin Scholl. – 2., rev. ed.
(Contributions to management science)
ISBN 978-3-7908-1180-3 ISBN 978-3-662-11223-6 (eBook)
DOI 10.1007/978-3-662-11223-6

Cover design: Erich Kirchner, Heidelberg

SPIN 10702921 88/2202-5 4 3 2 1 0 – Printed on acid-free paper

Preface

The book on hand describes decision problems which arise in flow-line production systems and proposes adequate solution methods. Flow-line production systems are present in different industrial environments and are utilized to manufacture a large variety of products. Usually, several models (versions) of a standardized product differing from each other by optional components and equipment have to be assembled jointly on (mixed-model) assembly lines. Two main types of decision problems have to be solved when such production systems are installed and operated. The *assembly line balancing problem* consists of partitioning the assembly work (of each model) among different stations of a line. This medium-term problem is connected with the short-term problem of finding an operating sequence of the models required to meet customer demands *(mixed-model sequencing problem)*. These problem types are discussed and classified in **Part I** of this work. Different versions of balancing and sequencing problems are defined and arranged within a hierarchical planning system.

Since the optimization problems mentioned above are NP-hard, a lot of research work has been done so far in order to develop procedures which provide acceptable solutions within an appropriate amount of computation time on a computer system. **Part II** of the book surveys existing procedures and presents a number of very promising new solution methods for several versions of both basic problem types. Comprehensive numerical investigations showing the effectiveness of the new methods are reported.

In case of the well-known simple assembly line balancing problem type 1 (minimize the number of stations for a given cycle time) considerable improvements are achieved by a new branch and bound procedure (SALOME-1) containing a so-called Local Lower Bound Method and a bidirectional branching strategy. For related problem types, adequate modifications of SALOME-1 are developed. These procedures are adapted to solve mixed-model balancing problems as well. Furthermore, heuristic improvement procedures utilizing the recent heuristic meta-strategy tabu search are investigated for several types of balancing problems.

For a particular mixed-model sequencing problem (minimize the total work overload), a branch and bound procedure as well as a hybrid heuristic procedure are developed. The latter approach is based on an analogy of the sequencing problem to the one-dimensional cutting stock problem and appears to be very successful even in solving large-scale problem instances which are intractable for branch and bound procedures.

In the present second edition decision problems of balancing U-shaped assembly lines, which are frequently used as components of modern just-in-time production systems, are additionally considered. By means of comprehensive computational results we show that they can efficiently be solved by modifications of our procedures for the simple assembly line balancing problem. Furthermore, it is demonstrated that organizing an assembly line as a U-line frequently increases the efficiency of the production system compared to traditional serial lines.

I am indebted to *Prof. Dr. Wolfgang Domschke* who supported the development of this work from the origin to the completion with valuable advice and numerous suggestions for improvements. Furthermore, I want to express my appreciation to *Prof. Dr. Hartmut Stadtler* for reading the manuscript and providing an excellent critique. I am grateful to my colleagues *Dipl.-Wirtsch.-Inf. Robert Klein*, *Dipl.-Math. Gabriela Krispin*, *Dr. Birgit Schildt*, and *Prof. Dr. Stefan Voß* for the productive co-operation and numerous inspiring discussions. Without their helpful comments, the book would not be on hand in the present form. Furthermore, thanks are due to my father *Karl Heinz Scholl* who read the manuscript and corrected some typing errors and inconsistencies. For assisting in programming some parts of solution procedures, I am grateful to the students *Christian Herrlich* and *Hans-Jürgen Undt*. I wish to express my thanks to *Prof. Thomas Hoffmann* and *Prof. Roger Johnson* for providing me with copies of their codes of EUREKA and FABLE, respectively.

The book is dedicated to my marvellous wife *Sonja*, my sons *Benjamin*, *Simon Eric*, and *Joachim* as well as my lovely daughter *Magdalena* ***Salome*** who gave the name to the procedure SALOME.

Darmstadt, September 1998 *Armin Scholl*

Contents

Part II Solution Methods

Notations

General Notations and Symbols

$x := y$	x is defined by the value of y
$\mathbb{N}$	set of natural numbers (= {1,2,3,...})
$\mathbb{R}$	set of real numbers
$\lceil x \rceil$	smallest integer $\geq x$
$\lfloor x \rfloor$	largest integer $\leq x$
$\|a\|$	absolute value of a number a
$\|A\|$	number of elements in a set A (cardinality of A)
ϕ	empty set; empty interval
∞	an infinite number
$i \in I$	i is element of the set I; i is in the interval I
$i \notin I$	i is not element of the set I; i is not in the interval I
$I \subseteq J$	I is a subset of J ($I \subset J$: I ist a proper subset of J)
$I \cup J$	union of the sets I and J
O(f(n))	order of a function f(n)
TU	time unit
LP	linear programming
LB	lower bound on the value of an objective function
UB	upper bound on the value of an objective function
SALBP	simple assembly line balancing problem
MALBP	mixed-model assembly line balancing problem
MSP	mixed-model sequencing problem
cf.	confer
p.	page
pp.	page and following pages

Notations for Single - Model Assembly Line Balancing

n	number of tasks; index: j
V	set of all tasks $(=\{1,\dots,n\})$
c	cycle time
p	production rate $(=1/c)$
m	number of stations; index: k
t_j	operation time (task time) of task j
p_j	station requirement of task j $(=t_j/c=t_j\cdot p)$
t_{max}	maximal task time $(=\max\{t_j \mid j=1,\dots,n\})$
t_{min}	minimal task time $(=\min\{t_j \mid j=1,\dots,n\})$
t_{sum}	sum of task times $(=\Sigma_j t_j)$
P_j (P_j^*)	set of immediate (and transitive) predecessors of task j
F_j (F_j^*)	set of immediate (and transitive) successors (followers) of task j
A	set of direct precedence relations $(=\{(i,j) \mid i\in V \text{ and } j\in F_i\})$
A^*	set of direct and transitive precedence relations $(=\{(i,j) \mid i\in V \text{ and } j\in F_i^*\})$
G	precedence graph $(=(V,A,t))$
G^*	transitive closure of the precedence graph $(=(V,A^*,t))$
OS	order strength $(=2\cdot \lvert A^*\rvert/(n\cdot(n-1)))$
S_k	station load, set of tasks assigned to station k
$t(S_k)$	station time of station k $(=\sum_{j\in S_k} t_j)$
E_j	earliest station of task j
L_j	latest station of task j
SI_j	station interval of task j $(=[E_j,L_j])$
B_k	set of tasks potentially assignable to station k $(=\{j \mid k\in SI_j\})$
E	line efficiency $(=t_{sum}/(m\cdot c))$
BD	balance delay time $(=m\cdot c-t_{sum})$
BR	balance delay ratio $(=1-E)$

Notations for Mixed - Model Assembly Line Balancing

M	number of models; index: $m = 1,...,M$
d_m	expected demand of model m during the planning period
D	total number of units required during the planning period $(=\Sigma_m d_m)$
J	number of operations (tasks), index: $j = 1,...,J$
PT	total time available for production during the planning period
c	(average) cycle time, launch interval $(c \leq \lfloor PT/D \rfloor)$
K	number of stations; index: $k = 1,...,K$
t_{jm}	operation time of task j for one unit of model m
$t_j^{\prime}$	cumulated operation time of task j for all required units $(=\Sigma_m d_m \cdot t_{jm})$
t_j	average operation time of task j per unit $(= t_j^{\prime}/D)$
S_k	station load, set of tasks assigned to station k
τ_{mk}	processing time per unit of model m in station k $(= \sum_{j \in S_k} t_{jm})$
$\tau_{mk}^{\prime}$	total operation time of model m in station k $(= d_m \cdot \tau_{mk})$
$\bar{t}_m$	average operation time per station and unit of model m $(= \Sigma_k \tau_{mk}/K)$
$\bar{t}_m^{\prime}$	average total operation time per station for model m $(= d_m \cdot \bar{t}_m)$
$\tau_k^{\prime}$	total operation time in station k $(= \Sigma_m \tau_{mk}^{\prime})$
τ_k	average operation time of station k per unit $(= \tau_k^{\prime}/D)$

Notations for Mixed - Model Sequencing

M	number of models; index: $m = 1,...,M$
d_m	demand for model m during the planning period
I	length of the sequence, total demand $(= \Sigma_m d_m)$; index: $i = 1,...,I$
c	fixed launch interval (cycle time, in TU)
K	number of stations; index: $k = 1,...,K$
l_k	length of station k (in TU; = physical length/conveyor speed)
τ_{mk}	processing time per unit of model m in station k $(\tau_{mk} \leq l_k)$
Δ_{mk}	displacement of the operator in station k by a unit of model m $(= \tau_{mk} - c)$
x_{mi}	binary variable, $= \begin{cases} 1 & \text{if a unit of model m is assigned to position i} \\ 0 & \text{otherwise} \end{cases}$
s_{ki}	start position of the operator in station k for the i-th unit
e_{ki}	end position of the operator in station k after processing the i-th unit
w_{ki}	utility work in station k to complete the i-th unit of the sequence
P	number of patterns to be sequenced
π_p	pattern; index: $p = 1,...,P$
L_p	length of pattern π_p
y_p	quantity of pattern π_p
Ψ	length of the pattern sequence $(= \Sigma_p y_p)$; index: $\tau = 1,...,\Psi$
β_{kp}	earliest end position after processing pattern π_p in station k
α_{kp}	latest start position for pattern π_p in station k
ω_p	reducible utility work of pattern π_p $(= \Sigma_k \beta_{kp})$
σ_p	reducible idle time of pattern π_p $(= \Sigma_k \alpha_{kp})$
MaxPatt	maximal pattern length

Part I: Decision Problems

1 Introduction

Assembly lines are special flow-line production systems which are typical in the industrial production of high quantity standardized commodities. In Section 1.1 we characterize flow-line production as a basic principle in production and discuss decision problems which arise in managing such systems. Among these, assembly line balancing problems are important tasks in medium-term production planning. Because of very different conditions in industrial manufacturing a large number of different problems of this type can be distinguished. After introducing some basic terms in Section 1.2, we provide a comprehensive classification in Section 1.3. A main characteristic is the number of products manufactured on an assembly line. Section 2 deals with single-model problems which concern the mass production of one homogeneous commodity. Mixed-model problems which occur when producing several versions (models) of a basic product on the same assembly line are treated in Section 3. In this case the balancing problem is connected with the short-term problem of finding an efficient model sequence.

1.1 Flow-Line Production

An important characteristic of industrial manufacturing is the *process pattern* (type of production). In a **job shop production system**, facilities which perform similar operations are spatially combined in workshops (e.g., all milling machines are combined in a milling shop). In contrast to this *job-oriented (functional)* organization, the facilities are arranged according to the technological sequence of operations in a **flow-line production system**. This *process-oriented* layout is well suited to *mass* or *large-scale series production* whereas job shop systems are mainly used in *production-to-order* and *small batch production*. Between these two extreme process patterns a number of hybrid systems can be identified which contain properties of both. In particular, group-oriented flexible manufacturing systems combine advantages of functional and process-oriented systems (cf., e.g., Tempelmeier and Kuhn (1993) as well as Scheer (1994)).

Advantages and disadvantages of flow-line layouts have intensively been discussed in the literature (cf., among others, Adamczyk (1969, pp. 34), Buffa (1983, pp. 203), Bhattacharjee and Sahu (1987), Zäpfel (1989, pp. 184)). The main **advantages** relative to job shop production systems are:

- The capacity utilization is usually high and throughput times are small.
- In-process-inventories (for unfinished jobs waiting in the system) are kept very low.
- The flow of material is regular and can simply be controlled.
- Only little manual material handling is needed, because workpieces are transferred by mechanical handling equipment like conveyor belts.
- The requirement of total floor space is small, because less space is needed for storage and material movement.
- Due to the strict division of labour, less skilled operators are needed who can be trained more quickly.

The following aspects are recognized as **disadvantages** of flow-line systems and therefore restrict their applicability to industrial manufacturing:

- The installation of the system causes high capital requirements, especially if automatic equipment is needed.
- The job satisfaction of workers employed at flow-lines is often low, because the high degree of labour division creates simple and monotonous job contents. This leads to high absenteeism and turnover rates of workers.
- Due to the high degree of specialization, flow-line systems are often inflexible. This is disadvantageous especially with respect to the steady shortening of product life cycles. Therefore, flexible transfer lines have become more important during the last years (cf. page 16).
- Maintenance and repairs are more critical issues than in job shop systems, because machine breakdowns may stop the complete system. The same is true if the requisite material is not available.
- The quality control must directly be included into the production process (inspection stations). When failures occur the complete line may be affected.

Summarizing the pros and cons, the following pre-conditions should have to be fulfilled for a successful installation of a flow-line production system: standardized products, high volume production, stable product demands, and continuous supply of material.

Due to its serial layout, a flow-line production system is often organized as an **assembly line** which consists of a number of **(work) stations** arranged along a conveyor belt or similar mechanical material handling equipment. The workpieces (jobs) are consecutively launched down the conveyor belt and are steadily moved from station to station. At each station, a certain part of the total work, necessary to manufacture the product, is performed. The decision problem of optimally partitioning (balancing) the assembly work among the stations (with respect to some objective) is known as the **assembly line balancing problem (ALBP)**.

In case of a *paced* assembly line, the time available for performing the allotted work on a workpiece is limited to the same maximum value at each station. This limit is called **cycle time** c, because a job arrives in a station and another one leaves after being processed every c units of time, respectively. Due to the cycle time restriction, paced assembly lines have a fixed *production rate* (production quantity per unit of time).

In the absence of a common cycle time, i.e., the stations may have different production rates, workpieces may have to wait before they can enter the next station. Therefore, *buffers* have to be installed between the stations which hold such *in-process-inventories*. In this case, the decision problem of positioning buffers and dimensioning **buffer capacities** arises. On the one hand, buffers tie up invested capital, on the other hand, jobs which cannot be stored in a buffer may cause an interruption of the production process.

Remark 1.1. Even though assembly line balancing problems have originally been defined for paced assembly lines, they occur in unpaced flow-line production systems as well (cf., e.g., Hahn et al. (1968) and Steffen (1977, p. 21)). In these systems, the output rates of the stations have to be balanced, and the cycle time can be interpreted as a mean time interval available for performing work on a job. If the distribution of work is not as equal as possible, expensive in-process-inventories are built in the buffers. Since buffer capacities are usually restricted, the stations cannot continue their work whenever they are not able to pass completed jobs to the buffer of the next station. Furthermore, stations may have to wait for work if the utilization of the preceding station is too high (cf. Section 1.3.2).

If several products are manufactured on the same assembly line, the efficiency of the line is influenced by the sequence of jobs, because the work time needed to assemble the products may differ considerably. Hence, a further decision problem consists of finding a sequence which is optimal with respect to some efficiency measure. This problem falls into the general class of **flow shop (sequencing) problems** (cf. Domschke et al. (1997, chapter 5.5)). In case of assembly line production, the type of sequencing problem depends on the variability of the products. A *mixed-model assembly line* is characterized by the fact that several related *models* of a basic pròduct are jointly manufactured on the line. Hence, the **mixed-model sequencing problem** consists of determining an intermixed sequence of the models. With a *multi-model assembly line*, a number of different products is assembled in batches. Due to a trade-off between set-up costs caused by product changes and inventory costs increasing with batch sizes, a **lot sizing and scheduling problem** has to be considered.

1.2 Basic Terms of Assembly Line Production

We define terms which are needed to describe assembly line production. Some of them have already been used intuitively in the previous section.

Assembly. Assembly is the process of collecting and fitting together various parts in order to create a *finished product* (end item). It is characterized by the used parts and the work necessary to combine them. Parts may be subdivided into (purchased) *components* and *sub-assemblies* (intermediate goods). The relationships of parts and the flow of material can be visualized by *assembly charts* (*gozinto charts*, cf., e.g., Buffa (1983), p. 196). The unfinished units of the product are called *work-pieces*.

Operation. An operation *(task)* is a portion of the total work content in an assembly process. The time necessary to perform an operation is called *operation (task) time*. Operations are considered indivisible, because they cannot be split into smaller work elements without creating unnecessary additional work.[1] For example, a task may consist of the following elemental motions: reaching for a screw driver, grasping it, taking the screw and putting it into the thread, tightening the screw, returning the tool.

Station. A (work) station is a segment of an assembly line where a certain amount of work (a number of operations) is performed. It is mainly characterized by its dimension, the machinery and equipment as well as the kind of assigned work. To this effect, stations can be subdivided into manual or automated stations depending on the subjects performing the work. Traditionally, work is manually done by *human operators* using a few (simple) tools or by semi-automated machines which have to be controlled by an operator. In the last years, flexible transfer lines with fully-automated machines (like CNC or DNC machines) are used in certain areas of assembly production. For ease of presentation we will use the term "operator" even

[1] Methods for analyzing work and measuring operation times are, e.g., described in Barnes (1980) as well as REFA (1975, 1992).

in the case of an automated production. The work content of a station (set of assigned tasks) is referred to as *station load*, the time necessary to perform the work as *station time*. A station is said to be *earlier* than another one if it is closer to the start of the line (workpieces flow from the early to the late stations).

Cycle Time. The cycle time represents the maximal amount of time a workpiece can be processed by a station of a paced assembly line. Since tasks are indivisible work elements, the cycle time can be no smaller than the largest operation time. In unpaced flow-line production systems (including mixed-model lines), the cycle time serves as maximal possible average station time. The time interval during which a workpiece is accessible to a station is called *tolerance time*. The *output rate* or *production rate* of the line equals the reciprocal of the cycle time. A positive difference between the cycle time and the station time is called *idle time*, because the operator is idle after performing the station load (or he works continuously at a slower pace). The sum of idle times for all stations of the line is called *balance delay time*.

Precedence Constraints. Due to technological restrictions, the ordering in which operations must be performed may partially be prespecified. This partial ordering of tasks can be illustrated by means of a *precedence diagram* (*precedence graph*; cf. Prenting and Battaglin (1964)) which contains nodes for all operations and arcs (i,j) if an operation i must precede an operation j. An example of a precedence diagram is given in Figure 1.1. The operation times are denoted as node weights. In this example, task 1 must be performed before tasks 2 and 5, task 3 prior to task 4, etc.

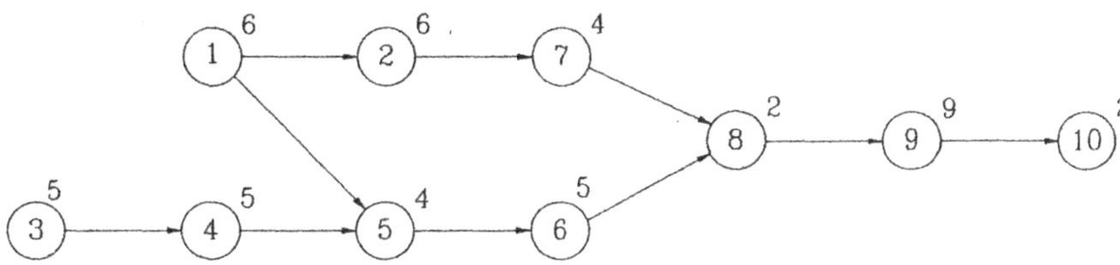

Figure 1.1. Precedence Graph

Line Balance. A line balance (feasible task assignment) represents a feasible solution of a balancing problem. A feasible solution is characterized by the following properties: Because of its indivisibility each task is assigned to exactly one station. The precedence constraints are fulfilled, i.e., no task j which must succeed a task i is assigned to an earlier station than i. The station times of all stations (or the average station times) do not exceed the cycle time.

For our example, a line balance with a cycle time of 11 and 5 stations is given by the station loads {1,3}, {2,4}, {5,6}, {7,8}, {9,10} of the stations 1 through 5 (tasks

1 and 3 are performed at station 1, tasks 2 and 4 at station 2, etc.). While no idle time occurs in stations 1, 2, and 5, the stations 3 and 4 show idle times of 2 and 5, respectively. Hence, the balance delay time is 7.

1.3 Classification of Assembly Line Balancing Problems

Assembly line production systems are present in different industrial environments and are utilized to manufacture a large variety of products. Especially, they are used to produce *consumer goods* such as cars, engines, domestic appliances, television sets, computers, and other electrical appliances. These products are rather different, and it is necessary to implement different production systems. Correspondingly, a wide range of balancing problems has to be considered. In this section, we provide characteristics and assumptions of balancing problems considered in the literature and give a classification scheme (cf., e.g., Wild (1972), Buxey et al. (1973), Aneke and Carrie (1984), Ghosh and Gagnon (1989), Shtub and Dar-El (1989)).

Figure 1.2 illustrates the main characteristics of assembly line balancing problems considered in the literature and their relationships. While continuous lines indicate that a particular combination of characteristics is typical, broken lines signify that it is unusual. The first two of these characteristics have already been mentioned in Section 1.1 and are discussed in greater detail in Sections 1.3.1 and 1.3.2, respectively. Subsequently, a variety of other criteria characterizing assembly line balancing problems is considered. For characteristics which will not be addressed in the Chapters 2 and 3, we give some explanations and references.

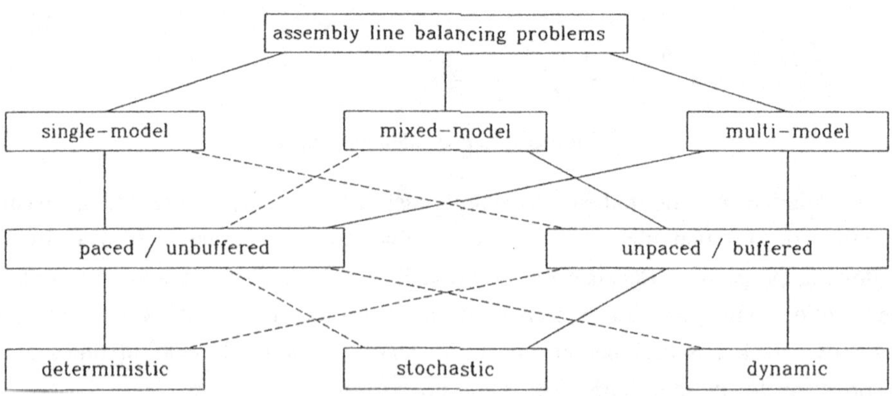

Figure 1.2. Classification of Assembly Line Balancing Problems

1.3.1 Number of Products

As already mentioned, the number of different products assembled on the same line(s) is an important characteristic of assembly line balancing problems.

Single-Model Lines. One homogeneous product is continuously manufactured in large quantities (mass production). All stations repeatedly have to perform the same tasks on identical workpieces. The work loads of all stations remain constant over time.

Mixed-Model Lines. Several models (versions) of a basic product are manufactured on the same line (e.g., versions of a personal computer). The production processes of the models are quite similar, because the models differ from the basic product only with respect to some attributes (like type of processor, type and capacity of hard disk) and optional features (e.g., CD-rom drive yes/no, additional floppy disk drive yes/no). Consequently, the same set of basic operations is necessary to produce each of the models. Some operations show different execution times, while others are present or absent depending on an optional part being installed or not. Due to the similarity of the production processes, no or only minor set-up activities arise when changing from one model to another. In order to satisfy up-to-date demands, the models are simultaneously produced on the line in an intermixed sequence. In this case, the balancing problem is connected with the short-term problem of determining an efficient model sequence (cf. Section 3.1).

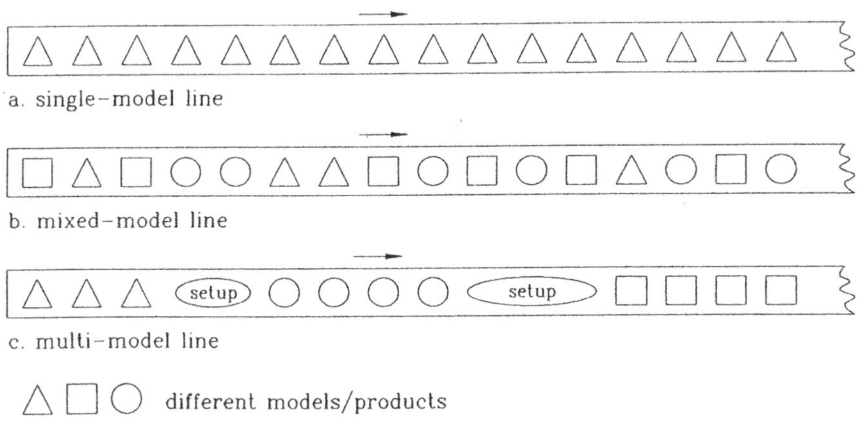

Figure 1.3. Assembly Lines for Single and Multiple Products

Multi-Model Lines. Several (similar) products are manufactured on one or several assembly lines. Because of significant differences in the production processes, rearrangements of the line equipment are required when product changes occur. Consequently, the products are assembled in separate batches in order to minimize set-up inefficiencies. While set-up costs are reduced by enlarging batch sizes, inventory costs are increased. This trade-off leads to the problem of determining economic lot sizes and production cycles of all products (cf. Domschke et al. (1997, chapter 3)). Simultaneously, the product batches have to be scheduled on the line. Hence, a combined lot sizing and scheduling problem has to be considered (cf., e.g., Seeland (1967), Hahn (1972, chapters 5 and 6), Burns and Daganzo (1987), Dobson and Yano (1994)).

Figure 1.3 illustrates the different line types. Note that the productive time is reduced by (sequence-dependent) set-up times in case of multi-model lines.

1.3.2 Paced and Unpaced Lines

A flow-line production system consists of an operating subsystem and a material handling subsystem. The latter one links the stations which build the operating subsystem and decides on the manner in which material flows through the line.

Paced Lines. Mechanical material handling equipment like conveyor belts couple the stations in an inflexible manner. Either the workpieces are steadily moved from station to station by the conveyor belt at constant speed, or they are intermittently transferred after being processed. In both cases, all stations have available the same amount of time (cycle time) to perform the assigned tasks on all workpieces.

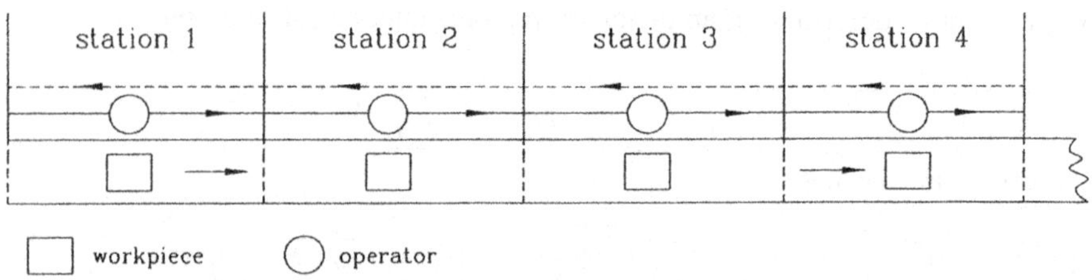

Figure 1.4. Layout of a Paced Assembly Line

Figure 1.4 shows an assembly line with a continuously moving conveyor belt. The workpieces are equi-spaced on the belt. While a workpiece passes a station, the operator moves in line with it, performs his work, and returns to the start of his station. In case of intermittent material movement, the workpieces stay in the stations and are simultaneously moved to succeeding stations after being processed. In order to have no significant transportation delays, the material movement has to be very fast, or some workpieces must be available on the material handling system between the stations. Then such a system serves as a limited buffer.

Unpaced Lines. In unpaced lines, the stations are decoupled by buffer stocks which hold workpieces when the succeeding station is still working on previous items

(Figure 1.5 a shows a typical layout). Since buffer capacities are normally restricted, a station may be *blocked* when the subsequent buffer is full. Then the considered station is idle until the succeeding station requires an item stored in the buffer. Another inefficiency, called *starvation*, occurs when the input buffer of a station is empty after terminating the current job. Then the station is idle until a workpiece enters the buffer. Starvation may be caused by a lower production rate or a *breakdown* of the preceding station (cf. Section 1.3.3). Note that a starvation cannot be avoided in paced systems whenever the total work content is not equally distributed among the stations.

Figure 1.5 b shows an alternative layout of an unpaced line. Here, the material handling equipment, which is a circular belt, serves as a common buffer for all stations. While the workpieces steadily circulate, the operators take one of the passing items as soon as they have finished their work on the current unit and have released it onto the belt. Obviously, such a layout is only practicable in small systems with manual stations. Furthermore, it is difficult to identify workpieces which have already been or are still to be processed in a station.

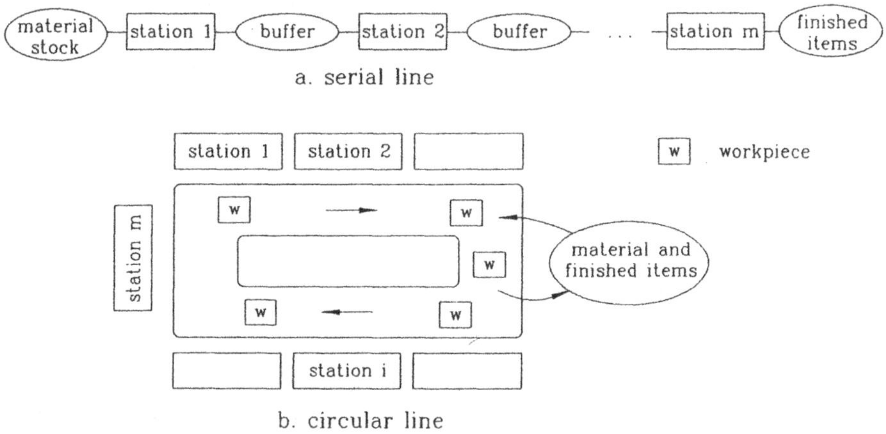

Figure 1.5. Different Layouts of Unpaced Lines

The buffer locations and sizes directly influence the production rate of the line. In the absence of buffers, breakdowns of stations may stop the complete line if no alternative equipment is available (cf. Davis (1966), Buzacott (1967, 1968), Suhail (1983), Baker et al. (1990)), while large buffers decouple the stations almost completely. Correlations between machine breakdowns and buffer sizes are investigated by, e.g., Koenigsberg (1959), Buzacott (1971), Yamashina and Okamura (1983) as well as Hillier and So (1991). While higher production rates (less down-time of the line) may be achieved, the storage space, material handling and inventory costs increase with larger buffer sizes. This trade-off between *down-time costs* and inventory related costs establishes the decision problem of dimensioning the buffer sizes. For such problems see Anderson and Moodie (1969), Basu (1977), Conway et al. (1988), Hopp et al. (1989), Pourbabai (1989), Szendrovits et al. (1990) as well as Malakooti (1994). Figure 1.6 shows the opposite trends of down-time costs and inventory costs for varying buffer sizes as well as the resultant total cost function on principle. In real-world situations it is very difficult to specify down-time costs which may represent tardiness penalties or lost sales.

Together with the buffer sizes, the number and the positions of the buffers have to be determined (cf., e.g., Jafari and Shanthikumar (1989), Zäpfel (1989, pp. 219), Hillier et al. (1993), Powell (1994)).

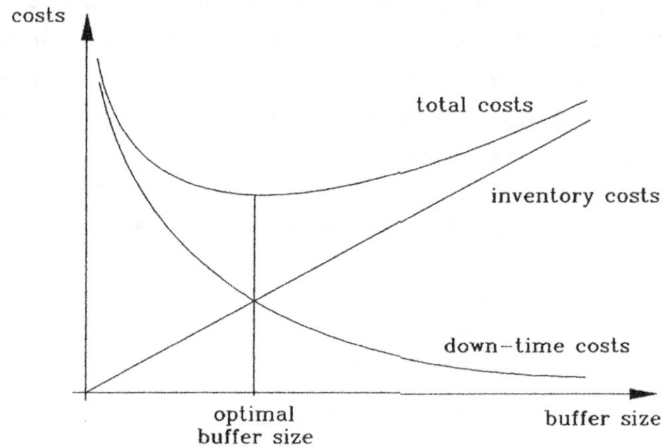

Figure 1.6. Influence of the Buffer Size on Costs

Mixed-model lines are most often unpaced, even in the case of a steadily moving conveyor belt, because the station times usually vary with the models. Then buffers may be used for temporarily balancing differing capacity supplies and demands (cf. Hiraki et al. (1977), Blumenfeld (1989), McCormick et al. (1989)). In fact, the conveyor belt itself serves as a buffer if the station lengths are larger than the distance of workpieces on the belt. Several workpieces are simultaneously accessible to the stations, and the tolerance time exceeds the cycle time.

Besides the effect of decoupling the stations, buffers allow *resequencing* of workpieces, i.e., workpieces may leave a buffer in another order than they have entered it. This may help to reduce sequence-dependent inefficiencies in mixed-model lines (cf. Bauer and Scholz (1984), Decker (1993 a, pp. 91, 1993 b)). Furthermore, buffers may be used to extract defective parts without stopping the line for an on-line repair. The parts are repaired at a separate repair station and return to the buffer afterwards (cf. Robinson et al. (1990)).

1.3.3 Variability of Operation Times

Depending on the nature of tasks and operators, the operation times may vary more or less. While the variance of operation times is expected to be very small for simple tasks, complex and failure sensitive operations may have considerably varying execution times. Especially in the case of human work, the task times are subject to physical, psychological or social influence factors.

Deterministic Operation Times. Most assembly line balancing models are based on fixed deterministic operation times. This assumption is justified whenever the expected task time variability is sufficiently small. In order to have almost steady task times in assembly line systems which mainly utilize human work, the production method must be sophisticated and robust while the workers have to be highly qualified and motivated. Advanced manufacturing technology reduces the task time variability by increased automation (cf., e.g., Johnson (1983)). Modern computer-controlled machines and robots used in flexible transfer lines are able to work permanently at a constant speed (as long as they are available).

Stochastic Operation Times. In case of human work, significant variations of task times may considerably influence the performance of the system. The operation times depend on the skills and the motivations of the employed workers, their current work rate, and other factors (cf. Buzacott (1990)). Excessive operation times result in inefficiencies such as starving of succeeding stations or blocking of preceding ones in unpaced systems. In case of paced assembly lines, workpieces cannot be processed completely whenever the station time exceeds the cycle time. Some of the possible consequences are that the complete line must stop until the work is completed, that additional (utility) workers have to be employed, or that incomplete units have to be reworked at additional off-line stations. The same problems arise when defective parts are produced (cf. Robinson et al. (1990)).

Even though the task times are almost constant in automated flow-line production systems, varying production rates of the stations may result from machine breakdowns. Whereas breakdowns cause the stoppage of the entire line in paced systems, the buffers of an unpaced system allow that in preceding and succeeding stations the production continues until blocking and starving occurs, respectively. In order to incorporate machine breakdowns into a balancing model, the (deterministic) operation times may be modified by adding a stochastic component reflecting the probability of machine breakdowns and the duration of repair processes. This may introduce high coefficients of variation into the operation times (cf. Chow (1987), Hillier and So (1991)). Since machine breakdowns may occur when a machine is working or also when it is inactive, one may distinguish between operation-dependent and time-dependent failures (cf. Buzacott and Hanifin (1978)).

Obviously, the line output rate depends on the distribution of task times. If the operation times are assumed to be exponentially distributed, the so-called *bowl-phenomenon* arises in unpaced systems. This phenomenon has been examined by Hillier and Boling (1966, 1979), Magazine and Silver (1978), Muth and Alkaff (1987), Hillier and So (1993) as well as Pike and Martin (1994). It means that the output rate may be increased by introducing some imbalance (in terms of the average station times) into the system. Empirical studies show that it may be better to build station loads in such a way that the stations in the middle are less utilized than those at the start or the end of the line (cf. El-Rayah (1979 a)). A typical Gantt-chart is shown in Figure 1.7. However, if the variation of operation times is much less than assumed by the exponential distribution, which is true in many real-world assembly line systems, balanced lines are optimal with respect to the line output (cf. Smunt and Perkins (1985)).

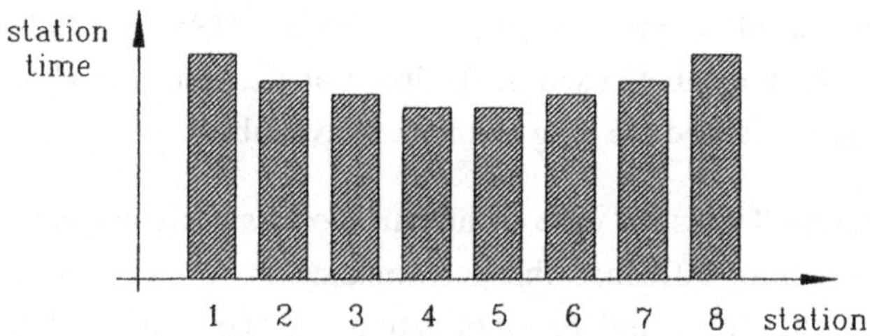

Figure 1.7. Bowl Phenomenon

Note that an equivalent bowl phenomenon exists for the distribution of buffer capacities along the line, because the output rate of the line may be increased when buffer sizes in the centre of the line are slightly larger than those at the beginning or the end of the line (cf. El-Rayah (1979 b), Hillier and So (1993)).

Dynamic Operation Times. Besides stochastic variations of the operation times, systematic reductions are possible due to *learning effects* or successive improvements of the production process. Especially when a new assembly line system is installed, the operators take a longer time to execute tasks than after becoming familiar with the work. The degree and the rate of learning related time reductions depend on the nature of the tasks, the flexibility of the production system, and the qualification levels of the operators (cf., e.g., Thomopoulos and Lehman (1969), Boucher (1987) as well as Chakravarty (1988)).

1.3.4 Assignment Restrictions

In addition to the precedence constraints, several types of conditions may restrict possible assignments of tasks to stations (cf. Lutz (1974, pp. 53)).

Task Related Constraints. In some situations, two tasks are desired at or must be assigned to the same station or section of the line. Such restrictions may, e.g., occur if a certain task needs the same state (temperature, moisture, etc.) of the workpieces as another one. This can be modelled by *maximum distances* (in terms of time or space; cf. Deckro (1989)).

In some cases, tasks are incompatible, i.e., they must not be performed at the same station or segment of the line (e.g., milling and measuring operations, painting and drilling operations; cf. Agnetis et al. (1995)). This can be expressed by *minimum distances* between the tasks. These restrictions are frequently called *zoning constraints* (cf. Ignall (1965)). Sometimes, this term is used to describe more general assignment restrictions (cf. Caruso (1965), Schofield (1979)).

Station Related Constraints. If special machines or tools which are needed to execute a certain task are only available in one or a few stations and cannot be moved without causing prohibitive costs, the task has to be assigned there (*must-do-tasks*, *fixed facility restrictions*; see, e.g., Kilbridge and Wester (1961 a)). A similar restriction occurs if material needed for some task is only available in a particular section of the line. This may be due to limited space for material stocks.

Position Related Constraints. Especially in the case of large and heavy workpieces (cars, washing machines, etc.), tasks may need a certain position of the workpieces (front or back, up or down). Since it may be neither possible nor economical to turn the workpieces too often, tasks which need the same position have to be grouped together in a segment of the line.

Workpieces which cannot be turned into another position at all are called *fixed items*, while *removable items* may be turned or removed from the conveyor (cf. Buxey and Sadjadi (1976), Wang and Wilson (1986)). For fixed items, tasks may be subdivided into left-hand side and right-hand side tasks as well as tasks which are possible at both sides of the (two-sided) line. Because of the difficulty of crossing the conveyor, the side specific tasks have to be done by different operators (cf. Section 1.3.5).

Operator Related Constraints. Depending on their complexity, tasks require different levels of skill. The requisite qualification of a worker is determined by the most complex operation assigned to the respective station. Therefore, it may be necessary to concentrate complicated tasks in a few stations (cf. Iskander and Chou (1990)). Furthermore, workers are usually paid with respect to the most complicated task they have to perform.

From the ergonomic point of view, it may be desirable to combine rigid and variable (non-rigid) tasks in the same station in order to reach a higher level of *job satisfaction* (cf., e.g., Gunther et al. (1983), Agrawal (1985), Downey and Leonhard (1992)). In this context of job design, concepts like *job enrichment* and *job enlargement* can be considered in the form of assignment constraints (cf., e.g., Zäpfel (1975, pp. 17), Globerson and Tamir (1980)).

1.3.5 Layout of the Production System

Because of the rigid process orientation, the layout of flow-line production systems is partially predetermined by the flow of materials. Nevertheless, some layout possibilities exist.

Serial Lines. This is the traditional layout of assembly lines. Single stations are arranged in a straight line along a conveyor belt.

U-shaped Lines. Traditional serial lines may have several disadvantages, e.g. the monotone and boring type of work mainly requiring low-level skilled operators, the sensibility and inflexibility of the production system concerning failures and changing demand rates. Negative consequences may be low-motivated operators, quality problems and large inventories due to rigid output rates.

In order to overcome these disadvantages, modern paradigms of production such as the *just-in-time principle* and *group technology* have been incorporated into assembly line production by many firms (see e.g. Hall 1983, Monden 1998). Such modern assembly lines are often organized as a "U-line" (see Figure 1.8).

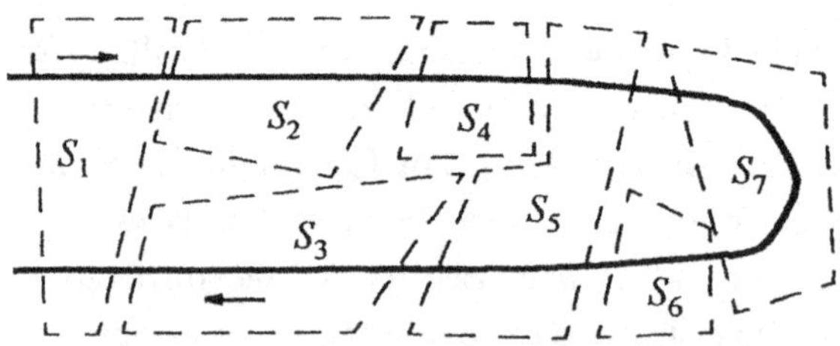

Figure 1.8. U-shaped Assembly Line

Both ends of the line are closely together forming a rather narrow "U". Stations may work at two segments of the line facing each other simultaneously.

For example, station 1 works at the beginning and the end of the line, i.e. it performs the first and the last tasks for every product unit. Operators are involved in different parts of the production process enlarging their horizons and skills. Stations are closer together such that visibility of the whole production process and communications between operators are improved. Furthermore, helping each other in cases of congestion and sharing machine capacities is facilitated. As a consequence of this job enrichment, workers acquire multiple skills leading to higher motivation, improved quality of products and increased flexibility. This flexibility is strongly required in just-in-time production systems, because demand rates of parts being produced for subsequent operations are dynamically changing from time to time. Then a re-balancing is necessary to match the output rate of the assembly line with the current demand rate. As the above discussion indicates this is much easier with U-lines than with traditional serial lines (see Monden 1998).

Parallel Lines. In order to increase flexibility and to decrease failure sensitivity of the system, it may be advantageous to install several parallel lines for manufacturing one or several products. Parallel lines enable the management to react to demand changes, because the number of working lines can be changed. The risk of machine breakdowns stopping the production is lowered. Furthermore, the use of parallel lines allows the enlargement of cycle time(s) which has several advantages. At first, better line balances can often be obtained, because more combinations of tasks exist. Secondly, horizontal job enlargement is possible, i.e., workers get more different tasks. These economical and social advantages of parallel production are utilized by a new production concept, the so-called *reflective production* which is successfully practiced in some Swedish plants (cf. Ellegard et al. (1992)). By the increased cycle times pacing aspects become less relevant and the operators are able to vary their individual working speed. Furthermore, the extent of labour division is relatively small so that only few operators are employed in every parallel unit. These units are organized as autonomous work teams.

In this context, the (strategic) decision problem of finding the *optimal number of lines* arises, because the installation of additional lines increases capital investment. Besides the line balancing problem, the assignment of models or different products as well as of work force to the installed lines builds a medium-term problem (cf. Lehman (1969), Geoffrion and Graves (1976), Globerson and Tamir (1980), Zäpfel (1989, pp. 212), Ahmadi et al. (1992)). With a flexible number of parallel lines it is possible to realize (optimal) cycle times which are not possible if only one line has to meet certain demands.

Parallel Stations. Even with a single line the advantages of parallelization can be utilized by installing parallel (multiple) stations, i.e., the workpieces are distributed among several operators who perform the same tasks. As is the case with parallel lines, the equipment has to be installed several times. Parallel stations allow the reduction of the (global) cycle time of the system if certain tasks have operation times longer than the desired cycle time. These operations may be assigned to parallel stations, the (local) cycle time of which is a multiple of the global cycle time. Models with parallel stations are considered by Freeman and Jucker (1967), Buxey (1974), Pinto et al. (1981), Sarker and Shanthikumar (1983), Bard (1989) as well as Daganzo and Blumenfeld (1994).

Another possibility of reducing the global cycle time below the largest task time is the concept of paralleled tasks (cf. Arcus (1966), Pinto et al. (1975)). Respective tasks are assigned to several stations of a serial line which cyclically perform them completely on different workpieces (cf. Domschke et al. (1997), chapter 4.3.6.1). A similar concept is described by Hu (1961), Nanda and Scher (1975, 1976) as well as Inman and Leon (1994).

Unpaced systems with parallel stations and buffers between the stages are referred to as *hybrid flow shop* and *flexible flow shop* problems in the literature, respectively (cf. Gupta (1988), Gupta and Tunc (1991, 1994) and Voß (1993 a)).

Two-Sided Lines. For the assembly of heavy workpieces it may be necessary to operate a two-sided line which consists of two serial lines in parallel. Instead of single stations, pairs of opposite stations on either side of the line (left-hand side and right-hand side stations) work in parallel, i.e., they work simultaneously at opposite sides of the same workpieces (cf. Hautsch et al. (1972), Bartholdi (1993)). Some tasks have to be performed on the right-hand and left-hand side, respectively, while others may be done on either side of the line. Most of the left-hand side tasks have a right-hand side mirror image, and vice versa (e.g., installing the left-hand/ right-hand side front wheel of a car). These pairs of tasks should be performed at the same pair of opposite stations.

Two-sided lines may be advantageous even for the assembly of small items, because the same number of stations as in a serial line needs a shorter physical line length and some kind of parallelism is included. However, two-sided lines require that two operators are able to work on the same workpiece simultaneously.

Feeder Lines and Supplementary Units. In flexible manufacturing systems, assembly lines are often used for the final assembly of products. Components (subassemblies) are produced by different supplementary units which may be organized as job shops, flexible manufacturing cells, or feeder lines. Then the balancing problem is connected with the problem of synchronizing the different production processes, i.e., the production rates of the supplementary units have to be determined (cf. Golenko-Ginzburg and Goldfeld (1991)).

1.3.6 Type of Station

Depending on the type of equipment and production control, stations may have different degrees of automation and flexibility. As a consequence of the physical line layout and the nature of the tasks, stations may be opened or closed.

Degree of Automation and Flexibility. Stations can be subdivided into manual, semi-automated, and automated stations. In *manual stations*, a human operator (or a group of operators) performs the tasks by means of rather simple tools or machines. *Semi-automated stations* are characterized by the fact that human interaction is restricted to control and material supply functions, while work is performed by automatic equipment. In (fully) *automated stations* even production control and material supply are performed by automatic equipment (cf. Zäpfel (1989), pp. 187).

In case of mass production of one homogeneous product, the stations of a semi-automated or fully automated line *(transfer line)* are equipped with *special-purpose machines* which are highly productive. The paced flow of workpieces through the working areas of the machines *(internal chaining)* is executed by automatic material handling equipment. In order to produce several products (models) on the same line, a higher degree of flexibility is needed. Consequently, the stations of a *flexible transfer line* are equipped with flexible *multi-purpose machines* (CNC machines) which can be re-configured for different products by automatically changing tools held in a direct-access tool magazine (cf. Tempelmeier and Kuhn (1993), chapter 1.1). If tools are not available in the tool magazine or large-scale re-arrangements of the line are necessary, the line has to be stopped while set-ups are performed (cf. multi-model lines in Section 1.3.1).

The degree of flexibility of manual lines mainly depends on the skills of the employed workers and the requisite tools. While varying work loads can be met by adapting the (machine) output rates in automated lines, this is only partly possible with manual lines. In these systems, the capacity of a station can be changed by varying the number of workers which is restricted by the fact that some tasks cannot be done by several people (cf. Sculli (1979), Akagi et al. (1983), Wilson (1986) as well as Lutz et al. (1994)).

Closed Stations - Open Stations. The extent of *closed stations* is fixed, and it is not possible for the operators to cross the station boundaries. Closed stations are necessary if the working area requires specific environmental conditions, for example, in the case of paint shops or heating chambers.

The boundaries of *open stations* can be crossed, but the extent of the crossing is usually restricted by, e.g., the limited range of power tools or the scope of the material handling system. If only one of both station boundaries can be crossed, the respective stations are called *open-to-the-right* and *open-to-the-left*, respectively. The areas of adjacent (open) stations may overlap, but usually the operators are not permitted to interfere with each other by simultaneously servicing a workpiece. Open stations allow short-term balancing of capacity requirements and supplies, especially in the case of mixed-model production, because it is possible to depart from the station to finish work on units of labour intensive models, while less expensive models allow returning to the actual station area (cf., e.g., Buxey (1979)).

Inspection Stations. Because of the process-orientation of assembly line production, quality control has to be included into the production process by installing inspection stations (cf. Chow (1990, pp. 140)). Then assembly line balancing is accompanied by the problem of determining the number and positions of inspection stations as well as their dimensions (cf. Zäpfel (1975, p. 18), Gunter and Swanson (1985)).

1.3.7 Launching Discipline

The intervals at which workpieces are launched down the line influence the efficiency of the system. Two main launching strategies can be distinguished (cf. Wester and Kilbridge (1964)).

Fixed Rate Launching. Consecutive units are started down the line in regular intervals (*launch* or *feed interval*). In paced systems, this interval is equal to the cycle time c. Every c time units, a workpiece becomes accessible to each station and another one leaves the station. The tolerance time, during which a workpiece is accessible to a station, depends on the length of the station and the speed of the conveyor belt. The workpieces are equi-spaced on the line.

Variable Rate Launching. A more flexible strategy avoids starving of the first station, because the next unit is launched after the first station has finished work on the current one. This results in different distances between consecutive workpieces. The number of units accessible to a station varies over time. If consecutive units have small work loads in the first station, their distance is small, and work congestion may occur in successive stations. If the first station loads are heavy, later stations may have idle times.

In order to avoid these difficulties, the launching strategy may concentrate on *bottleneck stations*. That is, the launching intervals should be chosen in such a way that one or more heavily loaded stations show a minimal amount of idle time, because such stations are most critical with respect to unused capacity.

1.3.8 Equipment Selection and Processing Alternatives

Most assembly line balancing models are based on the assumption that the equipment of the stations and the production process are fixed. But especially in automated production, a number of different possibilities to design stations and to perform tasks may exist. When assembly systems are to be installed, one has to decide how to equip the stations with machines and tools (cf. Kusiak (1987)). Considering the flexibility of the chosen equipment, the tasks have to be assigned to the stations. While *general-purpose machines* are able to perform many different tasks, *special-purpose machines* are restricted to a small number of operations. The latter ones lead to station related assignment restrictions (cf. Section 1.3.4). Because the selected equipment influences the task assignment, and vice versa, the balancing problem and the equipment selection problem have to be considered simultaneously. The combined problem is called an **assembly system design problem**. It consists of selecting the equipment of the stations and of assigning the tasks to the stations in such a way that the total system costs (fixed and variable) are minimized, while a given production rate is satisfied. Different types of assembly system design problems are considered by, e.g., Graves and Lamar (1983), Aneke and Carrie (1986), Graves and Redfield (1988), Miyazaki and Ohta (1990) as well as Rachamadugu and Shanthikumar (1991). These problems are complicated by the fact that the chosen equipment influences the production process, i.e., the necessary operations. Furthermore, the operation times of tasks depend on the equipment of the stations to which they are assigned (cf. Pinto et al. (1983)).

1.3.9 Volume of Production

Concerning the volume of production one may distinguish between *small-lot assembly* of large and very costly products and *mass-production* of standardized and relatively cheap products. In the first case which arises in heavy industry (e.g., assembly of aircrafts or heavy agricultural machinery), the production rate is small and the cycle time is long (up to several days or weeks; cf. Shtub (1984 b)). In some settings of *job-site production*, the "workpieces" may have fixed positions, while the "stations" (crews of workers) have to change their locations. Even this type of production can be interpreted as an assembly line, because the order in which the workpieces are visited by the crews is unique. Greatly varying operation times complicate balancing those "lines" mainly due to the slow pace of learning (cf. Chakravarty and Shtub (1986 a), Dar-El and Rabinovitch (1988), Dar-El and Rubinovitz (1991), Engström and Medbo (1992)). Furthermore, long cycle times and costly products require a careful planning of the part delivery dates and the launching discipline (cf. Sarin and Das (1987)).

In case of a mass-production of standardized products, the frequent repetition of operations and the high division of labour result in fast learning. The steady movement of workpieces requires a continuous part availability. The high production volume calls for carefully balancing the lines, because small inefficiencies per unit add up to large inefficiencies in total.

1.3.10 Objectives

The installation of an assembly line production system is a long-term decision and usually requires large capital investments. Therefore, it is important that such a system is designed and balanced so that it works as efficiently as possible. Besides balancing a new system, a running one has to be re-balanced periodically or after changes in the production process or the production program have taken place. Because of the long-term effect of balancing decisions, the used objective(s) has (have) to be carefully chosen considering the goals of the enterprise. Objectives may reflect capacity, cost or profit related aspects as well as organizational and social considerations, respectively. In the literature, usually only one objective is used, while other goals are formulated as constraints. Only few references deal with multiple objective assembly line balancing problems, namely, Gunther et al. (1983), Baybars (1985), Shtub and Dar-El (1990), Deckro and Rangachari (1990), Malakooti (1991, 1994).

Capacity Oriented Goals. Most commonly, assembly line balancing problems contain the objective of *maximizing the capacity utilization* of the line which is measured by the *line efficiency* (percentage of productive time in the line, cf. Section 2.1). In case of deterministic operation times and single-model production, the line efficiency only depends on the cycle time and the number of stations. Since this dependence is non-linear, simpler linear objectives are often considered:

- Minimize the number of stations for a given cycle time.
- Minimize the cycle time for a given number of stations.
- Minimize the weighted sum of the cycle time and the number of stations.

Other capacity or time related objectives include[2]

- minimizing the *flow time* (throughput time, interval of time between launching a workpiece down the line and removing the finished product from the line),
- equalizing the *levels of capacity utilization* of the stations,
- minimizing the *balance delay time* (sum of idle times) and the *balance delay (ratio)* (percentage of idle times) over all stations, and
- minimizing the *waiting times* of work pieces.

Cost Oriented Goals. An important objective is to minimize the total costs of the line including long-term investment costs and short-term operating costs. Assembly line production systems may concern the following *cost categories* which are mainly influenced by the cycle time and the number of stations (cf. Freeman (1968), Zäpfel (1975, pp. 31), Klenke (1977, pp. 25), Steffen (1977, pp. 65), and Ziegler (1990, pp. 71)).

[2] Most of the terms used are properly defined in Section 2.1.

Costs of Machinery and Tools. These costs are usually measured by machine-hour rates (including depreciations, interest charges, repair, maintenance and space costs) and depend on the number of stations as well as on the type of machines (special-purpose or general-purpose machines) and tools.

Wage Costs. The wages depend on the number of workers, their qualifications, and the work contents of the stations. They are usually paid according to the performance but contain fixed base rates (cf., e.g., Rosenberg and Ziegler (1992)). When regular work is not sufficient to meet the required production rate, an additional capacity has to be created by overtime and/or employment of additional workers. Short-term capacity shortages may be compensated by *utility workers* (floaters). Both overtime and utility work are more expensive than regular work because of larger wage rates.

Material Costs. The material needed is mainly determined by the output rate and, hence, by the cycle time. Furthermore, material costs are influenced by the production process and the wastage rates.

Idle Time Costs. Idle times represent unused (non-productive) capacities of machines and workers. They may result from the unequal division of labour, different production rates of the stations, or machine breakdowns. If the unused capacity might profitably be utilized for performing other work or for producing at a higher output rate, idle time costs may be used to measure the loss of profit, i.e., the alternative costs. The idle time and the respective alternative costs depend on the number of stations and the cycle time as well.

Demurrage/Penalties. If minimum production quantities are given due to supply commitments, penalties for non-fulfillment are to be paid, when the production is not able to meet the demand, i.e., the cycle time is too large or the shifts are too short.

Incompletion Costs. In case of stochastic task times, workpieces may leave stations before the assigned operations have been finished. This results in additional costs for stopping the line or for re-working the respective units off-line (cf., e.g., Sarin and Erel (1990)).

Set-up Costs. In a multi-model production, set-up activities are necessary to prepare machines and tools for producing another batch of a model. These activities reduce the productive time as idle times do (cf. idle time costs) and cause additional costs for skilled labour as well as materials and supplies such as cleaning agents and lubricants (cf. Chakravarty and Shtub (1986 b), Dar-El and Rabinovitch (1988)).

Inventory Costs. Whenever materials, sub-assemblies or finished products are stored in some type of non-productive inventory, capital is tied up and prevented from being invested in profitable alternative courses of action. The respective loss of profit may be measured by inventory holding costs (cf. Jacobson and Sadowski (1977)). Furthermore, stocks need expensive materials handling and floor space causing fixed and variable costs.

In order to reduce inventory levels of material stocks, the dates by which materials are required have to be known as exactly as possible. Then the materials can be delivered *just-in-time* (cf., e.g., Schonberger (1982, chapter 1), Monden (1998)). In case of just-in-time production, the objective is to minimize the amount by which production differs from strict proportion with demand in order to obtain levelled part usage rates (cf. Miltenburg (1989), Kubiak and Sethi (1991)).

Costs related to buffers holding in-process-inventories are reduced by synchronizing (balancing) the output rates of the work stations (cf. Malakooti (1994)).

Profit Oriented Goals. While cost oriented goals assume that the output volume is fixed and can completely be sold at a fixed price, profit oriented objectives require considering the production rate (i.e., the cycle time) and the prices as decision variables. The latter two are related, because the external demand (the marketable output) depends on the selling prices. The total profit of a line per period is defined as output volume times unit contribution margin (price minus variable costs per unit) minus fixed costs per period. The objective of maximizing the profit per time unit or period is considered by Klenke (1977, pp. 30), Rosenblatt and Carlson (1985) as well as Martin (1994), respectively. The maximization of the total contribution margin per shift is examined by Zäpfel (1975, pp. 36).

Social and Organizational Objectives. Goals which are related to social or organizational aspects are often considered as constraints rather than as objectives to be optimized. Such social constraints may concern *job enrichment*, i.e., to assign less boring combinations of operations to workers (cf. the operator related constraints in Section 1.3.4 as well as Gunther et al. (1983)). Furthermore, aspects of *job enlargement* may be included in order to restrict the degree of specialization by introducing parallelism and reducing the production rate, respectively.

From the organizational point of view it may be desirable or even necessary to avoid moving (heavy and expensive) equipment and changing the line layout when a re-balancing (re-engineering) of an assembly line takes place. Further organizational aspects may concern the flow of materials or the allocation of tools to stations (cf. Gunther et al. (1983)).

Most of the above mentioned objectives are mainly influenced by the cycle time and the number of stations, respectively. Therefore, most models considered in the literature use the (surrogate) objectives of minimizing the cycle time or the number of stations, respectively. Reflecting this fact, we intensively discuss models with these objectives in Chapter 2.

2 Single-Model Assembly Lines

In this chapter, we focus on single-model assembly line balancing problems. Section 2.1 presents some notations and definitions used throughout the work. In Section 2.2, we introduce the well-known Simple Assembly Line Balancing Problem (SALBP) and discuss several versions. Extensions of this fundamental problem are presented in Section 2.3.

2.1 Notations and Definitions

We give some definitions needed for a proper mathematical statement of single-model assembly line balancing problems. Figure 2.1 shows a summary of the terms and notations most frequently used. Some of the terms are already introduced in Section 1.2.

Definition 2.1. A *precedence graph* $G=(V,A,t)$ is a non-cyclical digraph with a set $V=\{1,...,n\}$ of n nodes and a set $A=\{(i,j) \mid i \in V \text{ and } j \in F_i\}$ of arcs. Nodes symbolize tasks, and arcs represent direct precedence relations between the tasks. A task i which must be completed before a task j can be performed is a *predecessor* of j, while j is a *successor* or *follower* of i. A precedence relation between two tasks i and j is *direct* (immediate) if none of the tasks succeeding i is a predecessor of j. The set of immediate successors of a task i is denoted by F_i. For the sake of simplicity, the precedence graph is assumed to be *sorted topologically,* i.e., the inequality $i<j$ holds for each arc (i,j).

The node weights t_j represent the operation times of the tasks. A node corresponding to a task without predecessors (successors) is called a *source (sink)* of G.

A *reverse precedence graph* G^r is obtained by reversing the orientation of all arcs of G. Sources become sinks, predecessors become successors, and vice versa.

Definition 2.2. The *transitive closure* $G^*=(V,A^*,t)$ of a precedence graph G includes arcs (i,j) for all pairs of tasks i and j connected by a path in G, i.e., $A^*=\{(i,j) \mid i \in V \text{ and } j \in F_i^*\}$, where F_i^* contains all successors of a task i. The difference set A^*-A consists of the *transitive* arcs representing *indirect precedence relations*.

The cardinality of A^* divided by the maximal number $n \cdot (n-1)/2$ of possible arcs defines the *order strength* of G^* and G as well (cf. Mastor (1970)).

n	number of tasks
V	set of all tasks $(= \{1,...,n\})$
j	index for the tasks $(j = 1,...,n)$
c	cycle time
p	production rate $(= 1/c)$
m	number of stations
k	index for the stations $(k = 1,...,m)$
t_j	operation time (task time) of task j [1]
p_j	station requirement of task j $(= t_j/c = t_j \cdot p)$
t_{max}	maximal task time $(= \max \{t_j \mid j = 1,...,n\})$
t_{min}	minimal task time $(= \min \{t_j \mid j = 1,...,n\})$
t_{sum}	sum of task times $(= \Sigma_j t_j)$
P_j	set of immediate predecessors of task j
F_j	set of immediate successors (followers) of task j
A	set of direct precedence relations $(= \{(i,j) \mid i \in V \text{ and } j \in F_i\})$
P^*_j	set of immediate and transitive predecessors of task j
F^*_j	set of immediate and transitive successors (followers) of task j
A^*	set of all precedence relations $(= \{(i,j) \mid i \in V \text{ and } j \in F^*_i\})$
S_k	station load, set of tasks assigned to station k
$t(S_k)$	station time of station k $(= \sum_{j \in S_k} t_j)$
c_r	realized (operating) cycle time $(= \max \{t(S_k) \mid k = 1,...,m\})$
E_j	earliest station of task j
L_j	latest station of task j

Figure 2.1. Notations

Definition 2.3. A partition of the set $V = \{1,...,n\}$ of all tasks into m disjoint subsets (station loads) S_k with $k = 1,...,m$ is called a *task assignment,* if $i \in S_h$ and $j \in S_k$ with $h \le k$ holds for each arc $(i,j) \in A$. That is, each task is assigned to exactly one station, and the successors of a task i must not be assigned to an earlier station than i.

A task assignment is *feasible* for a cycle time c if the station times $t(S_k)$ of all stations are lower than or equal to c. A feasible task assignment is called *line balance.* A necessary condition for the existence of a line balance is $c \ge t_{max}$ due to the indivisibility of the tasks.

---

1 Without a relevant loss of generality the task times t_j and the cycle time can be assumed to be integral. Non-integer parameter values of real-world problems may be transformed into integers by suitable scaling operations (cf. Hax and Candea (1984, p. 362)). Therefore, we assume the integrality of input parameters throughout this work. Furthermore, task times are assumed to be positive, whenever zero operation times are not explicitly stated.

Definition 2.4. Let $T = m \cdot c$ denote the total time available for the assembly of each workpiece on the line. Hence, T represents the *capacity supply* of the line. Obviously, T must be no smaller than the sum of task times t_{sum}. The *flow time (throughput time)* denotes the time interval between launching a workpiece down the line and removing the finished product from the line. Hence, it specifies the line length in terms of time and is at least as large as T.

Definition 2.5. The capacity utilization of the line is measured by the *line efficiency* $E = t_{sum}/T$. The unused (idle) capacity is reflected by the *balance delay time* $BD = T - t_{sum}$ and the *balance delay ratio* $BR = 1 - E$. The *idle time* of a station k with station load S_k is defined by $I_k = c - t(S_k)$. Due to $t_{sum} = \Sigma_k\, t(S_k)$, the balance delay time is equal to the sum of idle times, i.e., $BD = \sum_{k=1}^{m} I_k$.

Definition 2.6. The *smoothness index* SX measures the equality of the distribution of work among the stations and is defined through

$$SX = \sqrt{\sum_{k=1}^{m} (c_r - t(S_k))^2}\,,$$

where the realized cycle time c_r is determined by the maximal station time (cf. Moodie and Young (1965)). Note that the realized cycle time c_r may deviate from a desired (pre-specified) cycle time c.

2.2 The Simple Assembly Line Balancing Problem (SALBP)

In this section, we describe several versions of the well-known Simple Assembly Line Balancing Problem (SALBP) which has intensively been studied during the last four decades. This classical single-model problem contains the following main characteristics (cf. the classification scheme in Section 1.3):

- mass-production of one homogeneous product by performing n operations of a given production process
- paced line with fixed cycle time
- deterministic and integral operation times
- no assignment restrictions besides the precedence constraints
- serial line layout, one-sided stations
- all stations are equally equipped with respect to machines and workers
- fixed rate launching, launch interval = cycle time

Several versions of the problem arise from considering different capacity oriented objectives. In order to concentrate on objective-independent properties of the problem, we start our presentation with the feasibility problem (SALBP-F).

Subsequently, several objectives are discussed, namely, the minimization of the number of stations for a given cycle time (SALBP-1), the minimization of the cycle time for a given number of stations (SALBP-2), and the maximization of the line efficiency (SALBP-E).

2.2.1 The Feasibility Problem (SALBP-F)

In the absence of an objective function, SALBP reduces to a decision problem in the strong sense (cf. Section 2.2.1.5), which we call *feasibility problem* (SALBP-F for short). This problem consists of finding out whether or not a task assignment exists for a given number m of stations, which is feasible for a given cycle time c (cf. Definition 2.3). That is, SALBP-F has to decide if a line with m stations can be operated with a certain cycle time c. Furthermore, a respective line balance has to be provided for the given (m,c)-combination in the case of a positive answer.

Example. We consider a problem with $n=11$ tasks and the precedence graph of Figure 2.2. It is characterized by $t_{min}=1$, $t_{max}=6$, and $t_{sum}=45$. The successor and predecessor sets of, e.g., task 4 are $F_4=\{7\}$, $F_4^*=\{7,9,10,11\}$, $P_4=\{1,2\}$, and $P_4^*=\{1,2\}$.

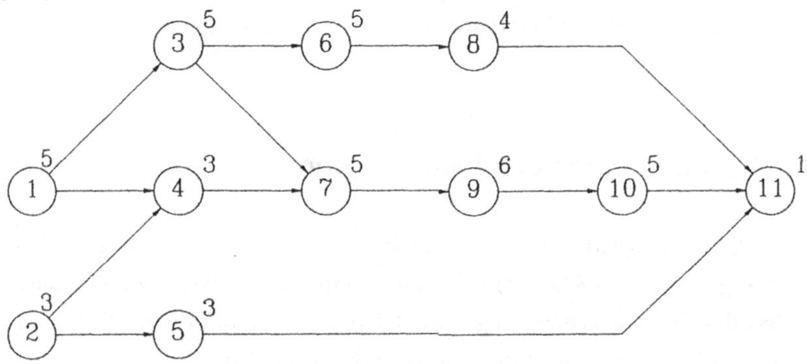

Figure 2.2. Example Precedence Graph

The task times, given as node weights, are assumed to be measured in minutes. Further, we assume that an output volume of 2,100 units should be realized within the planning horizon (1 month with 25 workdays and 2 daily shifts of 8 hours, each). The total available time for production is $25 \cdot 2 \cdot 8 \cdot 60 = 24{,}000$ minutes. Hence, the cycle time c should be no larger than $\lfloor 24{,}000/2{,}100 \rfloor = 11$ minutes per unit. With $c=11$ the production rate is $p=5.45$ units per hour. By assuming that an assembly line with 5 stations should be used, we obtain a SALBP-F instance with $m=5$ and $c=11$. A line balance for this parameter combination is presented in Figure 2.3. The operating cycle time of this solution is $c_r=t(S_3)=11$, and the line capacity is $T=5 \cdot 11=55$. Hence, the line efficiency is $E=45/55=0.82$ (82 %), the balance delay ratio is BR = 0.18 (18 %), and the balance delay time is BD = 55 - 45 = 10. The smoothness index is given by $SX = \sqrt{1^2 + 2^2 + 0^2 + 2^2 + 5^2} = 5.83$.

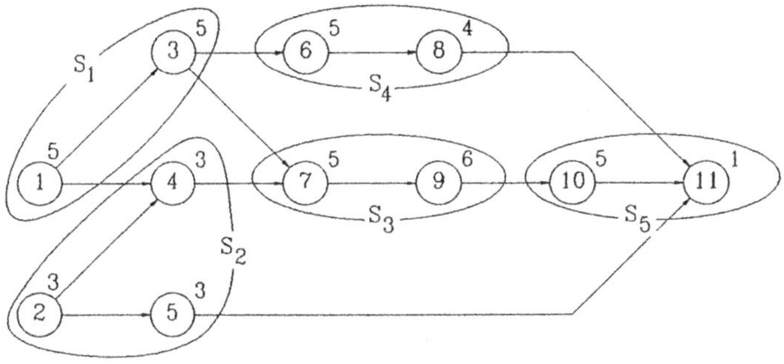

Figure 2.3. Solution for c = 11 and m = 5

Assume now that 2,400 units should be produced with the same assembly line within the planning horizon. Then the cycle time must not be larger than c = 10. Figure 2.4 shows a line balance for the corresponding SALBP-F instance with m=5 and c = 10. The realized cycle time is $c_r = 10$ (due to the station times of the stations 1, 3, and 4).

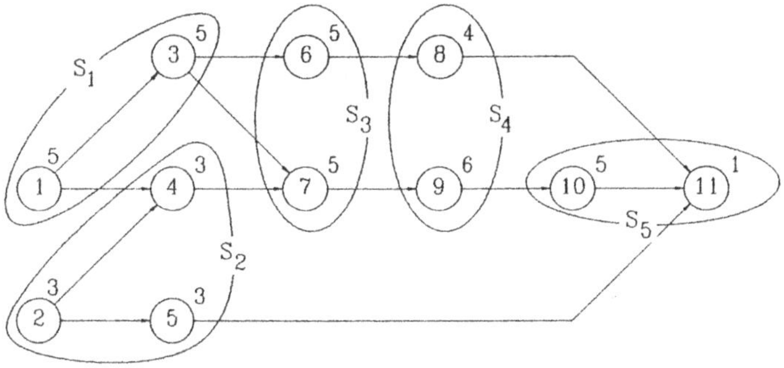

Figure 2.4. Solution for c = 10 and m = 5

This solution is characterized by E = 45/50 = 0.9 (90 %), BR = 0.1, and BD = 5. The production rate is p = 6 units per hour, and the smoothness index is computed by $SX = \sqrt{0^2 + 1^2 + 0^2 + 0^2 + 4^2} = 4.12$.

Obviously, the task assignment with c = 10 is superior to that with c = 11 with respect to all capacity and time related measures. The example shows the difficulty to pre-specify an appropriate cycle time, because it may be possible to operate a line with the same number of stations but with a smaller cycle time than the given one.

While the line with 5 stations can be operated with the cycle times 10 and 11, respectively, it is not possible to use smaller cycle times, because there is no feasible task assignment for SALBP-F with m = 5 and c = 9 (even though $T \geq t_{sum}$ holds).

2.2.1.1 Properties of SALBP-F

The precedence relations are utilized for restricting the number of stations to which a task can be assigned. For each task j, we obtain an **earliest station** E_j and a **latest station** L_j by the formulae (2.1) and (2.2).

$$E_j := \left\lceil (t_j + \sum_{h \in P_j^*} t_h)/c \right\rceil \qquad \text{for } j = 1,\dots,n \qquad (2.1)$$

$$L_j := m + 1 - \left\lceil (t_j + \sum_{h \in F_j^*} t_h)/c \right\rceil \qquad \text{for } j = 1,\dots,n \qquad (2.2)$$

The formulae (2.1) take into consideration that task j must not start before all preceding tasks are finished. A lower bound on the number of stations required for task j and the tasks of the predecessor set P_j^* is obtained by dividing the respective sum of operation times by the cycle time (which is the maximal possible station time). By rounding up the result to the nearest integer value, the index of the earliest station to which task j can be assigned in order to fulfill the precedence relations is obtained. Analogously, the formulae (2.2) define the index of the latest station to which task j can be assigned depending on the operation times of j and its successors as well as the given number m of stations.

The earliest and latest stations establish a *station interval* $SI_j = [E_j, L_j]$ of stations to which task j has to be assigned in order to find a line balance with cycle time c and m stations. Using these station intervals, we obtain the set B_k of tasks *potentially assignable* to station $k = 1,\dots,m$: $B_k := \{j \mid k \in SI_j\}$

Example. Table 2.1 contains the earliest and latest stations for our above mentioned example with m = 5 and c = 10.

Table 2.1. Earliest and Latest Stations for m = 5 and c = 10

j	1	2	3	4	5	6	7	8	9	10	11
t_j	5	3	5	3	3	5	5	4	6	5	1
E_j	1	1	1	2	1	2	3	2	3	4	5
L_j	2	3	2	4	5	5	4	5	4	5	5

The sets of potentially assignable tasks are $B_1 = \{1,2,3,5\}$, $B_2 = \{1,\dots,6,8\}$, $B_3 = \{2,4,\dots,9\}$, $B_4 = \{4,\dots,10\}$, and $B_5 = \{5,6,8,10,11\}$. Since task 11 is assignable to station 5 only, it must be assigned to this station in a feasible solution.

Remark 2.1. The above definitions of (2.1) and (2.2) require that all tasks representing sources and sinks of the precedence graph, respectively, have non-zero operation times to avoid $E_j = 0$ and $L_j = m+1$ for some task j. Thus, the formulae have to be complemented by $E_j \geq 1$ and $L_j \leq m$ in the absence of this assumption.

2.2.1.2 Mathematical Formulations of SALBP-F

A number of attempts has been made to develop mathematical formulations of simple assembly line balancing problems (especially for SALBP-1, see Section 2.2.2).

The first formulation of SALBP-1 is given by Salveson (1955). Since this model requires the pre-determination of all possible station loads, it is not practicable. A similar approach is described by Wedekind (1963).

Two different model formulations are proposed by Bowman (1960). The first model contains integer variables, which denote the number of time units devoted to an operation at a station, and zero-one variables, which ensure that a task is assigned to one station only. The second formulation contains integer variables expressing the clock time at which an operation starts.

An improvement of the first model of Bowman is given by White (1961) who eliminates the time related integer variables from the model. This model is further improved by Thangavelu and Shetty (1971) as well as Patterson and Albracht (1975) with respect to the objective function (which is not necessary for SALBP-F) and the way of expressing the precedence constraints.

We present the formulation of Patterson and Albracht (1975), because it contains the least number of constraints and variables. In order to formulate the feasibility problem SALBP-F which aims at finding a feasible assignment only, we omit the objective function.

Formulation 1 (Patterson and Albracht). It contains zero-one (binary) variables x_{jk} defined as follows:

$$x_{jk} = \begin{Bmatrix} 1 & \text{if task } j \text{ is assigned to station } k \\ 0 & \text{otherwise} \end{Bmatrix} \quad \text{for } j = 1,\dots,n \text{ and } k \in SI_j \tag{2.3}$$

The constraints (2.4) - (2.7) describe line balances for a given (m,c)-combination. If the model has a feasible solution, the respective SALBP-F instance has a positive answer. Otherwise, no line balance exists for the given combination of m and c.

$$\sum_{k \in SI_j} x_{jk} = 1 \qquad \text{for } j = 1,\dots,n \tag{2.4}$$

$$\sum_{j \in B_k} t_j \cdot x_{jk} \le c \qquad \text{for } k = 1,\dots,m \tag{2.5}$$

$$\sum_{k \in SI_h} k \cdot x_{hk} \le \sum_{k \in SI_j} k \cdot x_{jk} \qquad \text{for } (h,j) \in A \text{ and } L_h \ge E_j \tag{2.6}$$

$$x_{jk} \in \{0,1\} \qquad \text{for } j = 1,\dots,n \text{ and } k \in SI_j \tag{2.7}$$

The restrictions (2.4) are called *occurrence constraints*. Together with the *integrality constraints* (2.7), they ensure that each task is assigned to exactly one station of its station interval. The *cycle time constraints* (2.5) guarantee that the cycle time is not exceeded by the station time of any station. (2.6) represents the *precedence constraints* ensuring that no task is assigned to an earlier station than a predecessor.

The restrictions (2.6) may equivalently be replaced by (2.6') which increases the number of constraints by the factor m but requires only the coefficients 0 and 1 (cf. White (1961)).

$$x_{jk} \leq \sum_{q=1}^{k} x_{hq} \qquad \text{for } k=1,\dots,m,\ (h,j) \in A, \text{ and } L_h \geq E_j \qquad (2.6')$$

The formulation (2.4) - (2.7) contains up to $m + n + |A|$ restrictions and up to $n \cdot m$ binary variables. Because of the obvious inequality $m \leq n$ for reasonable problems, the model contains $O(n^2)$ constraints.[2] Patterson and Albracht (1975) propose possibilities to reduce the problem size, e.g., the variables x_{j,L_j} can be eliminated.

Formulation 2. We present a new formulation and use binary variables y_{jk} defined slightly different from the variables x_{jk} of formulation 1. The variables are related by $x_{jk} = y_{jk} - y_{j,k-1}$ with $y_{j,Ej-1} = 0$ for all j.

$$y_{jk} = \begin{cases} 1 & \text{if task } j \text{ is assigned to station } k \text{ or earlier} \\ 0 & \text{otherwise} \end{cases} \qquad \text{for } j=1,\dots,n \text{ and } k \in SI_j \quad (2.8)$$

Each vector $(y_{j,E_j},\dots,y_{jk},\dots,y_{j,L_j})$ contains a block of zeros and a consecutive block of ones. The switch to the second block occurs at the position for that station k to which task j is assigned. For instance, the vector (0,0,0,1,1) implies that a task is assigned to station 4, provided that it is potentially assignable to the stations 1 to 5.

SALBP-F is defined by the set of constraints (2.9) - (2.13).

$$y_{jk} - y_{j,k-1} \geq 0 \qquad \text{for } j=1,\dots,n \text{ and } k \in SI_j \qquad (2.9)$$

$$\sum_{j \in B_k} t_j \cdot (y_{jk} - y_{j,k-1}) \leq c \qquad \text{for } k=1,\dots,m \qquad (2.10)$$

$$y_{hk} - y_{jk} \geq 0 \qquad \text{for } (h,j) \in A \text{ and } L_h \geq E_j \qquad (2.11)$$

$$y_{j,E_j-1} = 0,\ y_{j,L_j} = 1 \qquad \text{for } j=1,\dots,n \qquad (2.12)$$

$$y_{jk} \in \{0,1\} \qquad \text{for } j=1,\dots,n \text{ and } k \in SI_j \qquad (2.13)$$

Together with (2.13), the constraints (2.9) ensure that each task is assigned to exactly one station. Restrictions (2.10) follow from (2.5) by the definition of the

[2] This O-notation is introduced in Section 2.2.1.5 for the time complexity of algorithms. It denotes the order of a function.

variables. The precedence relations are easily expressed by (2.11). The equations (2.12) guarantee that each task j is assigned to a station of its station interval.

The restrictions (2.12) can be eliminated by substituting the respective variables within the other constraints. Then formulation 2 contains up to $n \cdot m + m + |A|$ constraints and up to $n \cdot m$ variables. Hence, it is of the same order as formulation 1 but actually contains more restrictions. However, only the cycle time constraints (2.10) show coefficients unequal to 1, -1, and 0, respectively.

Formulation 3. We present a new mixed-integer formulation which is similar to a disjunctive model for the flow shop sequencing problem (cf. Domschke et al. (1997, pp. 372)). It combines ideas of Bowman (1960) and Talbot and Patterson (1984).

The formulation includes integer variables z_j denoting the number of the station to which task j is assigned. Continuous variables b_j are introduced for the clock time at which task j is started (measured in the time elapsed since entering the first station). Furthermore, binary variables w_{hj} are used which are defined as follows:

$$w_{hj} = \begin{cases} 1 & \text{if task h is performed before task j} \\ 0 & \text{otherwise} \end{cases} \quad \text{for } h < j \text{ and } (h,j) \notin A^* \text{ and } SI_h \cap SI_j \neq \phi \tag{2.14}$$

The set of constraints (2.15) - (2.22) defines SALBP-F, where $T = m \cdot c$.

$$b_j \geq c \cdot (z_j - 1) \qquad \text{for } j = 1,...,n \tag{2.15}$$

$$b_j + t_j \leq c \cdot z_j \qquad \text{for } j = 1,...,n \tag{2.16}$$

$$(1 - w_{hj}) \cdot T + b_j \geq b_h + t_h \qquad \text{for } h < j, (h,j) \notin A^* \text{ and } SI_h \cap SI_j \neq \phi \tag{2.17}$$

$$w_{hj} \cdot T + b_h \geq b_j + t_j \qquad \text{for } h < j, (h,j) \notin A^* \text{ and } SI_h \cap SI_j \neq \phi \tag{2.18}$$

$$b_h + t_h \leq b_j \qquad \text{for } (h,j) \in A \text{ and } L_h \geq E_j \tag{2.19}$$

$$E_j \leq z_j \qquad \text{for } j = 1,...,n \tag{2.20 a}$$

$$z_j \leq L_j \qquad \text{for } j = 1,...,n \tag{2.20 b}$$

$$z_j \geq 0 \text{ and integer, } b_j \geq 0 \qquad \text{for } j = 1,...,n \tag{2.21}$$

$$w_{hj} \in \{0,1\} \qquad \text{for } h < j, (h,j) \notin A^* \text{ and } SI_h \cap SI_j \neq \phi \tag{2.22}$$

The constraints (2.15) and (2.16) ensure that each task j is fully performed within one station, because j enters station k at time $c \cdot (k - 1)$ and leaves it at time $c \cdot k$ (provided that the tolerance times are equal to c and there is no space between neighbouring station areas). The disjunctive restrictions (2.17) and (2.18) guarantee that for each pair of tasks, which are not related by precedence and may interfere with each other, either task h is completely processed before task j is started, or vice versa. The restrictions (2.19) ensure the fulfillment of the precedence constraints, because a task j is not allowed to start before all predecessors are completed. The assignment of the tasks is restricted to the station intervals by (2.20 a) and (2.20 b), while (2.21) and (2.22) define the variables.

Formulation 3 contains up to $4 \cdot n + 2 \cdot (n \cdot (n-1)/2 - |A^*|) + |A|$ constraints, n integer variables, up to $n \cdot (n-1)/2 - |A^*|$ binary variables, and n continuous variables. Therefore, the constraints and the variables are of order $O(n^2)$ as well. However, the number of constraints and variables is usually much larger than in the other formulations. Formulation 3 becomes more attractive in the case of a high order strength, because the number of variables and restrictions mainly depends on the number of task pairs not related by precedence.

2.2.1.3 Different Versions of SALBP

As already mentioned, SALBP-F is related to the following optimization problems:

- SALBP-1: Minimize the number m of stations for a given cycle time c.
- SALBP-2: Minimize the cycle time c for a given number m of stations.
- SALBP-E: Maximize the line efficiency E or, equivalently, minimize $m \cdot c$.

The feasibility problem SALBP-F is to find out whether or not a certain combination (m,c) is feasible. In terms of feasibility, the optimization problems may be stated as follows (the asterisks indicate optimality):

- SALBP-1: $m^* = \min \{ m \mid (m,c) \text{ is feasible for given } c \}$
- SALBP-2: $c^* = \min \{ c \mid (m,c) \text{ is feasible for given } m \}$
- SALBP-E: $E^* = \max \{ t_{sum}/(m \cdot c) \mid (m,c) \text{ is feasible for possible values of } c \text{ and } m \}$

This description of the optimization problems shows that each of them requires the solving of at least one SALBP-F instance. That is, a certain feasible (m,c)-combination and a corresponding line balance have to be determined in each case.

SALBP-1 and SALBP-2 may be regarded as being *dual* to each other. SALBP-1 considers the cycle time as a restriction, while it is contained in the objective function of SALBP-2. By contrast, the number of stations restricts SALBP-2 and is minimized by SALBP-1. Note that this duality differs from those for linear programs.

Since the cycle time and, hence, the production rate have to be specified as fixed parameters, SALBP-1 is present when a new assembly line system has to be *installed* and the external demand can be well estimated. Opposite to this, SALBP-2 leads to the maximization of the production rate of an existing assembly line. This problem may arise when changes in the production process or the demand structure require a *re-engineering* of the production system (cf. Section 3.4 for a respective planning hierarchy in the mixed-model case). SALBP-E represents the generalization of both SALBP-1 and SALBP-2. It is important if the cycle time as well as the number of stations may vary in certain ranges. This is frequently the case in industrial problems.

While SALBP-2 minimizes the sum of idle times for a given number of stations, SALBP-1 always leads to additional idle times if a smaller cycle time than the given one causes the same number of stations, and vice versa (cf. Figures 2.3 and 2.4). This shows the difficulties of pre-specifying such parameters as c and m. Only SALBP-E avoids installing respective inefficiencies into the system and shows opportunities of improving the production process.

2.2.1.4 Related Combinatorial Problems

Simple assembly line balancing problems are members of the general class of combinatorial optimization problems (cf. Domschke and Drexl (1998, pp. 113)). Hence, SALBP (and its versions) is related to various assignment, sequencing, grouping, and selection problems. In this section, we discuss such relationships.

The most obvious correspondence results from omitting the precedence constraints. Then SALBP-1 reduces to the well-known *bin packing problem*, which is to pack a given collection of items into a minimum number of equal-sized bins (cf. Wee and Magazine (1982)). SALBP-2 passes into the problem of scheduling m *identical parallel machines* with the objective of minimizing the makespan (cf. Coffman et al. (1984)). These two problems show the same type of duality as SALBP-1 and SALBP-2 (cf. Hochbaum and Shmoys (1987)).

By its relationship to the bin packing problem, the *capacitated vehicle routing problem* is associated with SALBP-1. One version of this problem consists of minimizing the number of identical vehicles with given capacity which are needed to deliver goods from a depot to a number of customers (cf. Graves and Lamar (1983), Labbé et al. (1991)). A vehicle routing problem related to SALBP-2 arises if the number of vehicles is fixed and the distance of the longest tour is to be minimized. Similarly, SALBP-2 is associated with *bottleneck transportation problems* (cf. Domschke (1997, chapter 2.3)).

Since SALBP includes subdividing the set of tasks into a number of subsets, it is associated with the class of *set partitioning problems*. Within these problems, each subset is weighted with a cost value. The objective is to find a partition of the considered set into disjoint subsets such that the total weight of the chosen subsets is minimized (cf. Balas and Padberg (1979)).

The cycle time constraints connect SALBP to the *zero-one knapsack problem* (cf. Martello and Toth (1990 a, chapter 2)). This problem is to decide on the contents of a knapsack with given capacity. Among a collection of items, each of which is defined by a capacity requirement (e.g., weight) and an utility value, a subset with a maximal total utility value has to be chosen such that the capacity of the knapsack is not exceeded.

SALBP is related to *one-machine scheduling problems*, because in both problem types a sequence for performing operations has to be determined (the assembly line may be interpreted as a single machine). Precedence constraints are partly reflected by so-called *heads* and *tails* of the operations (cf. Carlier (1982)). Furthermore, SALBP may be interpreted as a *single resource constrained project scheduling problem* (cf. Drexl (1990), Domschke and Drexl (1991), Klein and Scholl (1999)).

2.2.1.5 Complexity of SALBP

In general, the *complexity* of a problem concerns the requirements of resources like computation time and storage space needed for solving the problem with a computer program (representing a particular algorithm). Most commonly, the computation time requirements are considered as the dominant factor deciding whether or not an algorithm is efficient enough to solve a problem.

In order to formalize aspects concerning the time complexity of problems, the *theory of NP-completeness* has been developed starting with the fundamental work of Cook (1971). This theory is excellently presented in Garey and Johnson (1979). Furthermore, see Papadimitriou and Steiglitz (1982, chapter 15) or Pulleyblank (1989, pp. 437). Hence, we give only a short summary of the essential perceptions and relate them to SALBP.

When discussing complexity related issues, we first have to distinguish between the problem itself (also called *problem type*) and actual *problem instances* which are obtained by specifying particular values for all problem *parameters* (input data, like the task times). Usually, the time requirements depend on the actually considered problem instance, which is mainly characterized by its *size* (*input length,* length of a suitable encoding of the input data). If a fixed number of digits is used to represent parameter values, the input length is sufficiently characterized by the number of parameters, e.g., the number of tasks or precedence relations, respectively.

The time requirements of algorithms for a problem type are measured by a *time complexity function,* which specifies an upper bound on the computation time for any possible instance size, i.e., the *worst case* solution time. Usually, one is only interested in the order of time complexity functions. A function f(n) is said to be of *order O(g(n))* if $f(n) \le c \cdot g(n)$ with c being an arbitrary positive constant.

Concerning time complexity functions, two classes of algorithms are distinguished:

- *Polynomial time algorithms:* The computation time is bounded by a polynomial function which depends on the input length n. That is, the time complexity function is of order $O(n^p)$, where p denotes a positive constant.
- *Exponential time algorithms:* The computation time increases exponentially with increasing problem size. For example, the time complexity function may be of order $O(2^n)$.

Complexity of Problem Types

As a matter of convenience, the problem types are subdivided into decision problems and optimization problems. *Decision problems* have only two possible solutions, the answers "yes" and "no". They arise from optimization problems by replacing the objective function by the question whether or not a feasible solution exists with an objective function value not exceeding a particular constant (in the case of a minimization problem). Since the term decision problem is also used in a more general context (describing all situations in which decisions have to be made), we prefer the unusual term *feasibility problem*.

As stated above, SALBP-F can be formulated as a feasibility problem with the following question:

"Does a feasible task assignment with m stations and cycle time c exist?"

Note that the question implicitly contains that the sum of idle times must not exceed the constant (balance delay time) $m \cdot c - t_{sum}$.

The theory of NP-completeness is restricted to feasibility problems and is based on the definition of the following classes.

Definition 2.7. The *problem class NP* consists of all feasibility problems, the solutions of which can be verified by a polynomial time algorithm. That is, it takes polynomial time to check whether or not the answer is correct for a given "yes" solution. The class NP contains two subclasses:

The *class P* consists of those problems within NP which are solvable by a polynomial time algorithm.

The *class of NP-complete problems* (NPC) contains those problems within NP, for which no algorithm solving all instances in polynomial time is known. Furthermore, all problems in NPC can be *transformed* to any other one within this class in polynomial time. That means, whenever a polynomial time algorithm would be found for any of those problems, all problems of this class would be solvable in polynomial time, too.

Note that the classes P and NPC are disjoint only under the not yet proved assumption that P ≠ NP. The class NPC contains the "hardest" problems within NP. A feasibility problem π is proved to be NP-complete (i.e., to be member of NPC) by showing that it is in NP and can be transformed to another NP-complete problem π' in polynomial time. A transformation of a problem π to another problem π' is called *reduction* if π' is a special case of π.

SALBP-F is contained in the class NP, because the feasibility of a given task assignment is easily verified. SALBP-F reduces to the problem *partition* which is to decide whether or not a set of positive integral weights can be subdivided into two sets having the same sum of weights (cf. Karp (1972)). The reduction is shown for the special problem instance of SALBP-F with $m=2$, $c=t_{sum}/2$, and no precedence constraints. In this case, the answer "yes" for SALBP-F immediately results in a "yes" answer for partition, too.

The above argumentation is restricted to feasibility problems which are members of NP. However, the results can be applied to other problems outside NP (including optimization problems) as well. This is based on the fact, that any problem to which a problem of NPC can polynomially be transformed cannot be solved in polynomial time unless P = NP. Such problems are called *NP-hard.*

Since the optimization problems SALBP-1, SALBP-2, and SALBP-E, respectively, contain SALBP-F as their underlying decision (feasibility) problem as illustrated in Section 2.2.1.3, they are NP-hard. By solving an instance of one of these problems, at least one related instance of SALBP-F is solved, too.

Complexity of Problem Instances

The above discussion deals with the worst case complexity of problem types contained in several classes. Even though problem types are NP-complete or NP-hard, respectively, reasonable algorithms usually solve a large number of corresponding instances in short computation times, while others are computationally intractable in terms of available computation time. Therefore, one is interested in measures which indicate whether or not a certain problem instance will probably take much computation time. Such information may help deciding on the solution procedure to be used, especially if exact or heuristic methods are more likely to obtain reasonable solutions with acceptable computation times. However, such measures cannot exactly predict the time requirement (complexity) of an instance, but they show tendencies. Furthermore, all complexity statements for problem instances refer to certain solution procedures, because it is possible that instances which take much computation time with one algorithm are quickly solved by another one.

In the sequel, we describe characteristics of SALBP and argue their *expected* influences on the complexity of problem instances. In order to quantify those complexity aspects, we present some numerical measures of problem characteristics and discuss their suitability as complexity indicators.

Problem Characteristics. We assume that problems are solved by some enumeration method as described in the literature which builds loads for the stations in the order k = 1,...,m (cf. the forward procedures in Chapter 4). The following argumentation concerns optimization problems rather than SALBP-F.

Generally, it can be expected that problem instances which have one or both of the following properties are favourable with respect to computation time requirements.

- The number of optimal solutions is large, and the optimal solutions are widely spread in the solution space.
- The objective function value of the optimal solution is equal to a known lower bound.

If a problem instance has the first property, the direction of the search is not as important as in a case where only few optimal solutions exist which are located in a small part of the solution space. Since only one optimal solution has to be found, the probability of finding it very soon increases with the number of existing optimal solutions. If an optimal objective function value is equal to a lower bound, a search procedure can stop as soon as a respective solution is found. Otherwise, it may take a large amount of computation time to prove a known solution to be optimal.

The following characteristics of problem data tend to have an influence on the complexity of an actual problem instance, especially with respect to the above properties.

Number n of Tasks. As mentioned in Section 2.2.1.3, SALBP falls into the general class of sequencing problems (cf. Baker (1974)). Ignoring precedence constraints, there are n! feasible task sequences. Therefore, the complexity is expected to grow exponentially with increasing n if the precedence constraints are not too restrictive (see below).

Task Times and Cycle Time. If task times are small compared with the cycle time, there should be more station loads with low idle times and probably more optimal solutions than in case of large operation times, because combining small items is usually easier than combining large ones. Furthermore, optimal solutions tend to have objective function values equal to a lower bound. This may be advantageous for optimum seeking procedures as stated above. Additionally, it can be expected that operations with significantly varying task times (including rather small values) can be combined to form station loads with low idle times better than such with almost identical operation times. This is especially true if these times are rather large relative to the cycle time. However, an instance with identical operation times for all tasks is simple.

Precedence Constraints. These constraints have an influence which is twofold. On the one hand, they restrict the number of feasible sequences which may lead to the assumption that the complexity decreases with increasing *number of precedence relations.* On the other hand, such constraints may complicate the process of finding optimal solutions, because the loads of the stations 1,...,k determine the tasks which are assignable to the stations k+1,k+2, ...,m. Therefore, the decision on a current station load is more sensitive than it is in the absence of precedence constraints.

As a consequence, not only the quantity but also the *structure of precedence relations* has an influence on the complexity. For enumeration procedures which load the stations in the order k=1,2,... (as assumed above), it should be advantageous that only few alternative station loads exist for the early stations, especially with respect to the sensitivity of assignment decisions. This requires that the precedence graph has only few sources and some type of *diverging* structure, i.e., the number of direct successors tends to increase when approaching the sinks of the graph. Note that a graph is *strictly* diverging if it is an out-tree (cf. Definition 2.9, p. 42). For such structures, there exist only few tasks potentially assignable to the first stations, while more tasks get assignable to later stations. That is, the sets B_k of potentially assignable tasks tend to become larger for the later stations. Since the sensitivity of assignment decisions decreases from early to late stations due to the decreasing number of available (unassigned) tasks, it can be expected that problem instances with diverging or "almost-diverging" precedence structures are less complex than those with other structures (cf. Definition 2.10, p. 42).

Furthermore, the discussion implies that the planning direction of a solution procedure may considerably influence the time complexity of an instance. Note that SALBP instances may also be solved by considering the reverse precedence graph which may have a more favourable structure (cf. Section 4.1.4.3).

Number m of Stations. The influence of the number m of stations on the complexity is not unique. The less stations are available, the more combinations of tasks have to be examined for each station (with fixed n). As a consequence, one may expect that the complexity increases if m decreases. However, the number of stations available for the assignment of each task (length of the station interval) decreases, too, especially if there are precedence constraints. Additionally, the probability that the objective function value of the optimal solution is equal to a lower bound is large, if many different tasks are potentially assignable to a station. Note that the extreme cases m = 1 and m = n are trivial.

Measures of Complexity. In the following, we present simple numerical measures (besides the number n of tasks) of the above mentioned problem characteristics and discuss their suitability as indicators for the complexity of problem instances (cf., additionally, Baybars (1986 a)).

- **Order strength** (cf. Definition 2.2 as well as Mastor (1970), Bhattacharjee and Sahu (1990)): The order strength OS measures the relative number of precedence relations in the precedence graph. Problems with a large order strength are basically expected to be more complex than such with small OS values. However, this assumption is critical as mentioned in the above discussion on the influence of precedence constraints. This is underlined by the trivial case $OS = 1$, where only one task sequence is feasible. On the other hand, SALBP-1 and SALBP-2 reduce to the NP-hard bin packing and parallel machine scheduling problems (cf. Section 2.2.1.4), respectively, for $OS = 0$.
- **Flexibility ratio** (Dar-El (1973)): The flexibility ratio $FR \in [0, 1]$ is defined as the number of zero entries in the (transitive) precedence matrix divided by the total number of entries. This measure is equivalent to the order strength because of $FR = 1 - OS$ (cf. Elmaghraby and Herroelen (1980)).
- **West ratio** (Dar-El (1973)): The west ratio $WR = n/m$ is the average number of tasks per station. In case of SALBP-1 and SALBP-E, respectively, this value depends on the optimum number of stations or even a lower bound. The above discussion on the influence of m on the complexity implies that problem instances with small values of WR tend to be more complex than such with larger values.
- **Time interval** (Wee and Magazine (1981 b)): The interval $TI = [t_{min}/c, t_{max}/c] \subseteq [0,1]$ is a measure of the interrelation between the cycle time and the task times. For SALBP-2 and SALBP-E, respectively, this measure depends on the minimal cycle time or a lower bound. A small length of TI indicates that the task times vary only in a small range. Furthermore, the position of TI in the interval $[0,1]$ shows if task times are large or small relative to the cycle time c. Following the above argumentation, problems are expected to be relatively complex if TI is small and has a position near to the right border of $[0,1]$.
- **Time variability ratio:** A measure similar to the time interval is defined by $TV = t_{max}/t_{min}$. In contrast to TI, only one value is used to characterize a problem instance. Furthermore, this measure is independent of the cycle time and, therefore, reflects the time structure of a precedence graph rather than single problem instances. Small values of TV indicate that the operation times vary only in a small range or that the minimal task time is remarkably large. Following the above discussion on the time interval, the complexity of problem instances is expected to grow with decreasing values of TV.

The above measures have the major drawback that each of them only controls a part of the problem characteristics which may influence the complexity. Therefore, in the literature, it is proposed to choose a combination of measures (e.g. number of tasks, order strength, west ratio; cf., e.g., Johnson (1988)). Nevertheless, the interrelations between several characteristics which may enforce or balance their influences on the complexity cannot be considered. Furthermore, the order strength is only a very rough measure for the precedence relations, because it does not reflect their structure. Hence, there is a need for further measures which utilize the information provided by the precedence structure more intensively.

Measures of Precedence Structures. A suitable measure should incorporate the structure of the precedence graph together with the task times. As a consequence, such measures cannot consist of single values as proposed above. A possible approach to get more versatile measures may be obtained from subdividing the graph into several stages (see Definition 2.8) and characterizing these parts separately.

Definition 2.8 (cf. Jackson (1956)). The *depth* d(j) of node (task) $j = 1,...,n$ is defined by the maximal number of arcs in a path from a source node to node j in the precedence graph G. Obviously, the depth of the source nodes is 0. The maximal depth of a node determines the *depth of G* which is denoted by D(G). The graph G is subdivided into *stages* $h = 0,...,D(G)$. Each stage ST_h consists of all tasks having depth h, i.e., $ST_h = \{j \mid d(j) = h\}$ for $h = 0,...,D(G)$.

Characteristics of Stages. Among others, the following values characterize the stages ST_h with $h = 0,...,D(G)$:

- number of elements (cardinality): $C(ST_h) = |ST_h|$
- average operation time: $\bar{t}(ST_h) = \sum_{j \in ST_h} t_j / C(ST_h)$
- total number of direct followers of tasks at stage h: $F(ST_h) = \sum_{j \in ST_h} |F_j|$

Example. The precedence graph of Figure 2.2 (see p. 26) has depth $D(G) = 5$, because the paths from the sources 1 and 2 to the sink 11 have at most 5 arcs. The stages are $ST_0 = \{1,2\}$, $ST_1 = \{3,4,5\}$, $ST_2 = \{6,7\}$, $ST_3 = \{8,9\}$, $ST_4 = \{10\}$, $ST_5 = \{11\}$. They are characterized by the values in Table 2.2.

Table 2.2. Characteristics of Stages

h	0	1	2	3	4	5
$C(ST_h)$	2	3	2	2	1	1
$\bar{t}(ST_h)$	4	3.67	5	5	5	1
$F(ST_h)$	4	4	2	2	1	0

Complexity Trends. In order to estimate the expected complexity of a problem instance, one should examine the stage characteristics reflecting their respective influence. For this purpose, the stages may be interpreted as parts of the graph which are successively considered by a solution procedure (following the above assumption that it builds station loads in the order 1,2,...), because tasks at stage h+1 do not become available before at least one task at stage h is assigned.

However, the stage-dependent tendency of values is more important than single values. Following basic conjectures may help to classify problem instances with respect to their complexity. If $C(ST_h)$ has the tendency to increase, i.e., the number of tasks per stage tends to be larger for later than for earlier stages, the graph is (almost-) diverging as mentioned above. Hence, it can be expected that problem instances with increasing trends of $C(ST_h)$ require less computation times than those with decreasing or more general trends. Furthermore, the structure of the graph is described by the behaviour of $F(ST_h)$ which reflects the distribution of direct precedence relations among the stages. Obviously, an increasing tendency indicates an (almost-) diverging structure and, hence, less complex problems.

In case the average operation times $\bar{t}(ST_h)$ have a decreasing trend, the tasks at the early stages tend to have larger operation times, and/or $C(ST_h)$ tends to grow. Since combining large tasks is more difficult than combining small ones, it seems to be advantageous if large tasks occur in the earlier stages where the solution procedure has more degrees of freedom for assignment decisions. Therefore, it is expected that instances with decreasing trends of $\bar{t}(ST_h)$ are less complex than such with increasing trends.

The above statements are rather vague but may help to construct solution procedures. The first observation is that the planning direction (the order in which stations are loaded) may play an important role with respect to the computation time an instance requires with a particular algorithm. Our discussion above refers to forward oriented procedures which load the stations in the order 1,2,... of the station numbers. If the reverse precedence graph has an (almost-) diverging structure, i.e., the original graph tends to *converge* from many sources to few sinks, the application of a backward oriented procedure (working from the late to the early stations) may be favourable. The precedence graph of our example is converging rather than diverging (cf. the trends in Table 2.2). Furthermore, a graph may be combined of diverging and converging parts such that it is advantageous to consider the stations in a forward and a backward direction simultaneously. In Section 4.1.4.3, we call this proceeding a *bidirectional planning*.

Stage-independent Measures. The structure of the precedence graph may be characterized by more general measures which are not based on the subdivision into stages. In the sequel, we define a degree which allows precising the above stated property of an "almost-diverging" precedence structure. Furthermore, such measures like the *bandwidth* or the *diameter* of a graph may be considered in this context (cf., e.g., Voß (1990, pp. 161) and Foulds (1992, p. 31)).

Definition 2.9. A directed graph $G=(V,E)$ is called *out-tree* if it has a single source or root $r \in V$, and each other node $j \in V - \{r\}$ is connected to r by a single (unique) directed path from r to j (cf. Domschke (1997, p. 4)). Analogously, a graph is called *in-tree* if the reverse graph is an out-tree.

Definition 2.10. The number δ_j^- of arcs entering a node j is called *in-degree* of j. A non-cyclical directed graph $G=(V,E)$ with a single source r is transformed to an out-tree by arbitrarily eliminating $\delta_j^- - 1$ of the entering arcs for each node $j \in V - \{r\}$. Therefore, we may define a positive *degree of divergence* for G as follows:

$$div(G) := 1 - \sum_{j \in V - \{r\}} (\delta_j^- - 1) / |E|$$

In the case of $div(G) = 1$, G is an out-tree, i.e., it is *strictly diverging*. If div(G) is smaller than (but near to) 1, we call the graph *almost-diverging with parameter div(G)* in conformity with the term *almost-tree* (cf. Voß (1990, pp. 161)).

An analogous *degree of convergence* conv(G) is achieved by considering the reverse precedence graph.

Note that a precedence graph can be transformed into a single-source graph by adding a fictitious source node with operation time 0 and arcs to the original sinks.

To follow the discussion above, SALBP instances with a high degree of divergence (convergence) are expected to be less complex than such with a low degree for forward (backward) procedures. In the case where both degrees are small, the respective instance is expected to be more complex than one which has a high degree of either type. These reflections are utilized for constructing solution procedures for SALBP-1 and SALBP-2 which allow changing the planning direction flexibly depending on the structure of the precedence graph (cf. the Sections 4.1.4.3 and 4.2.4.3).

2.2.2 Minimizing the Number of Stations (SALBP-1)

We consider the most famous version of SALBP (SALBP-1 for short) which is to minimize the number of stations for a given cycle time (cf., e.g., Baybars (1986 a), Gagnon and Ghosh (1991)). That is, SALBP-1 arises from SALBP-F by introducing the number m of stations as a decision variable and supplementing the objective of minimizing m.

In the following sections, we present methods for computing lower and upper bounds on the objective function value as well as modifications of the mathematical programming formulations of Section 2.2.1.2 for SALBP-1.

2.2.2.1 Lower Bounds of SALBP-1

We describe several approaches for determining lower bounds on the number m of stations. Even though bound arguments may have algorithmic character, we present them in this problem oriented part of the work for two reasons. First, bounds are useful to reduce the size of a problem formulation. Secondly, bounds depend on the structure of problems and are fundamentally independent of solution concepts.

For the description of the bounds, we use the following notations where J is a subset of the set $V = \{1,...,n\}$ of all tasks. In order to express the operation times in terms of stations, we transform the integer task times to fractional station requirements $p_j \in [0,1]$ defined as:

$$p_j := t_j / c \qquad \text{for } j = 1,...,n$$

Additionally, we define subsets of the set J containing operations, the station requirements of which fall into certain ranges:

$$J(x,y) := \{j \in J \mid x < p_j < y\} \qquad \text{for } J \subseteq V$$

$$J(x,y] := \{j \in J \mid x < p_j \leq y\} \qquad \text{for } J \subseteq V$$

$$J[x,y] := \{j \in J \mid x \leq p_j \leq y\} \qquad \text{for } J \subseteq V$$

For illustration purposes, we reconsider our example of Chapter 1 which is presented in Figure 2.5 and is characterized by $t_{sum} = 48$, $t_{max} = 9$, and $t_{min} = 2$. Furthermore, we assume that a cycle time $c = 10$ should be realized.

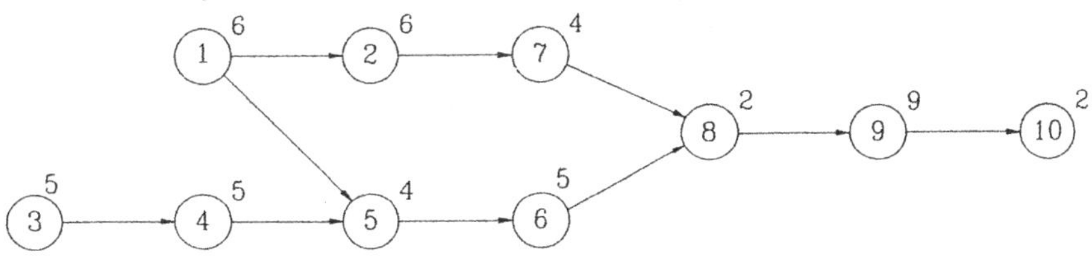

Figure 2.5. Example Precedence Graph (cf. Figure 1.1)

Bound LB_1. The most obvious bound follows from the inequality $m \cdot c \geq t_{sum}$, i.e., the total available time on the line must not be smaller than the total work content:

$$LB_1 := \lceil t_{sum} / c \rceil = \lceil \sum_{j=1}^{n} p_j \rceil \tag{2.23}$$

In the literature, LB_1 is called *theoretical minimum number of stations* (cf., e.g., Baybars (1986 a)).

Complexity. Computing LB_1 requires O(n) time due to the determination of t_{sum}.

Example. $LB_1 = \lceil 48/10 \rceil = 5.$

Two further bound arguments are based on counting tasks with certain operation times (cf. Johnson (1988)).

Bound LB_2. The number of tasks with their operation times exceeding c/2 (i.e., $p_j > 1/2$) is a lower bound on m, because all of those tasks have to be assigned to different stations. This bound can be strengthened by adding half of the number of tasks (rounded up to the next integer if necessary) with task time c/2 because two of them may share one station.

In terms of our above notation, LB_2 is computed for J = V as follows:

$$LB_2 := |J(\tfrac{1}{2},1]| + \left\lceil \tfrac{1}{2} \cdot |J[\tfrac{1}{2},\tfrac{1}{2}]| \right\rceil \tag{2.24}$$

Complexity. LB_2 has time complexity O(n), because each task is considered once.

Example. Three tasks exceed half of the cycle time while three further tasks have operation time c/2. The resulting value of LB_2 is $3 + \lceil 3/2 \rceil = 5$.

Bound LB_3. It generalizes LB_2 with respect to thirds of the cycle time and is computed by adding up weights which are determined as follows: All tasks j with $p_j > 2/3$ are given the weight 1, because they cannot be combined with any other one of the considered tasks. Operations with $p_j \in (1/3, 2/3)$ get the weight 1/2, because two of them may share a station. According to their station requirements, the tasks with $p_j = 1/3$ and $p_j = 2/3$ are weighted with 1/3 and 2/3, respectively.

With J = V the bound LB_3 is defined through

$$LB_3 := \left\lceil |J(\tfrac{2}{3},1]| + \tfrac{2}{3} \cdot |J[\tfrac{2}{3},\tfrac{2}{3}]| + \tfrac{1}{2} \cdot |J(\tfrac{1}{3},\tfrac{2}{3})| + \tfrac{1}{3} \cdot |J[\tfrac{1}{3},\tfrac{1}{3}]| \right\rceil . \tag{2.25}$$

Complexity. Obviously, LB_3 has a linear time complexity O(n), too.

Example. Only task 9 exceeds 2/3c, while seven tasks have operation times between 1/3c and 2/3c. Hence, $LB_3 = \lceil 1 \cdot 1 + 1/2 \cdot 7 \rceil = 5$.

Bound LB_4 (cf. Johnson (1988), Domschke et al. (1997, pp. 216)). The lower bound LB_4 is based on a relaxation of SALBP-1 to a special *one-machine scheduling problem* (cf. Section 2.2.1.4). Tasks are interpreted as jobs j = 1,...,n with "processing times" $p_j = t_j / c$ which have to be successively performed on one machine. After processing job j, a certain amount of time (which is called *tail* of j) is necessary before the job is terminated. The objective of the one-machine problem is to find a sequence of all jobs for which the *makespan*, i.e., the time interval from the start of the first to the termination of the last job, is minimized. An optimal solution for this problem is achieved by processing the jobs in the order of non-increasing tails. Assuming $<h_1,...,h_n]$ to be such a task ordering, the minimum makespan is computed by

$$\max \{p_{h_1}+n_{h_1}, p_{h_1}+p_{h_2}+n_{h_2},..., p_{h_1}+...+p_{h_n}+n_{h_n}\}. \tag{2.26}$$

In the case of SALBP-1, a tail n_j of task j is a lower bound on the number of stations (not necessarily integral) needed by its successors in F_j^*. Let us first assume that such tails n_j are given for each task. Then a lower bound LB_4 on m can be computed by solving the respective one-machine relaxation, i.e., by sorting the tasks according to non-increasing tails and computing (2.26). Since tails n_j are not immediately available, the bound LB_4 is determined through a recursive computation of tails.

For determining the tails, we use modifications of the bounds LB_1, LB_2, and LB_3, each defined for a subset J of tasks:

$$LB_1(J) := \sum_{h \in J} p_h \tag{2.27}$$

$$LB_2(J) := |J(\tfrac{1}{2},1]| + \tfrac{1}{2} \cdot |J[\tfrac{1}{2},\tfrac{1}{2}]| \tag{2.28}$$

$$LB_3(J) := |J(\tfrac{2}{3},1]| + \tfrac{2}{3} \cdot |J[\tfrac{2}{3},\tfrac{2}{3}]| + \tfrac{1}{2} \cdot |J(\tfrac{1}{3},\tfrac{2}{3})| + \tfrac{1}{3} \cdot |J[\tfrac{1}{3},\tfrac{1}{3}]| \tag{2.29}$$

Recursive Computation of Tails. The precedence graph is complemented by a fictitious *source node* representing a task 0 with operation time $t_0:=0$. This node is connected by arcs (0,j) to the nodes of all tasks j with $P_j = \phi$, i.e., to the original sources of the graph. The tails are determined in the order $j=n,n-1,...,1,0$. For all operations j without successors (i.e., $F_j = \phi$), n_j is set to 0. Otherwise, lower bounds n_{j1}, n_{j2}, and n_{j3} on the station requirement of F_j^* are derived as follows.

$$n_{j1} := LB_1(F_j^*) \tag{2.30}$$

$$n_{j2} := \begin{cases} LB_2(F_j^*) & \text{if } p_j \geq 1/2 \text{ or } LB_2(F_j^*) \notin \mathbb{N} \\ LB_2(F_j^*) - 1/2 & \text{otherwise} \end{cases} \tag{2.31}$$

$$n_{j3} := \begin{cases} LB_3(F_j^*) & \text{if } p_j \geq 2/3 \\ LB_3(F_j^*) - 1/3 & \text{otherwise} \end{cases} \tag{2.32}$$

The formulae for n_{j2} and n_{j3}, respectively, need some explanations. In some cases, the bounds LB_2 and LB_3 overestimate the station requirement of F_j^*. If task j additionally fits into the first of the stations at least occupied by its successors, the bounds have to be adjusted. For example, assume that a task has two successors h_1 and h_2 with $p_{h_1} = p_{h_2} = (1+\epsilon)/3$. Then, $LB_3(\{h_1,h_2\}) = 1$ for $0<\epsilon<1$. In this case, the actual station requirement is overestimated by $(1-2\epsilon)/3$. Hence, it is sufficient to reduce the value of LB_3 by 1/3 in the worst case. A similar argument holds for LB_2. Note that the adjustments given may be refined by examining further cases.

A further lower bound n_{j4} may be derived by solving the one-machine relaxation related with the set F_j^*. For this purpose, we use the already determined tails n_h of the tasks $h \in F_j^*$ (note that the tails are computed following a reverse topological ordering). By assuming $<h_1, h_2, ..., h_r]$ to be an ordered list of the $r := |F_j^*|$ successors of j according to non-increasing tails, we obtain n_{j4} from computing the minimum makespan of the related one-machine problem (cf. (2.26)):

$$n_{j4} := \max \{p_{h_1} + n_{h_1},\ p_{h_1} + p_{h_2} + n_{h_2}, ...,\ p_{h_1} + p_{h_2} + ... + p_{h_r} + n_{h_r}\} \tag{2.33}$$

Since n_{j1} through n_{j4} provide lower bounds on the station requirement of F_j^*, the tail of task j can be determined by

$$n_j := \max \{ n_{j1}, n_{j2}, n_{j3}, n_{j4} \}. \tag{2.34}$$

This value may be strengthened whenever task j does not additionally fit into the first of these stations, i.e., $p_j + n_j > \lceil n_j \rceil$. In this case, a non-integral n_j is rounded up to the nearest integer value. Finally, the bound LB_4 is computed by rounding up the tail n_0 of the fictitious source.

```
Procedure Tails;
begin
   for j := n downto 0 do
      if F_j = ∅ then n_j := 0
      else begin
         sort tasks of F*_j according to non-increasing tails to obtain
         the list <h_1, h_2,..., h_r] with r := | F*_j | elements;
         compute n_j1, n_j2, n_j3, and n_j4 by applying (2.30) through (2.33);
         n_j := max { n_j1, n_j2, n_j3, n_j4 };
         if n_j < ⌈n_j⌉ and p_j + n_j > ⌈n_j⌉ then n_j := ⌈n_j⌉
      end;
   end;
   LB_4 := ⌈n_0⌉;
end;
```

Figure 2.6. Computation of Tails

Figure 2.6 shows a pseudo-code formulation of the described procedure for computing the tails n_j as well as the lower bound LB_4.

Complexity. LB_4 takes time $O(n^2)$, because the computation of tails requires examining the set of successors for each task j.

Example. We consider the example of Figure 2.5 (p. 43) and demonstrate the computation of tails. Table 2.3 shows the computational steps and the interim results.

Table 2.3. Station Requirements and Tails of Tasks

j	10	9	8	7	6	5	4	3	2	1	0
p_j	0.2	0.9	0.2	0.4	0.5	0.4	0.5	0.5	0.6	0.6	0
n_{j1}	0	0.2	1.1	1.3	1.3	1.8	2.2	2.7	1.7	3.2	4.8
n_{j2}	0	0	0.5	0.5	1	1.5	1.5	2	1	2.5	4.5
n_{j3}	0	0	0.67	0.67	0.67	1.17	1.67	2.17	1.17	2.67	4.17
n_{j4}	0	0.2	1.9	2.2	2.2	2.7	3.4	3.9	2.6	4.1	5.7
n_j	0	0.2	1.9	2.2	2.2	2.7	3.4	3.9	2.6	4.1	5.7
rounded	0	1	2	2.2	2.2	3	3.4	4	3	4.1	(6)

We explain the computation of n_{14} and n_{04}. Since the successors of task 1 build the ordered list $<2,5,6,7,8,9,10]$, we obtain

$$n_{14} := \max\{0.6+3, 1+3, 1.5+2.2, 1.9+2.2, 2.1+2, 3+1, 3.2+0\} = 4.1 .$$

Due to the ordered list $<1,3,4,2,5,6,7,8,9,10]$ of all tasks, the tail of task 0 is

$$n_{04} := \max\{0.6+4.1, 1.1+4, 1.6+3.4, 2.2+3, 2.6+3, 3.1+2.2, 3.5+2.2, 3.7+2, 4.6+1, 4.8+0\} = 5.7 .$$

By rounding up n_0, we get the lower bound $LB_4 = 6$ for the entire problem.

Remark 2.2. The computation of LB_4 includes the bounds LB_1 through LB_3. Furthermore, the determination of n_{j1} is superfluous, because (2.33) contains the term for n_{j1}, and, hence, $n_{j4} \geq n_{j1}$ holds for all j. Even though the bounds LB_2 and LB_3 are not successful in our example, they may provide the best bounds for other problem instances, especially if more large tasks (with $p_j > 1/2$ and $p_j > 1/3$, respectively) are available.

Generalized Bound LB_4'. The bound argument LB_4 can be applied to the reverse precedence graph resulting in so-called *heads* a_j of the tasks $j = 1,...,n$. By analogy with tails, a head a_j is a lower bound on the number of stations required for the predecessors of task j (in the original graph). Note that tails of the reverse graph are heads of the original graph, and vice versa. Hence, the head a_{n+1} of a fictitious sink node n+1 with $t_{n+1}=0$ (connected with all original sinks) determines a lower bound $\lceil a_{n+1} \rceil$.

A further bound is obtained by $Z=\max\{a_j+p_j+n_j \mid j=1,...,n\}$. Note that Z is a simple lower bound on the minimum makespan of a related NP-hard one-machine scheduling problem with heads *and* tails (cf. Carlier (1982) as well as Section 2.2.1.4).

By means of this knowledge, a generalized bound LB_4' is defined through

$$LB_4' := \max\{\lceil a_{n+1} \rceil, \lceil n_0 \rceil, \lceil Z \rceil\} \tag{2.35}$$

Example. Table 2.4 shows the resulting heads and tails for all tasks of our example (the latter values being repeated from Table 2.3). The head $a_{11}=5.2$ of the fictitious sink node 11, the tail $n_0=5.7$ of the fictitious source node 0, and $Z=a_9+p_9+n_9=5.9$ lead to $LB_4'=6$.

Table 2.4. Heads and Tails of Tasks

j	0	1	2	3	4	5	6	7	8	9	10	11
p_j	0	0.6	0.6	0.5	0.5	0.4	0.5	0.4	0.2	0.9	0.2	0
a_j	-	0	1	0	0.5	1.6	2	1.6	3.5	4	5	5.2
n_j	5.7	4.1	3	4	3.4	3	2.2	2.2	2	1	0	-

Bound LB_5. A simple lower bound is obtained by means of earliest and latest stations (cf. (2.1) and (2.2) in Section 2.2.1.1). Since m is a variable in SALBP-1, we denote the latest stations defined in (2.2) by $L_j(m)$.

A necessary condition for the existence of a feasible solution with m stations is $E_j \le L_j(m)$ for each task j. Otherwise, at least one task cannot be assigned to any station. Hence,

$$LB_5 := \min\{m \mid L_j(m) \ge E_j \text{ for } j=1,...,n\} \tag{2.36}$$

defines a lower bound on the number of stations (cf. Saltzman and Baybars (1987)).

Complexity. The bound can easily be computed in linear time by using the equation $L_j(m+1)=L_j(m)+1$.

Table 2.5. Earliest and Latest Stations

j	1	2	3	4	5	6	7	8	9	10
t_j	6	6	5	5	4	5	4	2	9	2
E_j	1	2	1	1	2	3	2	4	5	5
$L_j(5)$	2	3	2	3	3	4	4	4	4	5
$L_j(6)$	3	4	3	4	4	5	5	5	5	6

Example. The bound LB_5 is computed starting with $m = LB_1 = 5$. Table 2.5 contains the resulting values for E_j and $L_j(5)$. Due to $E_9 > L_9(5)$, task 9 cannot be assigned to any of 5 stations, and m is increased to 6. Since the condition $L_j(6) \geq E_j$ holds for all tasks j, LB_5 gets the value 6.

Improved Earliest and Latest Stations. The computation of E_j and $L_j(m)$ as well as LB_5 may be strengthened by using further bounds (instead of LB_1) to estimate the station requirements of the reduced problems with tasks $\{j\} \cup P_j^*$ as well as $\{j\} \cup F_j^*$. Generally, this is achieved by using heads and tails of tasks which bound the number of stations required for P_j^* and F_j^*, respectively. Now, earliest and latest stations can be obtained by (2.37) and (2.38) using $p_j = t_j / c$.

$$E_j^! := \lceil a_j + p_j \rceil \qquad \text{for } j = 1,...,n \tag{2.37}$$

$$L_j^!(m) := m + 1 - \lceil p_j + n_j \rceil \qquad \text{for } j = 1,...,n \tag{2.38}$$

Complexity. The computational effort for determining (2.37) and (2.38) depends on the methods used for obtaining the heads and tails. If the bound $LB_4^!$ is computed anyway, only linear time is additionally needed.

Example. The resulting improved values for our example are presented in Table 2.6 (with a_j and n_j from Table 2.4). While only task 9 is restricted to one station by E_j and $L_j(6)$ in Table 2.5, the tasks 8, 9, and 10 are fixed to a single station by $E_j^!$ and $L_j^!(6)$. Furthermore, all tasks except task 9 show smaller station intervals.

Table 2.6. Improved Earliest and Latest Stations

j	1	2	3	4	5	6	7	8	9	10
E_j'	1	2	1	1	2	3	2	4	5	6
$L_j'(6)$	2	3	2	3	3	4	4	4	5	6

Bound LB_6. This bound argument combines and extends the logics of LB_2 and LB_3 (cf. Labbé et al. (1991) and Berger et al. (1992)). LB_6 is determined by the procedure in Figure 2.7.

The first three steps compute a lower bound on the number of stations required by the tasks in $X = J(1/2, 1]$ and $Y = J(1/3, 1/2]$.

Step 1 considers that each of the d_1 tasks of the first subset needs a station for its own, because these tasks cannot be combined with each other.

Step 2 computes the maximum number of tasks in the second subset, which can be combined with a task of the first subset. This problem may be formulated as a special matching problem, which consists of finding a matching of maximal cardinality in a bipartite graph defined as follows. The graph contains the two disjoint node sets X and Y and edges (h,j) with weight 1 for each combination of tasks $h \in X$ and $j \in Y$ which fulfills $p_h + p_j \le 1$. That is, the tasks h and j are connected if they can be combined in a station.

This bipartite matching problem with unit weights can be solved in linear time by step 2 because of the special structure of the underlying bipartite graph: Due to the definition of edges, all nodes of Y to which a node $h_1 \in X$ is connected have an edge in common with all nodes $h_2 \in X$ with $p_{h_2} \le p_{h_1}$, too.

Procedure LB_6:

Step 1. Assign the tasks in $J(\frac{1}{2}, 1]$ to the stations $1, \dots, d_1 := |J(\frac{1}{2}, 1]|$ in the order of non-increasing operation times.

Step 2. Consider the tasks in $J(\frac{1}{3}, \frac{1}{2}]$ in the order of non-decreasing operation times. Successively assign these tasks to the earliest of the first d_1 stations which shows enough idle time until all tasks are assigned or none of the d_1 stations remains.

Step 3. Let d be the number of tasks not assigned in Step 2. Then at least $d_2 := \left\lceil \frac{d}{2} \right\rceil$ stations are additionally needed for these tasks.

Step 4. Compute a bound d_3 on the station requirement for the tasks in $J[0, \frac{1}{3}]$:

$$d_3 := \max \{0, \max \{ d_3(q) \mid q \in \{p_j \mid j \in J[0, \tfrac{1}{3}]\} \} \}$$

Result. $LB_6 := \lceil d_1 + d_2 + d_3 \rceil$

Figure 2.7. Computation of LB_6

Step 3 computes a lower bound on the station requirement of those tasks of Y not assigned in step 2. At most two of them may share a station.

Step 4 considers all tasks with $p_j \le 1/3$. For each value $q \in [0, 1/3]$ realized by some p_j, a bound on the (additional) station requirement of $J[q, 1/3]$ is given by

$$d_3(q) := \sum_{j \in J[q, 1-q]} p_j - |J(\tfrac{1}{2}, 1-q]| - d_2, \qquad (2.39)$$

due to the equation $J[q, 1/3] = J[q, 1-q] - J(\frac{1}{2}, 1-q] - J(\frac{1}{3}, \frac{1}{2}]$.

Complexity. If the tasks are already ordered according to non-increasing operation times, bound LB_6 takes linear time (cf. Martello and Toth (1990 b)). The sorting of the tasks takes time $O(n \log n)$.

Example. We consider a problem with 9 tasks, cycle time $c = 10$, and task times 8, 8, 7, 6, 5, 5, 4, 3, 3. The structure of the precedence graph is arbitrary, because it is not exploited by LB_6. The sum of task times is 49, and $LB_1 = 5$. The value of both LB_2 and LB_3 is 5. LB_6 is computed as follows. Four tasks have operation times exceeding c/2 and cannot share a station with each other ($d_1 = 4$). Three of the remaining tasks exceed c/3. Among these, only that with task time 4 can be combined with the smallest of the first four tasks. The other two need at least one additional station ($d_2 = 1$). The tasks with task time 3 can only be combined with tasks not exceeding an operation time of 7. The sum of times of all such tasks is 33, which is equivalent to a station requirement of 33/10= 3.3. This includes 2 stations for the tasks with operation times 6 and 7 as well as one station for the tasks with task time 5 (cf. (2.39)). Thus, 0.3 stations remain ($d_3 = 0.3$), and $LB_6 = \lceil d_1 + d_2 + d_3 \rceil = 6$ follows.

Remark 2.3. The bounds LB_2, LB_3, and LB_6 are lower bounds for the related bin packing problem as well, because they do not utilize the precedence relations (cf. Section 2.2.1.4).

Bound LB_7. As discussed in Section 2.2.1.3, the problem versions SALBP-1 and SALBP-2 are in some sense *dual* to each other. Hence, one may use lower bounds for SALBP-2 in order to compute bounds for SALBP-1 within a search procedure (note that only the reverse direction is considered in the literature up to now).

For ease of presentation, we assume that tasks are numbered according to non-increasing operation times, i.e., $t_j \geq t_{j+1}$ for all $j = 1,...,n-1$. We start with the sharpest lower bound on the number of stations which is actually known, e.g., $m := \max \{LB_1,...,LB_6\}$. For this trial value of m, a lower bound on the cycle time is computed by a bound argument for SALBP-2 (see below). If this bound is larger than the given cycle time c, the value of m is increased by 1, and the process is repeated. Otherwise, the actual value of m provides a lower bound on the number of stations for SALBP-1.

Given m (< n) stations, the *lower bound on the cycle time* is determined as follows (cf. bound LC_2 in Section 2.2.3.1 for a more detailed discussion). Consider the m+1 largest tasks 1,...,m+1. A lower bound on the cycle time for this subproblem is $t_m + t_{m+1}$, which is the sum of the two smallest task times, because at least one station contains two tasks. In general, a lower bound on the cycle time is obtained by

$$c(m) := \max \{ \sum_{i=0}^{k} t_{k \cdot m+1-i} \mid k = 1,...,\lfloor (n-1)/m \rfloor \}. \tag{2.40}$$

By means of (2.40), the bound LB_7 for SALBP-1 can be formulated as

$$LB_7 := \min \{ m \mid c(m) \leq c \}. \tag{2.41}$$

Complexity. The computation of LB_7 may involve a number of linear time iterations on m to find c(m). Theoretically, the number of iterations is bounded by

$LB_1 - 1$ in the worst case (cf. Section 2.2.2.2, where theoretical upper bounds are described) but, usually, only few iterations are necessary. Furthermore, each iteration but the last results in an improvement of the lower bound.

Example. An example might be as follows. All of 50 tasks have an operation time of 3, and the cycle time is 10. No precedence constraints have to be observed. The values of LB_1 and LB_6 are $\lceil 150/10 \rceil = 15$, the bounds LB_2 and LB_3 are 0. For m = 15, a lower bound on the cycle time is

$$c(15) := \max\{t_{15}+t_{16}, t_{29}+t_{30}+t_{31}, t_{43}+t_{44}+t_{45}+t_{46}\} = 12 > 10.$$

The value of c(16) is also 12 due to $t_{46}+...+t_{49}$, whereas $c(17) = 9 < 10$. Hence, $LB_7 = 17$. The optimal solution consists of 16 stations with three and one station with two tasks.

2.2.2.2 Upper Bounds of SALBP-1

We discuss (theoretical) upper bounds on the number of stations for SALBP-1. These bounds are helpful in reducing the number of constraints and variables of model formulations and may be used as initial values for solution procedures.

Bound UB_1. The most simple upper bound is $UB_1 := n$, because assigning each task to a separate station results in a feasible solution, whenever the necessary condition for feasibility ($c \geq t_{max}$) is fulfilled.

Bound UB_2. A more sophisticated upper bound follows from the fact that in at least one optimal solution with m stations the loads of the first m - 1 stations are *maximal* (cf., e.g., Hackman et al. (1989) as well as Definition 4.3, p. 115). That is, the load of no station k < m can be extended by moving a task from a later station to k. Note that the following argumentation utilizes the integrality of the cycle time and the task times.

A necessary condition for the maximality of the first m - 1 station loads is:

$$t(S_k) + t_{max} > c, \text{ or equivalently, } t(S_k) \geq c + 1 - t_{max} \qquad \text{for } k = 1,...,m-1 \quad (2.42)$$

By summing up the inequalities (2.42), we obtain

$$\sum_{k=1}^{m-1} t(S_k) \geq (m-1) \cdot (c + 1 - t_{max}). \qquad (2.43)$$

Since m stations are needed, i.e., $t(S_m) \geq 1$, the inequality $t_{sum} \geq \sum_{k=1}^{m-1} t(S_k) + 1$ holds. Using this expression and the integrality of m, (2.43) can be transformed to the upper bound UB_2:

$$UB_2 := \lfloor (t_{sum} - 1)/(c + 1 - t_{max}) \rfloor + 1 \qquad (2.44)$$

Bound UB_3. A necessary condition for the optimality of a solution is:

$$t(S_k) + t(S_{k+1}) > c, \text{ or equivalently,}$$

$$t(S_k) + t(S_{k+1}) \geq c+1 \qquad \text{for } k = 1,\dots,m-1 \qquad (2.45)$$

The conditions (2.45) simply assure that no pair of adjacent stations k and k+1 can be combined to a single station. Summing up the inequalities results in:

$$t(S_1) + 2 \cdot \sum_{k=2}^{m-1} t(S_k) + t(S_m) \geq (m-1) \cdot (c+1) \qquad (2.46)$$

Since the left hand side of (2.46) is not larger than $2 \cdot t_{sum} - 2$, we obtain

$$UB_3 := \lfloor (2 \cdot t_{sum} - 2)/(c+1) \rfloor + 1. \qquad (2.47)$$

With UB_1, UB_2, and UB_3 we get the following theoretical upper bound on the number of stations:

$$UB := \min \{UB_1, UB_2, UB_3\} \qquad (2.48)$$

Complexity. All bounds can be computed in linear time.

Example. We consider the example of Figure 2.5 (p. 43) and obtain $UB = 9$ through $UB_1 = 10$, $UB_2 = 24$, and $UB_3 = 9$. Even though UB_2 gives an inferior bound in this case, it results in sharper bounds, whenever t_{max} is small relative to c. In the simple case of unit operation times, i.e., $t_{max} = 1$, it can be shown that UB_2 is equal to LB_1.

2.2.2.3 Mathematical Formulations of SALBP-1

We describe how models for SALBP-F have to be modified in order to formulate SALBP-1. First, m has to be introduced as a variable instead of a parameter. Secondly, an objective function expressing the minimization of m has to be added.

The first aspect requires an upper bound $\bar{m}$ on the number of stations, which can be determined by the methods described in Section 2.2.2.2. In all three formulations of Section 2.2.1.2, $L_j(\bar{m})$ has to be used instead of L_j. Thus, the value of $\bar{m}$ influences the number of the variables. Furthermore, $\bar{m}$ has to be used instead of m in the definition of the restrictions (2.5) and (2.10) in formulations 1 and 2, respectively.

Different types of objective functions are considered in the literature. Early approaches use a cost explosion to penalize additional stations, i.e., the coefficients in the objective function grow exponentially with increasing station numbers (see Bowman (1960), Thangavelu and Shetty (1971)). A simpler objective function is proposed by Patterson and Albracht (1975). It is based on introducing a fictitious sink into the precedence graph, whenever the graph contains several original sinks. Without loss of generality, we assume that task n is such a unique sink. Then the objective consists of minimizing the number of that station to which task n is assigned, because this is the latest used station. Hence, the following objective function can be supplemented to formulation 1 with the constraints (2.4) - (2.7):

$$\text{Minimize } F_1(x) = \sum_{k \in SI_n} k \cdot x_{nk} \tag{2.49}$$

In formulation 2 which is defined by the constraints (2.9) - (2.13), the following objective function can be used (note that $L_n = \bar{m}$):

$$\text{Minimize } F_2(y) = \bar{m} + 1 - \sum_{k \in SI_n} y_{nk} \tag{2.50}$$

Two alternative simple objective functions of formulation 3 which is defined by the constraints (2.15) - (2.22) are:

$$\text{Minimize } F_3(b,w,z) = z_n \quad \text{or} \quad \text{Minimize } F_3(b,w,z) = b_n \tag{2.51}$$

2.2.3 Minimizing the Cycle Time (SALBP-2)

In this section, we consider the version of SALBP which minimizes the cycle time for a given number of stations. This problem is called SALBP-2 and modifies SALBP-F by introducing the cycle time c as a decision variable and supplementing the objective of minimizing c.

In the following sections, we propose lower and upper bounds on c. Furthermore, we show how to modify mathematical formulations for SALBP-F. For illustration purposes, we take a slight modification of the above example. It is defined by the precedence graph in Figure 2.8 and a given number of $m = 5$ stations. The task times are characterized by $t_{sum} = 50$, $t_{min} = 2$, and $t_{max} = 9$.

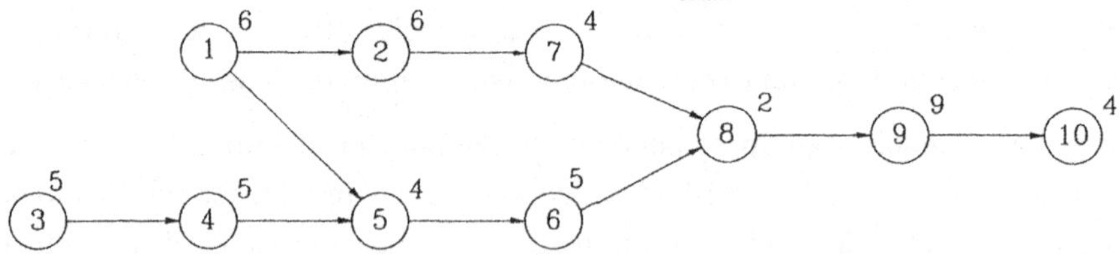

Figure 2.8. Example Precedence Graph

2.2.3.1 Lower Bounds of SALBP-2

In order to distinguish between bounds for SALBP-1 and SALBP-2, we use the abbreviation LC for lower bounds on the cycle time. As in the case of SALBP-1, bounds for SALBP-2 can be obtained by utilizing relationships to other combinatorial optimization problems (cf. Section 2.2.1.4). For example, SALBP-2 passes into the problem of scheduling jobs (tasks) on identical parallel machines (stations) with the objective of minimizing the makespan (cycle time) by omitting the precedence constraints. Since this parallel machine problem is NP-hard, too, only lower bounds for this problem are utilized to get SALBP-2 bounds.

Bound LC_1. By analogy with LB_1, a simple lower bound on the cycle time follows from the necessary feasibility condition $m \cdot c \geq t_{sum}$. Furthermore, the indivisibility of tasks requires that $c \geq t_{max}$. Hence, a lower bound is given by

$$LC_1 := \max\{t_{max}, \lceil t_{sum}/m \rceil\}. \tag{2.52}$$

This bound can be stated in terms of the related parallel machine scheduling problem as follows. LC_1 results from allowing job *preemption*, i.e., processing of a job may be interrupted and continued on another machine. Furthermore, jobs cannot be processed on two machines simultaneously. A corresponding solution of the preemptive parallel machine problem (which may be infeasible for SALBP-2) is generated by successively adding tasks (according to a topological ordering) to station 1, until the station time exceeds LC_1. In this case, the task which does not completely fit is split up between the machines 1 and 2 such that the first is filled exactly up to LC_1. The same procedure is applied to the stations 2,...,m (cf. McNaughton (1959)).

Complexity. The bound LC_1 is computed in linear time.

Example. $LC_1 = \max\{9, \lceil 50/5 \rceil\} = 10$. Figure 2.9 shows a corresponding preemptive solution which is obtained by considering the tasks in the order of their numbers. The solution is infeasible for SALBP-2, because it contains split tasks.

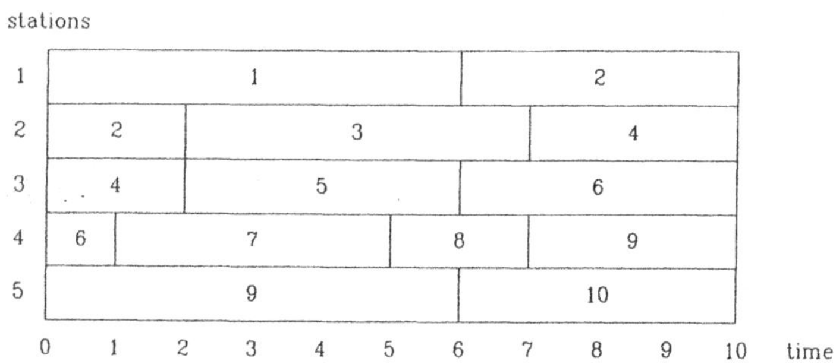

Figure 2.9. Infeasible Lower Bound Solution

Bound LC_2. Another lower bound for the related parallel machine problem is obtained as follows. In order to facilitate the presentation, we assume that the tasks are numbered according to non-increasing operation times, i.e., $t_j \geq t_{j+1}$ for $j = 1,...,n-1$. Consider the m+1 largest tasks 1,...,m+1. A lower bound on the cycle time for this reduced problem is $t_m + t_{m+1}$, the sum of the two smallest task times, because at least one station contains two tasks. For the $2m+1$ largest tasks, a lower bound is given by the sum of the three smallest operation times, $t_{2m-1} + t_{2m} + t_{2m+1}$.

In general, a lower bound LC_2 is obtained by

$$LC_2 := \max\{ \sum_{i=0}^{k} t_{k\cdot m+1-i} \mid k=1,...,\lfloor (n-1)/m \rfloor \}. \qquad (2.53)$$

Complexity. The bound LC_2 has the time complexity O(n log n) due to the required sorting of the tasks. For sorted tasks, the evaluation of (2.53) takes linear time.

Example. We get $LC_2 = 5+5 = 10$, because the six largest tasks are 9, 1, 2, 3, 4, and 6 with the operation times 9, 6, 6, 5, 5, and 5.

Earliest and Latest Stations. These values are defined for a *trial cycle time* c as follows (cf. (2.1) and (2.2) in Section 2.2.1.1, p. 28):

$$E_j(c) := \left\lceil (t_j + \sum_{h \in P_j^*} t_h)/c \right\rceil \qquad \text{for } j=1,...,n \qquad (2.54)$$

$$L_j(c) := m+1 - \left\lceil (t_j + \sum_{h \in F_j^*} t_h)/c \right\rceil \qquad \text{for } j=1,...,n \qquad (2.55)$$

By means of the *station intervals* $SI_j(c) := [E_j(c), L_j(c)]$, the set of tasks *potentially assignable* to station k = 1,...,m is defined by:

$$B_k(c) := \{ j \mid k \in SI_j(c) \} \qquad \text{for } k=1,...,m \qquad (2.56)$$

By analogy with (2.37) and (2.38) in Section 2.2.2.1 (p. 49), sharper values $E_j^!(c)$ and $L_j^!(c)$ are obtained by

$$E_j^!(c) := \lceil a_j(c) + p_j(c) \rceil \qquad \text{for } j=1,...,n \qquad (2.57)$$

$$L_j^!(c) := m+1 - \lceil p_j(c) + n_j(c) \rceil \qquad \text{for } j=1,...,n \qquad (2.58)$$

where $a_j(c)$, $n_j(c)$, and $p_j(c)$ denote the heads, the tails, and the station requirements of the tasks j = 1,...,n for a trial cycle time c, respectively.

Bound LC_3. A necessary condition for the existence of a feasible solution with m stations is $E_j^!(c) \le L_j^!(c)$ for each task j. Otherwise, at least one task cannot be assigned to any station. Thus,

$$LC_3 := \min\{ c \mid L_j^!(c) \ge E_j^!(c) \ \text{ for } j=1,...,n \} \qquad (2.59)$$

defines a lower bound on the cycle time (cf. the lower bound LB_5 for SALBP-1). Note that $E_j^!(c) \le E_j^!(c-1)$ and $L_j^!(c) \ge L_j^!(c-1)$.

Complexity. The bound is computed by performing some linear time iterations on the cycle time c, provided that the earliest and latest stations are given. Otherwise, the complexity depends on the way of computing heads and tails.

Example. We compute the earliest and latest stations of our example problem by (2.54) and (2.55) and start with $c = \max\{LC_1, LC_2\} = 10$. Table 2.7 contains the values for $E_j(10)$ and $L_j(10)$ which are derived from Table 2.5 in Section 2.2.2.1 (p. 48). Due to $E_9(10) > L_9(10)$, task 9 cannot be assigned to any of 5 stations, and c is increased to 11. Since the values of $E_j(11)$ and $L_j(11)$ are identical with those of $E_j(10)$ and $L_j(10)$, respectively, c is increased to 12. The values $E_j(12)$ and $L_j(12)$, which are contained in Table 2.7, too, indicate that each of the tasks 8, 9, and 10 can only be assigned to a single station in order to hold the cycle time c = 12. With (2.59) we obtain $LC_3 = 12$.

Table 2.7. Earliest and Latest Stations

j	1	2	3	4	5	6	7	8	9	10
$E_j(10)$	1	2	1	1	2	3	2	4	5	5
$L_j(10)$	2	3	2	3	3	4	4	4	4	5
$E_j(12)$	1	1	1	1	2	3	2	4	4	5
$L_j(12)$	2	3	3	3	4	4	4	4	4	5

Bound LC_4. By subdividing the interval [1,m] of all stations into two subintervals [1,i] and [i+1,m], we obtain two aggregate "stations". For a trial cycle time c, these "stations" have to observe the "cycle times" $i \cdot c$ and $(m-i) \cdot c$, respectively. This modified problem is solved as follows.

All tasks with $L_j(c) \leq i$ are assigned to the first "station" and those with $E_j(c) > i$ are assigned to the second one. The remaining tasks and the remaining idle times of the two aggregate "stations" constitute a residual problem. This problem can be formulated as the feasibility version of the bin packing problem with two bins and capacities equal to the remaining idle times of the two aggregate "stations" (2Bin-F for short). The question of this feasibility problem reads as: "Does there exist a partition of the remaining tasks (items) into at most two bins?"

Even though 2Bin-F is NP-complete, it can often be solved in a reasonable computation time, because the residual problem is usually much smaller than the original SALBP-2 instance and only two bins are considered. The problem may be solved by any exact procedure for the bin packing problem (cf., e.g., Martello and Toth (1990 a, chapter 8)). If the remaining idle times of the two "stations" are different, equal-sized bins are obtained by pre-assigning a fictitious item, the weight of which is equal to the idle time difference, to the smaller "station".

The transformation to the "two-station" problem can be used for determining a lower bound LC_4 as follows. A cycle time c is said to be *potentially feasible* for the underlying SALBP-2 problem, if the answer for 2Bin-F is "yes" for all possible subdivisions of [1,m] into two aggregate "stations" [1,i] and [i+1,m], i.e., for i = 1,...,m-1. The lower bound LC_4 is obtained by determining the smallest cycle time c which is potentially feasible for SALBP-2. This may be done by a *Lower Bound Search* which starts with a known lower bound on the cycle time and successively increases c by 1, until c becomes potentially feasible (cf. Section 4.2.2, p. 145).

Complexity. Since 2Bin-F is NP-complete, no polynomial time algorithm is known for this problem. For each trial cycle time (iteration), $m-1$ instances of 2Bin-F have to be solved. The number of iterations is data-dependent. However, all iterations but the last result in an improvement of the bound.

Example. We start with the trial cycle time $c = \max\{LB_1, LB_2, LB_3\} = 12$ and use the values $E_j(12)$ and $L_j(12)$ of Table 2.7.

For $i=1$ we get two "stations" with "cycle times" $c_1 = 12$ and $c_2 = 48$. The tasks 5 to 10 are assigned to "station" 2, i.e., $S_2^{\mathrm{I}} = \{5,\ldots,10\}$, while $S_1^{\mathrm{I}} = \{\}$. The two bins have capacities $\kappa_1 = 12$ and $\kappa_2 = 48 - t(S_2^{\mathrm{I}}) = 20$, and the tasks $1,\ldots,4$ remain. A feasible solution of 2Bin-F is given by the sets $S_1^{\mathrm{II}} = \{1,2\}$ and $S_2^{\mathrm{II}} = \{3,4\}$.

$i=2$: $c_1=24$, $c_2=36$; $S_1^{\mathrm{I}}=\{1\}$, $S_2^{\mathrm{I}}=\{6,8,9,10\}$; $\kappa_1=18$, $\kappa_2=16$; $S_1^{\mathrm{II}}=\{2,3,4\}$, $S_2^{\mathrm{II}}=\{5,7\}$

$i=3$: $c_1=36$, $c_2=24$; $S_1^{\mathrm{I}}=\{1,2,3,4\}$, $S_2^{\mathrm{I}}=\{8,9,10\}$; $\kappa_1=14$, $\kappa_2=9$; $S_1^{\mathrm{II}}=\{5,6\}$, $S_2^{\mathrm{II}}=\{7\}$

$i=4$: $c_1=48$, $c_2=12$; $S_1^{\mathrm{I}}=\{1,\ldots,9\}$, $S_2^{\mathrm{I}}=\{10\}$; no residual problem

Since 2Bin-F is answered with "yes" for $i=1,\ldots,4$, the value of LC_4 is 12.

Bound LC_5. This bound argument is based on computing lower bounds $LI_k(c)$ on the idle time of each station for trial cycle times c. Values for $LI_k(c)$ may be computed by considering a *special knapsack problem* with capacity c for each station $k=1,\ldots,m$. This problem is called *subset sum problem* (cf. Martello and Toth (1990 a, chapter 4)). It consists of finding a subset of items (tasks) such that the sum of weights (task times) is maximal without exceeding a given capacity c (cycle time). For each station, only the tasks $j \in B_k(c)$ are considered as items. The difference between c and the maximal sum of weights in an optimal solution of the subset sum problem for some station k provides a lower bound $LI_k(c)$ on the idle time of that station. By means of $LI_k(c)$, a lower bound LC_5 is obtained by

$$LC_5 := \min\{c \mid \sum_{k=1}^{m} LI_k(c) \leq BD(c)\}, \tag{2.60}$$

where the balance delay time (sum of idle times) is defined by $BD(c) = m \cdot c - t_{sum}$ (a similar bound argument is given for SALBP-1 by Talbot and Patterson (1984)).

Remark 2.4. A simpler way to compute lower bounds on the idle time consists of determining tasks which cannot be combined with any other task in the same station. A sufficient condition for the incompatibility of a task j with any other task h is:

$$t_j + \min\{t_h \mid h=1,\ldots,n,\ h \neq j,\ SI_j(c) \cap SI_h(c) \neq \phi\} > c \qquad \text{for } j=1,\ldots,n \tag{2.61}$$

If (2.61) is true for a task j, an idle time of $c - t_j$ cannot be avoided in one of the stations in $SI_j(c)$, provided that the cycle time c is to be realized.

Complexity. For each trial cycle time, m subset sum problems have to be solved. This may be achieved by some exact algorithm for the zero-one knapsack problem or a specialized algorithm for the subset sum problem (cf., e.g., Martello and Toth (1990 a, chapter 4)). Even though the subset sum problem is NP-hard, it is usually solved very fast by such algorithms, because it is highly structured. If a particular problem instance cannot be solved in a reasonable amount of computation time, the lower bound $LI_k(c)$ is set to zero for the corresponding station.

As is the case with LC_4, some iterations on several cycle times have to be performed in order to find the minimum value of (2.60).

Example. We start with the trial cycle time $c := \max\{LC_1, ..., LC_4\} = 12$. With $E_j(12)$ and $L_j(12)$ from Table 2.7 we obtain $B_1(12) = \{1,2,3,4\}$, $B_2(12) = \{1,...,5,7\}$, $B_3(12) = \{2,...,7\}$, $B_4(12) = \{5,...,9\}$, and $B_5(12) = \{10\}$. Optimal solutions for the related subset sum problems are given by the sets $S_1' = \{1,2\}$, $S_2' = \{1,2\}$, $S_3' = \{2,3\}$, $S_4' = \{8,9\}$, and $S_5' = \{10\}$. The corresponding lower bounds on the idle times are $LI_1(12) = LI_2(12) = 0$, $LI_3(12) = LI_4(12) = 1$, and $LI_5(12) = 8$. Since the sum of these values is equal to $BD(12) = 10$, the lower bound cannot be improved by LC_5. However, the computation has shown in which stations idle time must occur in order to hold the cycle time 12.

Bound LC_6. Further bounds may result from the close relationship to SALBP-1. The smallest cycle time, for which a corresponding lower bound of SALBP-1 does not exceed the given number m of stations, provides a lower bound of SALBP-2 (cf. Section 2.2.2.1 for a survey of SALBP-1 bounds). This is expressed by (2.62), where LB(c) denotes an arbitrary lower bound of SALBP-1 for a cycle time c:

$$LC_6 := \min \{ c \mid LB(c) \leq m \} \tag{2.62}$$

Complexity. Obviously, the complexity depends on the used bound argument(s) for SALBP-1. As in case of LC_3 to LC_5, some iterations on several values of c have to be performed in order to find the minimum of (2.62).

Table 2.8. Heads and Tails for c = 12

j	0	1	2	3	4	5	6	7	8	9	10	11
$p_j(12)$	0	0.5	0.5	0.42	0.42	0.33	0.42	0.33	0.17	0.75	0.33	0
$a_j(12)$	-	0	0.5	0	0.42	1.33	2	1	3	3.17	4	4.33
$n_j(12)$	4.92	3.58	2.33	3.42	3	2.42	2	2	1.75	1	0	-

Example. We use the generalized bound LB_4' for computing LB(c) with c = 12. Table 2.8 contains the corresponding heads and tails (rounded to two digits after the point). With $a_{11}(12) = 4.33$, $n_0(12) = 4.92$, and $Z(12) = \max\{a_j(12)+p_j(12)+n_j(12) \mid j = 1,...,n\} = 4.92$, we obtain $LB(12) = 5$ (cf. (2.35), p. 48). Due to $LB(11) = 6$, we get $LC_6 = 12$.

2.2.3.2 Upper Bounds of SALBP-2

We present (theoretical) upper bounds on the cycle time which may be helpful as initial values for solution procedures.

Bound UC_1. A simple upper bound on the cycle time is given by:

$$UC_1 := \max\{t_{max}, 2 \cdot \lfloor t_{sum}/m \rfloor\} \tag{2.63}$$

Since UC_1 does not utilize the precedence relations, it is an upper bound for the related parallel machine scheduling problem as well (cf. Coffman et al. (1978)). We do not give a proof of the correctness of this bound, but refer to Hackman et al. (1989) who present a slightly improved version.

Bound UC_2. If $LC_1 > t_{max}$, another upper bound is obtained by:

$$UC_2 := \lceil t_{sum}/m \rceil + t_{max} - 1 = LC_1 + t_{max} - 1 \tag{2.64}$$

This bound is based on the infeasible preemptive solution associated with LC_1 (see p. 55). Each task split up between two stations k and k+1 can completely be processed at station k if the cycle time UC_2 is realized, because its operation time at station k+1 is no larger than $t_{max} - 1$.

Complexity. Obviously, the computation of both upper bounds takes linear time.

Example. We obtain $UC_1 = 20$ from the second term of (2.63), and $UC_2 = 18$.

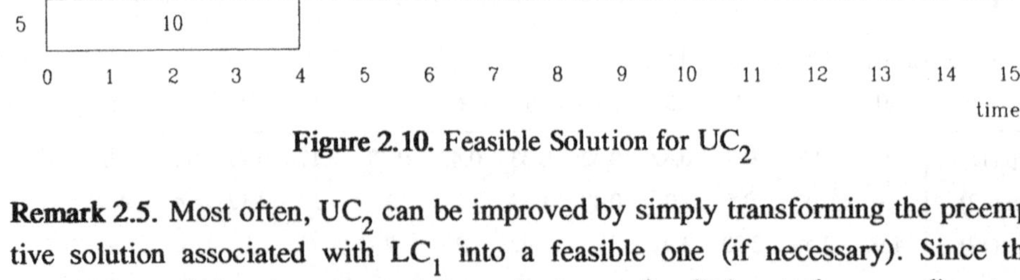

Figure 2.10. Feasible Solution for UC_2

Remark 2.5. Most often, UC_2 can be improved by simply transforming the preemptive solution associated with LC_1 into a feasible one (if necessary). Since the preemptive solution is generated by assigning and splitting tasks according to a topological ordering, we obtain a feasible solution by assigning each split task completely to its first station. For our example, the infeasible solution of Figure 2.9 is transformed into the feasible one of Figure 2.10 with the improved upper bound 15. In the worst case (a task with operation time t_{max} has been split up into parts of "length" 1 and t_{max} - 1), the resulting cycle time is equal to UC_2. Obviously, alternative ways of transforming the preemptive solution into a feasible one are possible.

2.2.3.3 Mathematical Formulations of SALBP-2

The mathematical formulations of SALBP-F are easily modified for SALBP-2 by introducing the cycle time c as a variable and an objective function for the minimization of c. For determining the earliest and latest stations we use an upper bound $\bar{c}$ on the cycle time (see Section 2.2.3.2). That is, $E_j(\bar{c})$, $L_j(\bar{c})$, $SI_j(\bar{c})$, and $B_k(\bar{c})$ are used instead of E_j, L_j, SI_j, and B_k in the formulations of Section 2.2.1.2.

Formulation 1. The model contains the restrictions (2.4)-(2.7) and the objective:

$$\text{Minimize } F_1(x,c) = c \tag{2.65}$$

An equivalent formulation, which eliminates the variable c, includes the restrictions (2.4), (2.6), and (2.7) as well as the mini-max objective function:

$$\text{Minimize } F_1(x) = \max \{ \sum_{j \in B_k(\bar{c})} t_j \cdot x_{jk} \mid k = 1,...,m \} \tag{2.66}$$

Formulation 2. The modification of formulation 2 is essentially the same as that of formulation 1, i.e., the model is defined by the restrictions (2.9) - (2.13) and the objective function:

$$\text{Minimize } F_2(y,c) = c \tag{2.67}$$

Formulation 3. By introducing c as a variable, the restrictions (2.15) and (2.16) become non-linear. The objective function reads as:

$$\text{Minimize } F_3(b,c,w,z) = c \tag{2.68}$$

2.2.4 Maximizing the Line Efficiency (SALBP-E)

We discuss the most general version of SALBP which is the problem of finding a combination of the number m of stations and the cycle time c as well as a respective line balance such that the line efficiency is maximized (SALBP-E for short). Hence, it is to find a feasible SALBP-F instance (m,c) with maximal value of $E = t_{sum}/(m \cdot c)$.

We illustrate the problem by means of an example with the precedence graph of Figure 2.11.

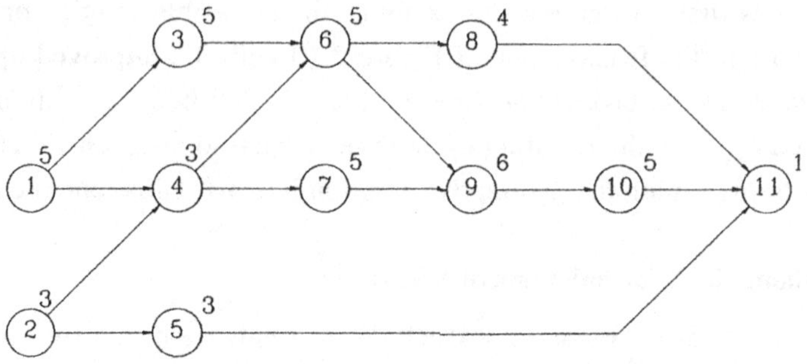

Figure 2.11. Precedence Graph (cf. Figure 2.2)

2.2.4.1 Equivalence of Objectives

In Definition 2.4 and Definition 2.5, capacity oriented measures for the line performance are introduced. The line efficiency E is defined as the quotient of the sum of operation times and the total time available at the line.

Since E only depends on the constant t_{sum} and the non-linear term $m \cdot c$, the line efficiency is maximized by minimizing $m \cdot c$. Therefore, the following capacity oriented objectives are equivalent to the maximization of E:

- Minimize the line capacity $T = m \cdot c$.
- Minimize the balance delay time $BD = m \cdot c - t_{sum}$.
- Minimize the balance delay ratio $BR = 1 - E$.

The equivalence of the objectives is illustrated by Figure 2.12, which contains a line balance for the precedence graph in Figure 2.11 with $m = 4$ and $c = 12$. Each workpiece has to pass m stations in each of which it is available for c time units. Hence, the line capacity is $T = 48$. The balance delay time $BD = 48 - 45 = 3$ is indicated by the non-hatched regions. The line efficiency $E = 45/48 = 0.9375$ and the balance delay ratio $BR = 1 - 0.9375 = 0.0625$ correspond to the proportion of hatched and non-hatched regions, respectively.

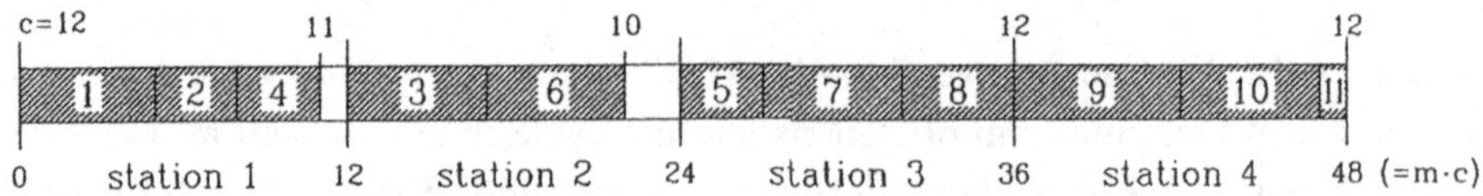

Figure 2.12. Equivalence of Objectives

2.2.4.2 Statement of the Problem

SALBP-E is an optimization problem only if a lower bound m_{min} on the number of stations has to be observed, because a line efficiency of 1 is obtained by taking $m = 1$ and $c = t_{sum}$. Such a minimum number of stations may result from a desired degree of the division of labour. Note that this degree influences the determination of task times during the work measurement. Furthermore, an upper bound c_{max} on the cycle time may be imposed by a desired production rate. Due to the feasibility condition $m \cdot c \geq t_{sum}$, a lower bound on the number of stations (besides m_{min}) is given by $\lceil t_{sum}/c_{max} \rceil$.

SALBP-E may be further restricted by an upper bound m_{max} on m, which may result from space limitations. A lower bound c_{min} on the cycle time follows from the feasibility condition $c \geq t_{max}$ and, possibly, from external sales limits. By analogy with c_{max}, an additional lower bound $\lceil t_{sum}/m_{max} \rceil$ on the cycle time is derived from m_{max}.

Summarizing these restrictions, SALBP-E *instances* are defined by two intervals $[m_{min}, m_{max}]$ and $[c_{min}, c_{max}]$. In accordance with the above bounds, we assume that for each value of m and c at least one value of the respective other variable exists, such that the necessary feasibility condition is fulfilled, i.e., $m_{min} \cdot c_{max} \geq t_{sum}$ and $m_{max} \cdot c_{min} \geq t_{sum}$.

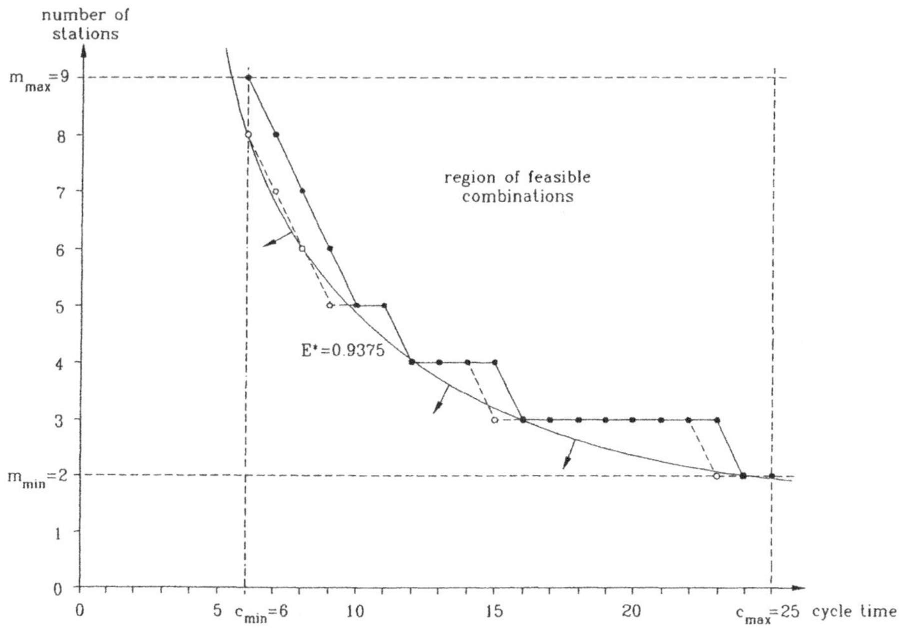

Figure 2.13. Feasible Region for SALBP-E

Example. In order to specify the instance with the precedence graph of Figure 2.11 completely, we assume that at least 1080 units of a product have to be assembled on the line within each shift of duration 7.5 hours. This results in a maximal cycle time of $c_{max} = 25$ seconds/unit. Due to $t_{sum} = 45$, at least $m_{min} = 2$ stations are required. Since we have no sales limits, a lower bound on the cycle time is given by $c_{min} = t_{max} = 6$ seconds/unit. Space limitations prevent from installing more than $m_{max} = 9$ stations. Hence, the problem is defined by $m \in [2,9]$ and $c \in [6,25]$.

Figure 2.13 shows the region of all feasible combinations of m and c. This region contains only points with integral coordinates and is restricted by the lower and upper bounds on m and c, respectively, as well as the necessary feasibility condition $m \cdot c \geq t_{sum}$. The bold points indicate the minimal numbers of stations for all possible values of c, i.e., the optimal combinations for fixed c. The non-filled points show combinations for which no line balance exists, even though they fulfill the necessary feasibility condition. The optimal line efficiency is 0.9375 (93.75 %) and is realized by the (m,c)-combinations (4,12), (3,16), and (2,24). The corresponding *efficiency isoquant* is shown in Figure 2.13, too.

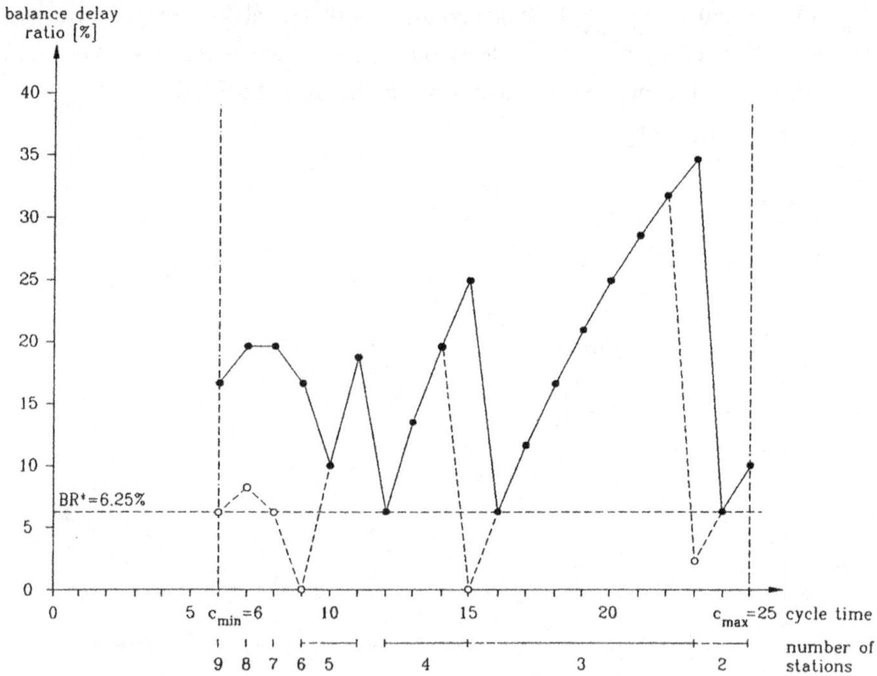

Figure 2.14. Balance Delay Function

Figure 2.14 shows the relationship between all possible cycle times $c \in [6,25]$ and the respective optimal objective function value (bold points). In order to facilitate the presentation, we use the balance delay ratio BR (in %) instead of the line efficiency E. Furthermore, the discrete points of the resultant function are connected by straight lines (cf. Kilbridge and Wester (1961 a) for this type of diagram). This *balance delay function* can be explained as follows. While the cycle time increases, the balance delay ratio is increased, too, until the required number of stations can be reduced. This reduction leads to a decrease of the balance delay ratio, i.e., a local minimum of the function is achieved. As in Figure 2.13, the non-filled points indicate infeasible combinations.

Remark 2.6. For reasonable problem instances, the interval $[m_{min}, m_{max}]$ of possible station numbers usually contains significantly less values than the cycle time interval $[c_{min}, c_{max}]$ (cf. the footnote on p. 274). As a consequence, for most values of m, several cycle times are possible. Only those cycle times which correspond to a local minimum of the balance delay function are relevant for the optimization of SALBP-E. Speaking in terms of restricted problems, only optimal solutions for SALBP-2 (with $m \in [m_{min}, m_{max}]$) are candidates for optimal SALBP-E solutions, while a lot of the optimal SALBP-1 solutions (with $c \in [c_{min}, c_{max}]$) are inferior. That is, SALBP-1 institutionalizes unnecessary idle times, whenever the given cycle time is too large (cf. Section 2.2.1.3).

2.2.4.3 Upper and Lower Bounds of SALBP-E

An *upper bound* UE on the line efficiency may be obtained in two ways. First, a bound UE_1 results from lower bounds LB(c) of SALBP-1 for all possible cycle times (cf. Section 2.2.2.1):

$$UE_1 := \max \{ t_{sum} / (LB(c) \cdot c) \mid c \in [c_{min}, c_{max}] \} \tag{2.69}$$

Analogously, a bound UE_2 may be computed by means of lower bounds LC(m) of SALBP-2 for all possible values of m (cf. Section 2.2.3.1):

$$UE_2 := \max \{ t_{sum} / (m \cdot LC(m)) \mid m \in [m_{min}, m_{max}] \} \tag{2.70}$$

Complexity. The time complexity of computing the bounds mainly depends on the used bound arguments for SALBP-1 or SALBP-2, respectively, and the size of the intervals $[m_{min}, m_{max}]$ and $[c_{min}, c_{max}]$. With respect to Remark 2.6, a SALBP-2 based determination of bounds may cause less computations than SALBP-1 based bounds.

Example. For $m = 5$ and $m = 3$, we obtain $UE_2 = 1$ by applying the simple bound LC_1 for SALBP-2 to compute LC(m). The corresponding (m,c)-combinations are (5,9) and (3,15) without balance delay time (cf. the non-filled circles on the x-axis of Figure 2.14).

Remark 2.7. The application of the simple bounds LB_1 for SALBP-1 and LC_1 for SALBP-2, respectively, just eliminates combinations (m,c) which do not fulfill the necessary feasibility condition $m \cdot c \geq t_{sum}$.

Lower bounds LE on the line efficiency representing feasible solutions are derived from any line balance for a possible (m,c)-combination. By analogy with upper bounds UE, we may use upper bounds $UB(c) \leq m_{max}$ on m for all possible cycle times c or upper bounds $UC(m) \le c_{max}$ on c for all possible values of m to obtain lower bounds on E (cf. Section 2.2.2.2 and 2.2.3.2).

2.2.4.4 Mathematical Formulations of SALBP-E

The formulations of SALBP-F in Section 2.2.1.2 may be extended to SALBP-E by introducing both m and c as variables and adding the non-linear objective function of minimizing $m \cdot c$ (cf. Section 2.2.4.1). Note that m is implicitly contained as the number of the station to which the (fictitious) source node n is assigned. The earliest and latest stations have to be determined using c_{max} and m_{max} instead of c and m in (2.1) and (2.2).

Formulation 1. The constraint set is defined by (2.4) - (2.7), where m is substituted by m_{max} in (2.5). The non-linear objective function reads as:

$$\text{Minimize } F_1(x,c) = c \cdot \sum_{k \in SI_n} k \cdot x_{nk} \tag{2.71}$$

Formulation 2. The modification is essentially the same as that for formulation 1. That is, the constraints are (2.9) - (2.13), where (2.10) is modified by using m_{max} instead of m. Due to $L_n = m_{max}$, the objective function is expressed by:

$$\text{Minimize } F_2(y,c) = c \cdot (m_{max} + 1 - \sum_{k \in SI_n} y_{nk}) \tag{2.72}$$

Formulation 3. The constraint set contains restrictions (2.15) - (2.22), where (2.15) and (2.16) are non-linear. We may use the objective function:

$$\text{Minimize } F_3(b,c,w,z) = c \cdot z_n \tag{2.73}$$

If the intervals on m and c defining SALBP-E instances are not rather small, the numbers of variables and constraints are very large in all formulations.

2.3 Generalized Single - Model Problems

In this section, we consider possible generalizations of SALBP with respect to different characteristics of assembly line production systems (cf. Section 1.3). Such more general problems often arise from SALBP by relaxing one or more of its rather restrictive assumptions. Therefore, solution methods for respective models are usually based on SALBP algorithms. For this reason, we restrict the discussion of generalized problems to their additional characteristics and show correspondencies to SALBP. Furthermore, we concentrate on those generalizations most frequently considered in the literature and provide respective references. Additionally, see Domschke et al. (1997, chapter 4.3.6).

2.3.1 Stochastic Task Times

As described in Section 1.3.3, operation times often contain some stochastic component caused by, e.g., variability in human work pace, defective parts or machine breakdowns. In paced systems, varying task times may cause the station times exceeding the cycle time, while blocking and starving of succeeding and preceding stations, respectively, may occur in unpaced assembly lines. Therefore, we distinguish between both basic operating strategies.

2.3.1.1 Paced Assembly Systems

Several time or cost oriented models of stochastic paced assembly lines are considered in the literature. All of them consider operation times as being random variables following a particular statistical distribution.

In most models, task times are assumed to be *independent normal variates*, i.e.,

$$t_j \sim N(\mu_j, \sigma_j) \qquad \text{for } j = 1,\dots,n, \tag{2.74}$$

where μ_j denotes the *mean value*, and σ_j (σ_j^2) the *standard deviation (variance)* of t_j. This assumption is considered to be realistic in most cases of human work (cf., e.g., Moodie and Young (1965), Mansoor and Ben-Tuvia (1966), Brennecke (1968)). Because of the special properties of the normal distribution, station times $t(S_k)$ also become normally distributed random variables with mean value MV_k and variance Var_k:

$$MV_k := \sum_{j \in S_k} \mu_j \qquad \text{for } k = 1,\dots,m \tag{2.75}$$

$$Var_k := \sum_{j \in S_k} \sigma_j^2 \qquad \text{for } k = 1,\dots,m \tag{2.76}$$

By transforming the normal distributions of the station time variables $t(S_k)$ into the standardized normal distribution with $MV_k = 0$ and $Var_k = 1$ for $k = 1,\dots,m$, we obtain the following condition, which ensures that the probability of not exceeding a particular station time is at least $1 - \beta$ *(confidence level)*.

$$\text{Prob}\left[\, t(S_k) \le MV_k + y_{1-\beta} \cdot \sqrt{Var_k}\,\right] \ge 1 - \beta \qquad \text{for } k = 1,\dots,m \tag{2.77}$$

The parameter $y_{1-\beta}$ denotes the $(1-\beta)$-quantile of the standardized normal distribution, i.e., that value for which the distribution function provides the probability $1 - \beta$.

By means of (2.77), the cycle time constraints may be formulated as follows:

$$MV_k + y_{1-\beta} \cdot \sqrt{Var_k} \le c \qquad \text{for } k = 1,\dots,m \tag{2.78}$$

The restrictions (2.78) guarantee for each station that its station time does not exceed the cycle time with probability $1 - \beta$.

Model formulations for any version of SALBP (cf. Section 2.2.1.2, 2.2.2.3, 2.2.3.3, 2.2.4.4) are easily modified for the case of normally distributed task times by replacing their cycle time constraints with (2.78), where the mean value and the variance have to be expressed in terms of the respective variables.

Figure 2.15 illustrates (2.78) by showing the density function $f(t(S_k))$ of a normally distributed station time variable $t(S_k)$ for some station k. The hatched area indicates the probability β of exceeding the cycle time.

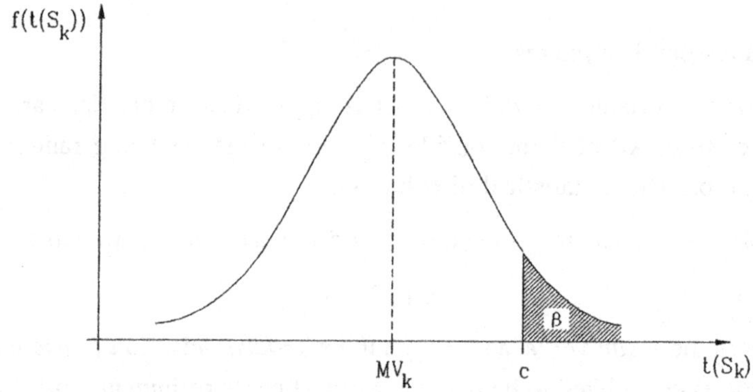

Figure 2.15. Density Function of Station Time Variables

This model may be transformed to a relaxed deterministic problem by using the following modified operation times:

$$t_j^! := \mu_j + y_{1-\beta} \cdot \sigma_j \qquad \text{for } j=1,...,n \qquad (2.79)$$

The "task time" $t_j^!$ contains a safety factor such that the realized (stochastic) operation time of task j does not exceed this value with a probability of at least $1-\beta$. Note that in this case the probability of not exceeding the cycle time depends on the number of tasks in a station. Furthermore, the tasks may get different safety factors depending on their importance for the production process.

Sometimes, the assumption of normally distributed operation times is not realistic. Other distributions, like the Poisson and the gamma distribution, are considered by, e.g., Ramsing and Downing (1970), Kao (1976, 1979), and Sniedovich (1981).

Fundamentally, stochastic models may contain the same objectives as deterministic problems (cf. Henig (1986), Carraway (1989)). In case of time related objectives, the stochastic model may be transformed to one of the SALBP versions by using the modified task times $t_j^!$. Sphicas and Silverman (1976) show that for certain distributions it is sufficient to solve a deterministic model with the cycle time being reduced to a certain proportion of its value. However, most procedures for SALBP can be used for directly solving stochastic models with minor changes. Instead of simply comparing the cycle time with the sum of task times in a station, one has to check the cycle time constraints (2.78) when assigning tasks to stations (cf. Henig (1986)).

Specific objectives of stochastic models are the minimization of the probability of exceeding the cycle time in any station (cf. Reeve and Thomas (1973)) or the variations of station times (cf. Raouf and Tsui (1982)). Usually, time related objectives do not consider the consequences of exceeding the cycle time. Since it is generally not possible to use a cycle time large enough to obtain sufficiently small excess probabilities, one has to consider effects of incompleted operations explicitly:

- The conveyor is stopped until the incompleted or missed operations have been performed *on-line* (cf. Silverman and Carter (1986), Lau and Shtub (1987)).
- The process continues, while the incompleted tasks and all their successors in the precedence graph are left out. These operations are performed at special *off-line stations* after the incompleted unit has left the last station (cf. Kottas and Lau (1981) as well as Carter and Silverman (1984)).
- Stations which tend to exceed the cycle time get an additional capacity by employing *additional operators* which increase the work pace (cf. Shtub (1984 a)).

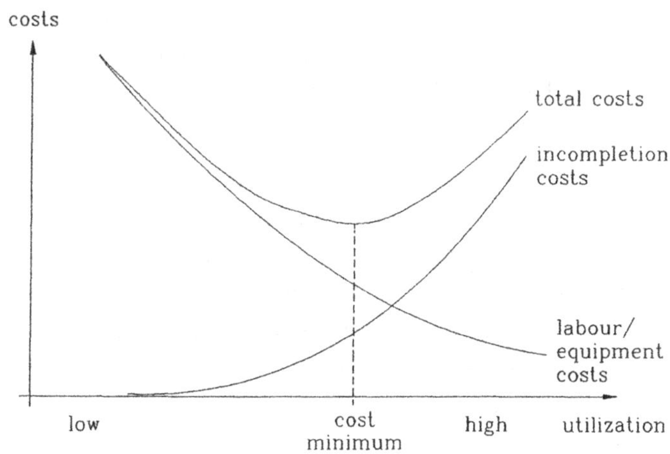

Figure 2.16. Costs of Stochastic Assembly Lines

Regardless of how incompletions are dealt with, they cause additional costs. These *incompletion costs* are reduced by decreasing the station utilizations. This can be done by increasing the number of stations or the cycle time. However, this raises the *labour and equipment costs*. The *cost trade-off* between incompletion costs and labour/equipment costs is illustrated by Figure 2.16 (cf. Kottas and Lau (1973)). Furthermore, the ordering of tasks inside the stations influences the incompletion costs with respect to their value and probability (cf. Kottas and Lau (1976)). The higher the potential incompletion costs of a task are, the more idle time has to be introduced into a station in order to avoid actual incompletions. Therefore, the idle time should be concentrated in early stations of the line, because the incompletion costs depend on the number of affected successor tasks.

2.3.1.2 Unpaced Assembly Systems

As mentioned in Section 1.3.3, stochastic task times may cause blocking and starving of stations in unpaced systems. While small variations of operation times can usually be compensated by buffers, operation times with large coefficients of variation may introduce considerable inefficiencies into the system. In the first case, it is often sufficient to balance the line as a paced one (see Section 2.3.1.1). For the latter case which may result from machine failures and repairs, it has empirically been shown that slightly unbalancing the line may increase the production rate. Since this imbalance should have the shape of a bowl, this effect is called *bowl-phenomenon* (see Figure 1.7, p. 12; cf. Hillier and Boling (1966)).

Whenever the degree of imbalance required is a priori known, the bowl-phenomenon can be included in a model by defining different (local) cycle times. The stations in the centre of the line get the smallest local cycle times and those at the start and the end of the line receive the largest ones. If the stochastic task times are transformed to deterministic ones, as described in Section 2.3.1.1, such models may be solved by procedures for SALBP (different cycle times can be realized by pre-assigning fictitious tasks to the stations with the smaller desired station times).

If the degree of imbalance and its distribution among the stations is not known, one may check several "bowl types" by repeatedly determining task assignments and computing the resulting output rates (cf. Remark 5.1, p. 187). Guidelines for constructing bowl-shaped work load distributions are given by Smunt and Perkins (1985) and So (1989). The output rates may be estimated by means of simulation or queueing theory (cf., e.g., Muth (1973), Warnecke et al. (1978), Taylor and Russell (1982), Buzacott and Shanthikumar (1993, chapters 5 and 6), Papadopoulos and O'Kelly (1993)).

2.3.2 Assignment Restrictions

Assignment restrictions reduce the number of possible task-station-combinations (cf. Section 1.3.4). Therefore, they have a similar influence on the complexity of the problem as precedence constraints.

Task related constraints can be incorporated into model formulations as maximum or minimum distances between pairs (h,j) of tasks (cf. Deckro (1989)). While in the formulations 1 and 2 of Section 2.2.1.2 additional binary variables are needed, such constraints can easily be introduced into formulation 3 by defining respective differences between the numbers z_h and z_j of the stations to which two tasks h and j are assigned. For example, two tasks h and j, not related by precedence, are prevented from sharing the same station by the following disjunctive conditions:

$$\left.\begin{array}{l} w_{hj} \cdot T + z_h - z_j \geq 1 \\ (1 - w_{hj}) \cdot T + z_j - z_h \geq 1 \end{array}\right\} \begin{array}{l} \text{for some } h < j \text{ and} \\ (h,j) \notin A^* \text{ and } SI_h \cap SI_j \neq \phi \end{array} \qquad (2.80)$$

The restrictions (2.80) express that task h is performed in a station which is either later or earlier than that to which task j is assigned. Such task related restrictions can be included in any solution procedure for SALBP, but may require modifying some bounding mechanisms or logical tests (cf. Remark 4.5, p. 154).

Station related constraints restrict the station intervals of particular tasks. Therefore, the number of variables and restrictions is reduced. If a task is fixed to one station, the respective variables and restrictions can be eliminated from the model. These constraints can easiliy be observed in most procedures for SALBP (cf. Johnson (1983)).

Position related constraints may be included in a model by distinguishing two station types (left-hand side and right-hand side stations). For tasks restricted to either side of the line, only the respective stations are available. This results in a reduction of variables and restrictions as in case of station related constraints.

Operator related constraints which concern qualification considerations may be included as follows. The tasks are grouped into several classes depending on the required skills of the workers. Different types of stations are defined to each of which only tasks of certain skill classes can be assigned (cf. Johnson (1983)).

The operational specification of job design principles, which aim at combining monotonous tasks with demanding ones, is more difficult (cf. Gunther et al. (1983)). One possibility is to build *work teams* which operate a segment of the line, consisting of a number of stations, in a semi-autonomous manner. Then balancing is performed on the basis of work teams instead of stations. Such problems arise in *lean production systems* which include a decentralized organization based on work teams (cf., e.g., Krafcik (1988), Hentze and Kammel (1994)). The resulting problems may be solved by means of modified SALBP procedures (cf., e.g., Johnson (1991)).

2.3.3 Parallel Lines and Stations

In a situation, where identical *parallel lines* (cf. Section 1.3.5) should be installed for single-model production, the balancing problem is related with an investment problem. Zäpfel (1989, pp. 212) gives a respective non-linear optimization model which simultaneously decides on the number of lines, the number of stations, and the cycle time of each line, and maximizes the *capital asset*. The cycle time and the number of stations are assumed to be identical for all lines.

The influence of the number of lines (degree of parallelism) becomes obvious by considering a single line. In this case, the output rate only depends on the cycle time which is bounded from below by the maximum operation time of a task. But small cycle times tend to produce worse line balances and increase the failure sensitivity of the system. Furthermore, they are considered as being disadvantageous with respect to job satisfaction. These drawbacks may be removed by operating parallel lines, because they obtain the same output rate by using much larger cycle times and less stations per line. Besides, the system becomes more flexible, because the number of lines actually operating may be changed in the short-run. While additional lines allow better line balances in terms of utilization and job satisfaction as well as increased flexibility, they cause high capital investments.

If the number of lines is known, the above problem reduces to SALBP-E with a maximal cycle time resulting from the desired output volume per period and the given number of lines. In case the number of lines is not known, the problem may be treated by repeatedly solving SALBP-E instances for several settings.

Other possibilities for introducing parallelism into an assembly line production system are *parallel stations* or *paralleled tasks* (cf. Section 1.3.5). In both cases, the cycle time of the system can be reduced below the maximal operation time of a task without requiring the installation of complete additional lines.

Stations which are in parallel with each other perform the same work content on different workpieces. For example, if two parallel stations are available, consecutive workpieces are alternately fed into one of both stations. If the line operates with the *(global) cycle time* c, each of both stations has the *local cycle time* $2 \cdot c$ available for performing the work on each unit. Thus, parallel stations provide similar advantages as parallel lines. They are likewise connected with additional investments for duplicated equipment and more complicated material handling devices. Hence, cost oriented models have to be considered.

Pinto et al. (1981) describe a model which includes fixed costs for the installation of station duplicates and labour costs which depend on the number of stations and the cycle time. A desired cycle time is derived from a given output volume. By introducing overtime work, which is more expensive than regular work, an increased cycle time may be realized. The objective consists of minimizing the total costs per period (for a detailed description see Domschke et al. (1997, pp. 233)).

A similar problem including fixed costs for station duplication and task dependent costs for paralleling tasks (see below) is stated by Bard (1989). Since overtime work is not allowed, the cycle time is fixed. Again, the objective is the minimization of the total costs per period.

Both approaches only allow a single duplicate of each station. Hence, a solution is described by the number of serial stations, a binary vector which indicates the duplicated stations, and the task assignment. Given the line layout, the problem of Pinto et al. reduces to a variation of SALBP-E which considers a labour cost objective (see Section 2.3.5). Due to the fixed cycle time, the problem of Bard reduces to SALBP-1 if no paralleled tasks are allowed. Thus, both problems may be optimized by iteratively solving SALBP instances for several possible line layouts.

Buxey (1974) proposes a model with parallel stations and the objective of minimizing the balance delay time. The possible number of duplicates of a station depends on the types of the assigned tasks. Parallel stations are only introduced when a (serial) station cannot be loaded without idle time (cf., furthermore, Sarker and Shanthikumar (1983)).

Since parallel stations require completely equipped additional stations, it may be beneficial to utilize the concept of paralleled tasks. It consists of performing identical tasks at two or more stations of a serial line but on different workpieces (cf. Pinto et al. (1975)). Whenever the tasks to be paralleled are known (because they exceed the desired cycle time), the problem can be transformed to SALBP by splitting the respective tasks. If the paralleled tasks have to be done by two stations, the tasks are split up into two smaller tasks, each of which gets half of the operation time and the same precedence relations. Even though such an artificial task is assigned to different stations, the original task is completely performed by one of the stations on each workpiece in an alternating manner. Since this leads to varying station times which alternately exceed and fall below the cycle time, the tolerance times of concerned stations must be sufficiently large relative to the cycle time (cf. Johnson (1983)).

2.3.4 The U-line Assembly Line Balancing Problem (UALBP)

The U-line assembly line balancing problem (UALBP) is an extension of SALBP (Section 2.2) which is based on a U-shaped assembly line instead of a serial line (see Section 1.3.5). As is the case with SALBP, we can also define three problem versions of UALBP (cf. Miltenburg and Wijngaard (1994) as well as Scholl and Klein (1997 b)):

- UALBP-1: Given the cycle time c, minimize the number m of stations.
- UALBP-2: Given the number m of stations, minimize the cycle time c.

• UALBP-E: Maximize the line efficiency E for c and m being variable.

Since models for UALBP differ from those for SALBP only with respect to the precedence constraints, we concentrate on this difference. In SALBP all (direct and indirect) predecessors of a task j performed at a station k must be assigned to one of the stations 1,...,k. A task can share a station k with an indirect predecessor or successor only if all intermediate tasks defining this precedence relationship are also in k.

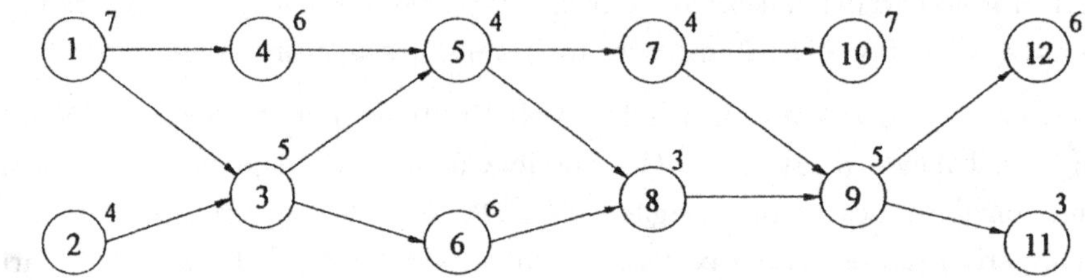

Figure 2.17. Example Precedence Graph

With the precedence graph given in Figure 2.17, task 1 can be performed in the same station as task 5 only if the tasks 3 and 4 are also in this station. In UALBP, each task in principle can share a station with any of its predecessors or successors. However, **all** predecessors or (and) **all** successors of a task j performed at a station k must be assigned to one of the stations 1,...,k. Due to these relaxed precedence constraints the optimal line efficiency of a SALBP instance is a lower bound on the optimal line efficiency of the corresponding UALBP instance. In many cases, a higher efficiency is possible with UALBP. Note that increasing the line efficiency has the further positive effect of smoothing the levels of station utilization, i.e., the stations get more equally loaded.

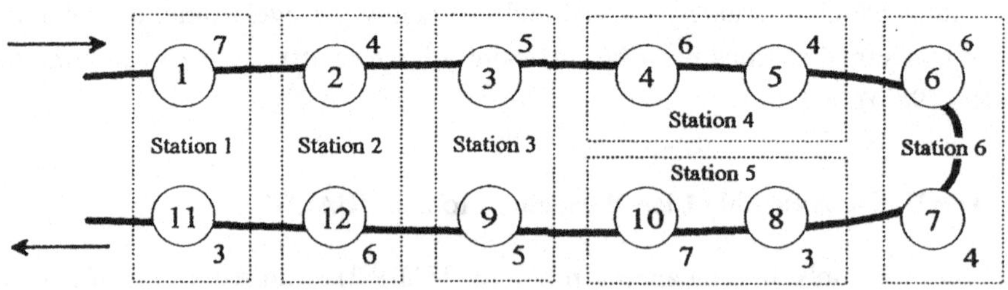

Figure 2.18. Optimal UALBP-1 Solution with c = 10

UALBP-1. Consider the example problem of Figure 2.17 and assume the cycle time to be c = 10. Figure 2.18 illustrates an optimal solution of UALBP-1 with six stations. As a convention, stations are numbered consecutively from left to right. Let us examine station 3 which performs the tasks 3 and 9, i.e. $S_3 = \{3,9\}$. Task 3 is executed whenever a product unit crosses the station for the first time (from left to right) after its predecessors have been performed in station 1 and 2, respectively. When the product unit returns to station 3 (from right to left) all predecessors of task 9 have already been done (in stations 1 to 6) and task 9 can be performed. Afterwards, the successors of 9 are executed in the stations 2 and 1.

An optimal solution to the corresponding SALBP-1 instance requires seven stations: $S_1=\{1\}$, $S_2=\{2,4\}$, $S_3=\{3,5\}$, $S_4=\{6,7\}$, $S_5=\{8,10\}$, $S_6=\{9,11\}$, $S_7=\{12\}$. Due to $t_{sum}=60$, the line efficiency of the U-line is 100% whereas the line efficiency of the straight line is only 85.7%.

All lower bounds for SALBP-1 which do not utilize precedence relations are also valid for UALBP-1. That is, LB_1, LB_2, LB_3, LB_6, and LB_7 are lower bounds for UALBP-1, too, whereas LB_4 and LB_5 cannot be applied (cf. Section 2.2.2.1). In particular, the concept of earliest and latest stations is not transferable to UALBP.

UALBP-2. This problem type arises when an existing U-line with a particular number m of stations has to be re-balanced which frequently takes place in a just-in-time production system. As in the case of UALBP-1, lower bounds for SALBP-2 which do not rely on precedence relations are applicable to UALBP-2. That is, LC_1, LC_2, and LC_6 can be used without changes while the other bound arguments of Section 2.2.3.1, which are based on earliest and latest stations, cannot be applied.

With $m=6$ stations given, we get $LC_1=10$. For UALBP-2, this lower bound cycle time can be realized (see above), whereas the optimal cycle time of SALBP-2 is $c=11$ with the line efficiency being $E=90.91\%$ instead of 100%.

UALBP-E. The more general problem type UALBP-E is present whenever the production system is flexible enough to change the number of stations and the cycle time as well. However, it will never be possible to vary those parameters arbitrarily. By the demand rate a maximal cycle time, which is the reciprocal of the production rate, is defined. A lower bound on the cycle time is given by t_{max}. The space available may set an upper limit on the number of stations, while a lower limit follows from the maximal cycle time and the necessary condition for feasibility $m \cdot c \geq t_{sum}$ (cf. Section 2.2.4).

Let us assume that $m \in \{3,4,5,6\}$ is possible in our example. Then the optimal line efficiency is $E=100\%$ for UALBP-E with $(m,c)=(3,20)$ or $(m,c)=(6,10)$. The optimal value $E=93.75\%$ of SALBP-E is obtained for the combination $(m,c)=(4,16)$.

2.3.5 Cost and Profit Oriented Objectives

As mentioned in Section 1.3.10, most cost and profit oriented goals mainly depend on the cycle time and the number of stations. Hence, respective models are strongly related to SALBP-E. The latter problem is usually solved by iterating on SALBP-1 or SALBP-2 instances, respectively (see Section 4.3.2). Thus, the same procedures can likewise be used for cost or profit oriented objectives on principle.

Zäpfel (1975, pp. 31) proposes a model with the objective of maximizing the *total contribution margin* per shift. It considers operating expenses, idle time costs, material costs, and advertising expenses as well as constant selling prices.

Problems with the objective of maximizing the *profit* are considered in some references. The model of Klenke (1977, pp. 30) includes fixed selling prices, material costs as well as wage and equipment costs (cf., additionally, Dick (1966, pp. 121)). A similar model which considers individual fixed costs for the different stations is proposed by Rosenblatt and Carlson (1985). This model is extended by Martin (1994) for the case of unpaced lines with buffers. Hence, inventory related cost components are included.

2.3.6 A Critique of SALBP

The assumptions of SALBP are very restrictive. Therefore, this simple type of an assembly line balancing problem is not realistic for most real-world assembly line production systems. Nevertheless, our above discussion shows that many extensions required for modelling industrial situations can be included in SALBP models. Furthermore, procedures for SALBP can often be modified for generalized problems. Consequently, it is absolutely necessary to have on hand efficient procedures for these basic problems in order to develop acceptable solution methods for more complex problems. Especially, mixed-model problems, which arise in today's industrial manufacturing frequently (cf. Chapter 3), require efficient solution concepts (cf. Gagnon and Ghosh (1992)). As stated in Section 6.1, such problems can be solved by SALBP methods with minor changes.

Besides its importance for generalized assembly line balancing problems, SALBP and corresponding procedures have applications in other fields, like, e.g., the assignment of installment lending operations to employees in commercial banks (cf. Levary and Renfro (1991)) or the scheduling of operations in vector computers (cf. Hu (1961)).

As a consequence of the above discussion, the present work mainly concentrates on developing efficient procedures for the different versions of SALBP (see Chapter 4 and Chapter 5). Even though considerable effort has been made to obtain such algorithms in the last four decades, most of them do not sufficiently utilize the structure of the problem.

3 Mixed-Model Assembly Lines

In this chapter, we concentrate on balancing and sequencing problems which arise in the context of installing and operating mixed-model assembly lines. In Section 3.1, we characterize the mixed-model assembly line production and discuss relationships of the problems. More detailed descriptions of the two problem types are given in the Sections 3.2 and 3.3, respectively. A hierarchical approach which attempts to integrate these problems into a common planning environment is discussed in Section 3.4.

3.1 Characterization of Mixed-Model Production

As stated in Section 1.3.1, several models (versions) of a standardized commodity are jointly assembled on a flexible assembly line. Models may differ from each other with respect to size, colour, used material, or equipment. Since all versions contain identical basic parts, the production processes are quite similar, i.e., a large percentage of the operations is common to most of the models. Another part of the tasks is present or absent depending on whether or not optional components are included in a particular model.

Mixed-model production systems are mainly used due to the following advantages. They provide a continuous flow of materials, reduce the inventory levels of final items, and are very flexible with respect to model changes (cf. Wester and Kilbridge (1964)). However, this flexibility requires expensive equipment which reduces or even eliminates delays due to set-up activities (cf., e.g., Schonberger (1982, chapter 6)).

Several decision problems with different planning horizons arise in managing mixed-model lines. *Medium-term (or long-term)* decisions concern the installation of the line and the division of work among the stations. This scope includes the determination of the line length (number of stations, lengths of the stations), the production rate or, equivalently, the cycle time as well as the work loads of the stations. Most of these decisions are part of the **mixed-model assembly line balancing problem (MALBP)**. As in the case of single-model production, it is the problem of finding a number of stations and a cycle time as well as a respective assignment of tasks to the stations such that a certain objective is optimized. Corresponding to the problem versions SALBP-1 and SALBP-2, the cycle time or the number of stations may be given, respectively. However, the problem is more difficult than in the single-model case, because the station utilizations have to be balanced with respect to each model. Hence, it is not sufficient to consider the simple capacity or cost oriented goals of single-model problems. In order to have a continuous flow of materials, to make use of learning effects, and to avoid installing equipment in several copies, tasks which are common for several models have to be assigned to the same station. Nevertheless, the utilization of a station may significantly vary because of the different production processes of the models (which include different tasks and task times). In contrast to the single-model case, the cycle time is usually not interpreted as an upper bound on the station time, but it corresponds to the *average* production rate. That is, station times may exceed the cycle time for particular models, whenever other models have station times sufficiently smaller than the cycle time.

The results of the medium-term balancing process determine the division of labour and the desired production rate. A *short-term* decision problem is called **mixed-model sequencing problem (MSP)**. It is to find a sequence of model units which meets the demands of all models given by the production program of a short-term planning period and optimizes some objective. Such objectives mostly correspond to inefficiencies arising from the variation of the station utilizations with respect to the models. Usually, the demands result from individual orders with certain delivery dates. Therefore, the sequencing problem arises per shift, day or week with particular demands for all models. The combination of demands for all models is referred to as *model mix*, and the respective planning horizon is called a *period.*

The two problem types outlined above (MALBP and MSP) are strongly related to each other even though they arise in different planning horizons. On the one hand, the short-term sequencing problem uses the solution of the medium-term balancing problem as its input data. Hence, the quality of sequencing decisions directly depends on the quality of the work load balancing. In the case where the line is almost perfectly balanced with respect to stations and models, the importance of the sequencing problem is minor, whereas it may not be possible to find acceptable sequences when the balancing solution leaves considerable imbalances. On the other hand, the quality of balancing decisions strongly depends on the expected model mix and the possible sequences which may occur when operating the line. Unfortunately, these informations are usually not available when the line has to be balanced. Furthermore, a single balancing solution should be efficient for different short-term situations each of which is characterized by a particular model mix.

Due to their different planning horizons, both problem types have to be solved separately. However, one should try to observe as many conditions of the short-term problems as adequate when balancing decisions are made. This may be done within the framework of a hierarchical planning system (see Section 3.4).

3.2 Mixed-Model Line Balancing

In this section, we give a precise statement of the mixed-model assembly line balancing problem (MALBP) and discuss versions of this problem especially with respect to the objective function used.

3.2.1 Description of the Problem

Partly, MALBP relies on the same basic assumptions as SALBP (deterministic operation times, no assignment restrictions, serial line-layout, fixed rate launching; cf. Section 2.2). Additional and modified characteristics which result from the joint assembly of several products are:

- The assembly of each model requires to perform a set of tasks which are connected by precedence relations (precedence graph for each model).
- A subset of tasks is common to all models; the precedence graphs of all models can be combined to a non-cyclical *joint precedence graph*. [1]
- Tasks which are common to several models are performed by the same station but may have different operation times; zero operation times indicate that a task is not required for a model.
- Fixed total time available for the production during the planning period (given by the number of shifts and the shift durations).
- Expected demands for all models (expected model mix) during the planning period.

Figure 3.1 contains notations some of which are already given in Figure 2.1 (see p. 24), while others are additionally introduced. Note that the character m is no longer used for the number of stations but as an index for models. The number of stations is denoted by K and indexed by k. Furthermore, the number of tasks is denoted by J instead of n. These changes are due to the large number of parameters and variables needed in the context of mixed-model problems.

---

[1] When contradicting precedence relations exist, several nodes have to be introduced for the concerned tasks in order to resolve these conflicts.

M	number of models, index: m = 1,...,M
d_m	expected demand for model m during the planning period
D	total number of units required during the planning period $(= \Sigma_m d_m)$
J	number of operations (tasks), index: j = 1,...,J
PT	total time available for production during the planning period
c	(average) cycle time, launch interval $(c \leq \lfloor PT/D \rfloor)$
t_{jm}	operation time of task j for one unit of model m
t'_j	cumulated time of task j for all required units $(= \Sigma_m d_m \cdot t_{jm})$
$\bar{t}_j$	average operation time of task j per unit $(= t'_j/D)$
K	number of stations, index: k = 1,...,K
S_k	station load, set of tasks assigned to station k
τ_{mk}	processing time per unit of model m in station k $(= \Sigma_{j \in S_k} t_{jm})$
τ'_{mk}	total operation time of model m in station k $(= d_m \cdot \tau_{mk})$
$\bar{t}_m$	average operation time per station and unit of model m $(= \Sigma_k \tau_{mk}/K)$
$\bar{t}'_m$	average total operation time per station for model m $(= d_m \cdot \bar{t}_m)$
τ'_k	total operation time in station k $(= \Sigma_m \tau'_{mk})$
τ_k	average operation time of station k per unit $(= \tau'_k/D)$

Figure 3.1. Notations for Mixed-Model Balancing

For illustration purposes, we use a small mixed-model problem with M = 3 models and J = 12 tasks. We assume that the available time during the planning period is PT = 77 time units. The demands of the three models are $d_1 = 4$, $d_2 = 2$, and $d_3 = 1$. Hence, the total number of units to be produced during the planning period is D = 7. In order to fulfill the demands, the cycle time must not be larger than c = 11, i.e., the average production rate has to be at least 1/11 units of any model per time unit. The precedence graphs of the models as well as the joint precedence graph with the cumulated operation times t'_j as node weights are shown in Figure 3.2. The operation times are summarized in Table 3.1.

Table 3.1. Operation Times of the Example Problem

j	1	2	3	4	5	6	7	8	9	10	11	12	$\Sigma_j t_{jm}$
t_{j1}	6	7	4	5	5	2	4	7	3	1	10	1	55
t_{j2}	5	11	6	5	3	1	1	4	4	1	11	0	52
t_{j3}	8	13	0	5	2	4	3	13	1	1	8	3	61
t'_j	42	63	28	35	28	14	21	49	21	7	70	7	385

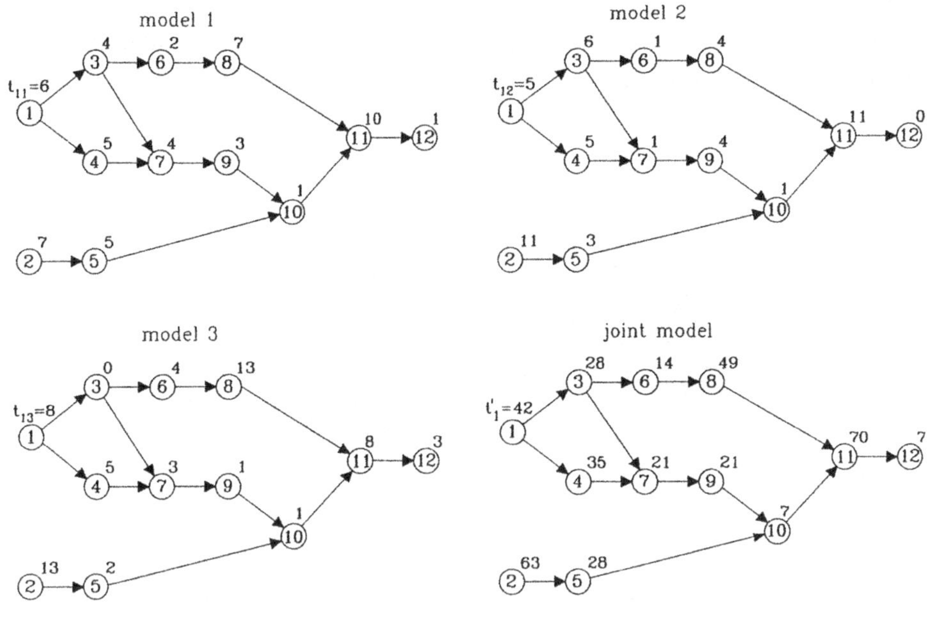

Figure 3.2. Precedence Graphs of the Models

A possible assignment of the tasks to m = 6 stations is given by $S_1 = \{1,4\}$, $S_2 = \{2\}$, $S_3 = \{3,5,7\}$, $S_4 = \{6,9,10\}$, $S_5 = \{8\}$, $S_6 = \{11,12\}$. The resulting operation times per model and station are summarized in Table 3.2. (Note that the terms operation time, station time and processing time are used synonymously.) This assignment fulfills the joint precedence constraints and ensures that the average station times per unit (τ_k) do not exceed the cycle time c = 11.

Table 3.2. Operation Times of Models and Stations

m \ k	1	2	3	4	5	6	$\bar{t}_m$ ($\bar{t}'_m$)
τ_{1k} (τ'_{1k})	11 (44)	7 (28)	13 (52)	6 (24)	7 (28)	11 (44)	9.17 (36.67)
τ_{2k} (τ'_{2k})	10 (20)	11 (22)	10 (20)	6 (12)	4 (8)	11 (22)	8.67 (17.33)
τ_{3k} (τ'_{3k})	13 (13)	13 (13)	5 (5)	6 (6)	13 (13)	11 (11)	10.17 (10.17)
τ_k (τ'_k)	11 (77)	9 (63)	11 (77)	6 (42)	7 (49)	11 (77)	

The spreads of the station times are illustrated by the *utilization diagram* in Figure 3.3, where hatched areas represent the intervals of the station times per unit of all models. The broken lines indicate the average station times τ_k.

The utilization diagram and Table 3.2 show that the distribution of work among the stations is rather unequal with respect to the average station times as well as their model-dependent variations. While the utilization of station 4 is constant at level 54.5 %, the station times for the stations 1, 2, 3, and 5 exceed the cycle time for one model, in each case. Only station 6 shows neither idle capacity nor work overload.

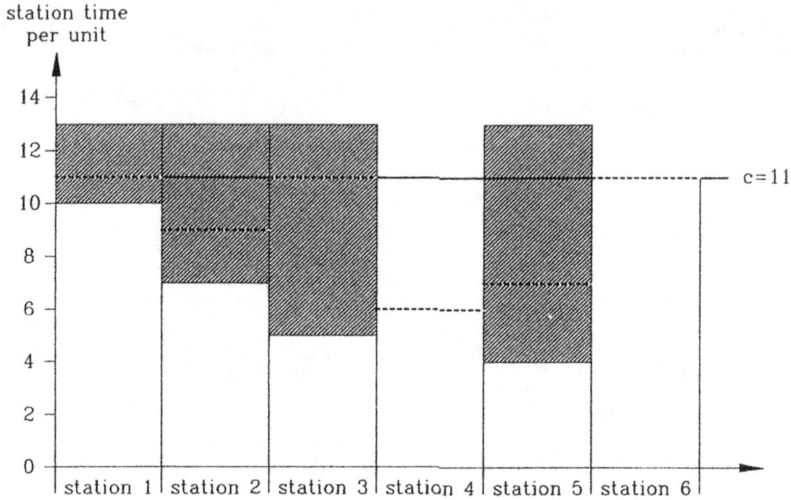

Figure 3.3. Utilization Diagram

3.2.2 Different Versions of MALBP

According to the single-model case, several versions of the mixed-model assembly line balancing problem (MALBP) may be distinguished. Hence, we first describe how to adapt the versions of SALBP to the mixed-model case. Furthermore, we examine the additional aspects of model-dependent work load variations and discuss objectives which attempt to equalize such variations.

3.2.2.1 Relationship of MALBP to Single-Model Problems

We describe mixed-model equivalents of the different versions of SALBP by means of the following definitions.

Definition 3.1. In a *mixed-model assignment*, each task of the joint precedence graph is assigned to exactly one station and the joint precedence constraints are fulfilled. A mixed-model assignment (with K stations) is *feasible* for a cycle time c if the average station times τ_k do not exceed c for all stations k = 1,...,K.

The definition includes that tasks which are common to several models are assigned to the same station, because those tasks are represented by the same node in the joint precedence graph. The feasibility condition ensures that the total operation times τ_k^t of the stations do not exceed the total available time PT. According to single-model problems, we use the term *line balance* for a feasible assignment.

Definition 3.2. An *average model* is defined by the joint precedence graph and the average task times t_j. A *joint (aggregate) model* includes the joint precedence graph with the operation times $t_j^!$.

With the average and the aggregate model, equivalent balancing problems are defined. While the problem is considered on a cycle time basis in the first case, it refers to the whole planning horizon in the latter case.

Feasibility Problem (MALBP-F). By analogy with SALBP-F, we assume that a cycle time c and a number K of stations are given. The cycle time determines the *launch interval,* i.e., every c time units a model unit is launched down the line. For a given combination (K,c), MALBP-F is to decide whether or not a feasible mixed-model assignment with K stations exists.

The given definition of MALBP-F corresponds to a transformation of the mixed-model problem to SALBP-F by using the average model. Following this transformation, each MALBP-F instance with M models can be formulated as a SALBP-F instance for the average model. That is, the solution spaces of both problem types are identical. Hence, the same problem properties, including the NP-completeness of MALBP-F, and model formulations are valid (see Section 2.2.1). Furthermore, generalized problems equivalent to SALBP-1, SALBP-2, and SALBP-E can be distinguished.

MALBP-1. For a given cycle time c, the number of stations is to be minimized. Usually, the cycle time is derived from the total available time PT and the desired output volume D, i.e., $c = \lfloor PT/D \rfloor$. An equivalent single-model problem (SALBP-1) is obtained by considering the average model with cycle time c or the aggregate model with "cycle time" PT (cf. Thomopoulos (1970)).

MALBP-2. For a given number K of stations, the cycle time c has to be minimized. Equivalently, the production rate, i.e., the output volume D of the planning period is maximized. Due to the variability of D, only expected demand proportions of the models *(relative model mix)* can be specified instead of numbers of required units. An equivalent SALBP-2 problem is obtained for the average model.

MALBP-E. According to SALBP-E, the cycle time c (the production rate) as well as the number K of stations may vary in certain ranges. The objective is to maximize the line efficiency or, equivalently, to minimize $c \cdot K$.

In the literature, only MALBP-1 has been considered so far. According to SALBP-1, this problem type has its applications mainly in the context of installing new mixed-model lines. Nevertheless, the number of stations may be fixed in a running system which has to be *re-balanced* due to changes of the demand structure or the production process. Then MALBP-2 becomes relevant (cf. Section 3.4). In order to avoid installing inefficiencies into the system due to inadequate pre-specified parameters c or K, respectively, MALBP-E may be used to examine different parameter settings. Furthermore, this problem type may help to decide on alternatives in the production system such as parallel lines or parallel stations (cf. Section 1.3.5 and Section 2.3.3).

3.2.2.2 Variations of Station Utilization

The objectives included in the versions of MALBP discussed above are based on considering the balancing problem with respect to *average* station utilizations. Even in the case of an optimal solution for the average model, considerable inefficiencies may occur when operating the line. This is due to the variations in the station times of the models. As shown by our example in Section 3.2.1, these variations may be rather large. In the following, we discuss the influence of work load variations on the performance of mixed-model lines.

Work Overload. Whenever the operator of a station is not able to complete the assigned operations on a workpiece, *work overload* occurs. It is measured in terms of the remaining operation time to complete the operations. Usually, work overload is compensated by the additional temporary employment of *utility workers (floaters)*. These workers have to be available whether or not work overload situations actually occur and must be highly qualified in order to perform a large set of different operations. Accordingly, this type of work is rather expensive and the (maximal) requirement of utility work should be as small as possible. In the absence of a sufficient floater capacity, work overload may be dealt with by stopping the line until the remaining work is completed or by off-line repair (see p. 69). However, work overload is inefficient and expensive and should be minimized. Unfortunately, the amount of work overload which really occurs cannot be computed for a solution of the balancing problem directly, because it depends on the unknown short-term production programs and corresponding production sequences. Therefore, it is essential to obtain balances in which *potential* work overload situations are minimized (see the discussion of cycle time violations below).

Tolerance Time. The interval of time during which a workpiece is accessible to a station is called *tolerance time* (cf. Section 1.2). It is an upper bound on the time available for performing operations in that station. Obviously, the tolerance time should not be smaller than the cycle time in order to avoid unnecessary idle times. In case of a continuously moving conveyor belt, the tolerance times are determined by the physical station lengths and the speed of the belt.

Work overload can only occur for feasible assignments if the tolerance times are restricted and some station times exceed the cycle time (cf. Dar-El (1978)). This is always the case in real-world problems, e.g., due to a restricted floor space or limited reachs of machines. Nevertheless, tolerance times are usually variable in certain ranges, especially when a new line is installed. Then they have to be determined in connection with balancing the line. If the tolerance times are fixed, they serve as restrictions of a balancing problem. Obviously, station times exceeding the tolerance times always result in work overload. Hence, it is reasonable to avoid such excess times by imposing respective upper bounds on the station times τ_{mk} of the models.

We illustrate the influence of tolerance times by assuming that they are equal to the cycle time in all stations (strictly paced line; cf. Deutsch (1971)). All station times of any model which are in excess of the cycle time (i.e. $\tau_{mk} > c$) result in incomplete operations (whatever sequence of models is used) and a corresponding work overload. If the tolerance times exceed the cycle time, the sequence in which model units are launched down the line influences the amount of work overload. For our example, this is explained by assuming that two units of model 1 are consecutively produced within a sequence. Furthermore, we assume that each workpiece is accessible to the stations for 13 time units, i.e., the tolerance time is 13 for all stations. For the task assignment given in Table 3.2, we recognize that in all stations but station 3 the operation times τ_{mk} do not exceed the cycle time $c=11$. Hence, in these stations, no work overload can occur for the considered two units of model 1. In station 3, the first unit requires a processing time of 13 time units. Since the launch interval is equal to the cycle time (fixed rate launching), the second unit is already accessible to the station for 2 time units when the first unit is finished. Therefore, only 11 time units remain for processing the second unit of model 1 before it leaves the station. Thus, a work overload of 2 time units has to be compensated by utility work in station 3. No work overload occurs in that station (but possibly in other ones) if other pairs of consecutive model units are considered.

Idle Time. As in the single-model case, idle time occurs when a station has completed its work on a unit and has to wait for the next unit arriving at the station. The idle times per unit are constant if only one model is produced. If several models are assembled, the idle times differ and depend on the sequence. Only for strictly paced lines, they are independent of the sequence (see above).

Cycle Time Violation. As mentioned above, work overload can only occur if some station times of models exceed the cycle time. In order to quantify such *cycle time violations,* we define:

$$\Delta^{+}_{mk} := \max\{0, \tau_{mk} - c\} \qquad \text{for } k=1,\dots,K \text{ and } m=1,\dots,M \tag{3.1}$$

According to the origin of cycle time violations, we distinguish three types:

Operation-dependent Cycle Time Violations. A violation of the cycle time cannot be avoided by balancing if the operation time t_{jm} of a single task j of model m exceeds the cycle time.

Model-dependent Cycle Time Violations. If the average operation time $\bar{t}_m$ per unit of model m exceeds the cycle time, cycle time violations cannot be prevented in one or more stations unless the cycle time c or the number K of stations is increased (note that $\bar{t}_m = \Sigma_j t_{jm}/K$).

Assignment-dependent Cycle Time Violations. Cycle time violations which are neither operation-dependent nor model-dependent are caused by the assignment of tasks to stations. Hence, these violations may be influenced by balancing decisions even for fixed c and K.

Slack Time. Station times which are smaller than the cycle time may cause idle times. These *potential* idle times are called *slack times* and are defined by:

$$\Delta^-_{mk} := \max\{0, c - \tau_{mk}\} \qquad \text{for } k = 1,...,K \text{ and } m = 1,...,M \qquad (3.2)$$

Like cycle time violations, slack times may have different reasons:

Model-dependent Slack Times. In case the average operation time $\bar{t}_m$ per unit of model m is smaller than the cycle time, slack times cannot be avoided in one or more stations unless the cycle time c or the number K of stations is decreased (if possible).

Assignment-dependent Slack Times. Slack times which are not model-dependent are caused by the assignment of the tasks to the stations and, therefore, they may be influenced by balancing.

While slack times tend to produce idle times, they give possibilities to equalize cycle time violations of other models. Therefore, it is not always the best decision to choose a task assignment which minimizes slack times.

The cycle time violations Δ^+_{mk} and the slack times Δ^-_{mk} for the solution of our example (cf. Table 3.2, p. 81) are presented in Table 3.3. While model 2 shows no cycle time violations, the cycle time is exceeded by model 1 in one station and by model 3 in three stations. That is, an arbitrary large number of units of model 2 can consecutively be processed without a work overload. On the contrary, the consecutive production of several units of model 3 will cause work overload of 6 time units per workpiece if the tolerance times are equal to the cycle time. In case the tolerance times are longer, a part of the work overload may be avoided by subsequently producing a number of model units with positive slack times. For example, the cycle time violation of model 3 in station 1 may be equalized by two units of model 2 and that in station 2 by one unit of model 1.

Table 3.3. Cycle Time Violations and Slack Times

m \ k	1	2	3	4	5	6	Σ
$\Delta^+_{1k}/\Delta^-_{1k}$	0/0	0/4	2/0	0/5	0/4	0/0	2/13
$\Delta^+_{2k}/\Delta^-_{2k}$	0/1	0/0	0/1	0/5	0/7	0/0	0/14
$\Delta^+_{3k}/\Delta^-_{3k}$	2/0	2/0	0/6	0/5	2/0	0/0	6/11
Σ	2/1	2/4	2/7	0/15	2/11	0/0	8/38

3.2.2.3 Objectives for Smoothed Station Loads

In this section, we present and discuss objectives which are intended to equalize work load variations and may be added to any of the problems of Section 3.2.2.1. As described above, these objectives should prefer line balances which avoid potential inefficiencies such as work overload and idle time. A large number of corresponding objectives are possible. We present and compare two objectives, which have been discussed in the literature, as well as a new one.

First of all, we assume that a number K of stations and a cycle time c are given (MALBP-F). While all feasible solutions of a SALBP-F instance are equivalent with respect to the idle time of the line, this is not necessarily true for MALBP-F instances, because different idle times and work overloads may occur depending on the distribution of work (note that work overload cannot arise in deterministic single-model lines). Therefore, additional objective functions have to be used to evaluate and order the feasible solutions for the given (K,c)-combination with respect to their work load variations and their expected influence on work overload and idle time as well.

Objective 1. Thomopoulos (1970) proposes the following objective function which minimizes the sum of absolute differences between the average total station times $\bar{t}'_m$ and the total operation times τ'_{mk} of all models in all stations:

$$\text{Minimize } \Psi'_1 = \sum_{k=1}^{K} \sum_{m=1}^{M} | \tau'_{mk} - \bar{t}'_m | \tag{3.3}$$

A modified version of this objective, which is independent of a particular model mix, is given by (3.4).

$$\text{Minimize } \Psi_1 = \sum_{k=1}^{K} \sum_{m=1}^{M} |\tau_{mk} - \bar{t}_m| \tag{3.4}$$

Objective 2. The following model mix-independent objective function minimizes the maximal deviation of an operation time of any model from the average station time per unit (cf. Link (1992)):

$$\text{Minimize } \Psi_2 = \max\{ |\tau_{mk} - \bar{t}_m| \text{ with } k=1,...,K \text{ and } m=1,...,M \} \tag{3.5}$$

By analogy with objective 1, a modified function Ψ_2' may be defined by using the total times τ'_{mk} and $\bar{t}'_m$.

Objective 3. This simple new objective function minimizes the sum of cycle time violations of all models in all stations (cf. Domschke et al. (1996 a)):

$$\text{Minimize } \Psi_3 = \sum_{k=1}^{K} \sum_{m=1}^{M} \max\{0, \tau_{mk} - c\} = \sum_{k=1}^{K} \sum_{m=1}^{M} \Delta^+_{mk} \tag{3.6}$$

Obviously, a large number of further objectives is possible. For example, one may use squared deviations (cf. the smoothness index in Definition 2.6, p. 25) or minimize the difference between the largest and the smallest operation times. To the best of the author's knowledge, there is no comprehensive discussion available in the open literature which compares such objective functions with respect to their ability of equalizing work load variations (some aspects are considered by Decker (1993 a, chapter 2)).

We compare the objectives with respect to our example and discuss structures of the solutions preferred as well as their influences on the short-term sequencing problem.

Discussion of Objective 1. If objective 1 is added to the MALBP-F instance defined by K = 6 and c = 11, we obtain the following solution. The optimal objective function value is $\Psi_1' = 70.33$.

$$S_1 = \{1,3\}, S_2 = \{4,7\}, S_3 = \{2\}, S_4 = \{6,8\}, S_5 = \{5,9,10\}, S_6 = \{11,12\}$$

Note that eight further optimal solutions are available which differ from the considered one only by the order of stations. The corresponding operation times are given in Table 3.4 and are illustrated by the utilization diagram of Figure 3.4.

The solution shows an acceptable average work load distribution, because all stations have similar average station times τ_k. The same is true for model 1, while the station loads for model 2 and, in particular, for model 3 vary remarkably. Both model 1 and model 2 show no cycle time violations. The assignment contains large slack times ($\Delta^-_{35} = 7$) and cycle time violations ($\Delta^+_{34} = 6$) for model 3. The cycle time violation in station 4 is mainly assignment-dependent, because task 6 is assigned in spite of the operation-dependent violation by task 8. The large station times of model 3 require large tolerance times (long stations) to maintain the possibility of equalizing these violations by units of other models.

Table 3.4. Operation Times of Models and Stations for Objective 1

m \ k	1	2	3	4	5	6	$\bar{t}_m$ ($\bar{t}'_m$)
τ_{1k} (τ'_{1k})	10 (40)	9 (36)	7 (28)	9 (36)	9 (36)	11 (44)	9.17 (36.67)
τ_{2k} (τ'_{2k})	11 (22)	6 (12)	11 (22)	5 (10)	8 (16)	11 (22)	8.67 (17.33)
τ_{3k} (τ'_{3k})	8 (8)	8 (8)	13 (13)	17 (17)	4 (4)	11 (11)	10.17 (10.17)
τ_k (τ'_k)	10 (70)	8 (56)	9 (63)	9 (63)	8 (56)	11 (77)	

The stated situation is mainly due to the influence of the model mix which is taken into consideration by objective 1, because variations of the models with large demands are penalized harder than those of the models which are required less frequently. Furthermore, the large deviations for model 3 are compensated by many small deviations for the models 1 and 2.

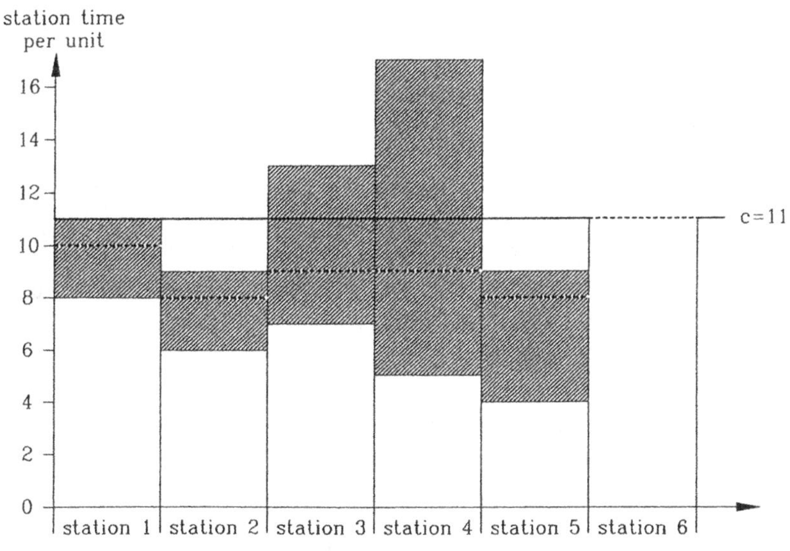

Figure 3.4. Utilization Diagram for Objective 1

For the given model mix, the sequence <3,2,1,2,1,1,1] of model units does not cause any work overload if the tolerance times (station lengths in terms of time) are set to the maximal station time in each station, i.e., $l_k := \max\{\tau_{mk} \mid m = 1,...,M\}$ for k = 1,...,K. In the case where more than two units of model 3 are required (for the given total demand D = 7), work overload cannot be avoided, because the remaining units of either model 1 or model 2 are not sufficient to equalize the cycle time violations.

The above observations are summarized as follows. Objective 1 tends to prefer solutions which are acceptable for the given model mix, but may result in rather large cycle time violations as well as slack times for models required less frequently. If the short-term model mix is not almost constant over time, such a line balance may lead to considerable inefficiencies. This disadvantage is partly removed by using the demand-independent objective function Ψ_1. However, this function still has the drawback of compensating large deviations by several small ones.

Discussion of Objective 2. With objective 2, we obtain the following solution for K = 6 and c = 11 with the optimal objective function value $\Psi_2 = 3.83$:

$$S_1 = \{1,3\},\ S_2 = \{2\},\ S_3 = \{4,7\},\ S_4 = \{5,6,9\},\ S_5 = \{8,10\},\ S_6 = \{11,12\}$$

The chosen solution is one out of five equivalent optimal line balances. The model-dependent and average station times are given in Table 3.5 and are illustrated by the utilization diagram of Figure 3.5.

Table 3.5. Operation Times of Models and Stations for Objective 2

m \ k	1	2	3	4	5	6	$\bar{t}_m$ $(\bar{t}'_m)$
τ_{1k} (τ'_{1k})	10 (40)	7 (28)	9 (36)	10 (40)	8 (32)	11 (44)	9.17 (36.67)
τ_{2k} (τ'_{2k})	11 (22)	11 (22)	6 (12)	8 (16)	5 (10)	11 (22)	8.67 (17.33)
τ_{3k} (τ'_{3k})	8 (8)	13 (13)	8 (8)	7 (7)	14 (14)	11 (11)	10.17 (10.17)
τ_k (τ'_k)	10 (70)	9 (63)	8 (56)	9 (63)	8 (56)	11 (77)	

The solution shows the same quality of the average work load distribution as the solution for objective 1. While the variations for the models 1 and 2 are similar, that of model 3 is considerably better. Nevertheless, the cycle time violation of model 3 in station 5 is partly assignment-dependent.

Due to the smaller maximal station times, the tolerance times can be reduced relative to the solution for objective 1. Several sequences allow the production of the given model mix without work overload. At most three units of model 3 can be produced (with D = 7) in order to avoid work overload.

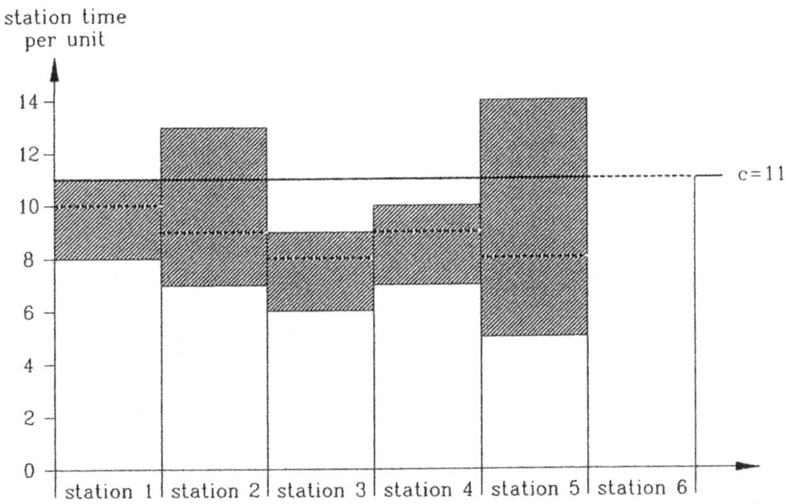

Figure 3.5. Utilization Diagram for Objective 2

Even though objective 2 attempts to prevent large variations of station times, it has major shortcomings, because only a single deviation determines the objective function value. If there is a model which requires only a subset of the tasks, it may be unavoidable that station times of this model are zero or very small. Then the optimal value Ψ_2^* of objective 2 is determined by a corresponding deviation, while all other station times are allowed to vary within the (large) intervals $[\bar{t}_m - \Psi_2^*, \bar{t}_m + \Psi_2^*]$ without influencing the objective function value. As a consequence, a further objective function is needed as a tie breaker in order to find reasonable solutions, whenever Ψ_2 is trapped by some deviations which cannot be removed. Link (1992) proposes the function Ψ_1 as tie breaking goal. However, this function has disadvantages, too (see above). In the case of our example, all solutions optimal for Ψ_2 have the same value of Ψ_1.

Discussion of Objective 3. This objective results in the following solution for $K=6$ and $c=11$ with the optimal objective function value $\Psi_3=4$:

$$S_1=\{2\},\ S_2=\{1,5\},\ S_3=\{3,4\},\ S_4=\{6,7,9,10\},\ S_5=\{8\},\ S_6=\{11,12\}$$

Besides the solution chosen, five equivalent optimal line balances exist. The station times are summarized in Table 3.6 and are illustrated by the utilization diagram of Figure 3.6. The given assignment does not contain assignment-dependent cycle time violations, because the violations of model 3 in the stations 1 and 5 are operation-dependent and cannot be avoided. The average distribution of work is similar to that of the solutions for objectives 1 and 2, respectively, but the average utilization of station 5 is rather low even though model 3 violates the cycle time.

This is due to the small station time of model 2 accepted by objective 3 which does not consider slack times.

Table 3.6. Operation Times of Models and Stations for Objective 3

m \ k	1	2	3	4	5	6	$\bar{t}_m$ ($\bar{t}'_m$)
τ_{1k} (τ'_{1k})	7 (28)	11 (44)	9 (36)	10 (40)	7 (28)	11 (44)	9.17 (36.67)
τ_{2k} (τ'_{2k})	11 (22)	8 (16)	11 (22)	7 (14)	4 (8)	11 (22)	8.67 (17.33)
τ_{3k} (τ'_{3k})	13 (13)	10 (10)	5 (5)	9 (9)	13 (13)	11 (11)	10.17 (10.17)
τ_k (τ'_k)	9 (63)	10 (70)	9 (63)	9 (63)	7 (49)	11 (77)	

With respect to possible production sequences, this solution is superior to those of the other objective functions. All sequences which include at least one unit of model 1 between every two units of model 3 do not generate work overload (for reasonable tolerance times), because model 1 equalizes the cycle time violations of model 3 in the stations 1 and 5.

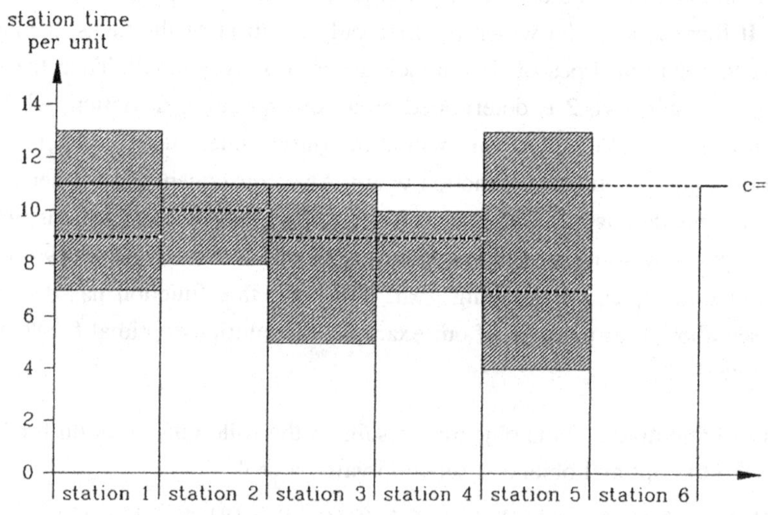

Figure 3.6. Utilization Diagram for Objective 3

The above discussion shows several advantages of objective 3 relative to the other objectives which are confirmed by numerical experiments of Domschke et al. (1996 a):

- Cycle time violations are avoided whenever possible. This is illustrated by a slight modification of our example. Assume the operation times $t_{23} = t_{83} = 13$ being reduced to 11 in order to eliminate operation-dependent violations. Then the solution of Table 3.6 represents a feasible solution without cycle time violations. Obviously, this solution is optimal with respect to objective 3, and all possible sequences for any model mix do not contain work overload. Even though such solutions exist, the objectives 1 and 2 prefer assignments with maximal station times of 15 and 12, respectively. That is, unnecessary cycle time violations are allowed which may result in work overload. No sequencing problem arises for the considered problem instance in case of objective 3 (provided that the total work overload has to be minimized), whereas some effort is necessary to find efficient sequences for the balances of the other objectives. Furthermore, work overload may be unavoidable for particular short-term model mixes.

- Slack times are not explicitly considered by objective 3, whereas they influence the objectives 1 and 2. As mentioned above, large unavoidable slack times are responsible for rather critical cycle time violations being accepted by these objectives. Furthermore, slack times may be desired due to their ability of avoiding work overload by compensating (unavoidable) cycle time violations. In spite of accepting a few large slack times, the minimization of the sum of cycle time violations results in a minimal sum of slack times, too.

Combination of Objectives. The above discussion is based on the assumption that a number K of stations and a cycle time c are given (MALBP-F plus objective). However, the objectives can be added to the optimizing versions of MALBP, too.

In the literature, the objective of minimizing the number of stations (MALBP-1) is taken as primary goal, while an additional objective for equalizing work load variations is considered as secondary goal (cf. Thomopoulos (1970), Link (1992)). That is, among all solutions with the minimal number of stations, the secondary goal is utilized to choose one task assignment which seems to be superior with respect to efficiently operating the line. The same approach can be used for objective 3. However, it may be advantageous to accept one or more additional stations if a much better division of work is possible.

Objective 3 is generalized for MALBP-1 by using the objective function

$$\text{Minimize } \Psi_3' = \sum_{k=1}^{K} \sum_{m=1}^{M} \max\{\tau_{mk}, c\}, \tag{3.7}$$

which considers cycle time violations, but excludes slack times in order to avoid compensation effects, too. The following transformation of the objective function shows that it aims at simultaneously minimizing the (weighted) number of stations and the sum of cycle time violations:

$$\text{Minimize } \Psi_3' = K \cdot M \cdot c + \sum_{k=1}^{K} \sum_{m=1}^{M} \Delta^{+}_{mk} \tag{3.7'}$$

The objective function (3.7') weights the primary goal such that an additional station is used if the sum of cycle time violations can be reduced by more than $M \cdot c$ at the same time. That is, a further station is accepted if the average sum of cycle time violations per unit ($\Sigma_k \Sigma_m \Delta^+_{mk} / M$) exceeds the cycle time c, which is the average capacity of the station per unit. This is based on the fact that cycle time violations represent potential work overloads which are usually more expensive than regular work.

Remark 3.1. Instead of the capacity oriented objective function (3.7'), we may consider a cost oriented function by using other weights for the two contained goals. For example, the number of stations may be weighted by an operating cost factor (including equipment and wage costs), while the sum of cycle time violations may be multiplied by a cost factor for the employment of floaters.

The objective (3.7') is applicable to the two other problem versions MALBP-2 and MALBP-E as well, because the first term contains both the cycle time and the number of stations. In all versions of MALBP, it helps to incorporate (expected) short-term consequences into medium-term decisions. Therefore, the application of such objectives is an important component of a hierarchical planning approach for mixed-model assembly lines (cf. Section 3.4).

3.3 Mixed-Model Sequencing

In this section, we consider the mixed-model sequencing problem (MSP). As already outlined in Section 3.1, this short-term decision problem is to find an intermixed sequence of model units such that the demands of all models are satisfied and some time or cost oriented objective is optimized.

We give a more precise statement of the problem (and its numerous versions) by describing possible operating characteristics of such flexible mixed-model lines and discuss several objectives. For a special version of the problem, we present a mathematical formulation. Surveys and problem classifications are given, among others, by Koether (1986), Yano and Bolat (1989), Ziegler (1990), Bard et al. (1992) as well as Domschke et al. (1997, chapter 4.4.2).

3.3.1 Operating Characteristics and Problem Statement

In what follows, we describe some of the numerous operational aspects of mixed-model lines. The short-term sequencing problem has to incorporate such operating characteristics as station types (open, closed), launching discipline, conveyor speed, and operator inference (cf., e.g., Dar-El (1978) and Bard et al. (1992)). Since real-world lines may incorporate any combination of such properties, it is not possible to give a general statement of the mixed-model sequencing problem, and a number of simplifying assumptions has to be made. In the following, we briefly discuss these assumptions, a part of which is already relevant for the balancing problem.

Line Balance and Layout. As mentioned in Section 3.1, the short-term sequencing problem is based on a given line balance, i.e., a solution of a particular MALBP instance, which determines the division of work among the stations of the line for all models. Furthermore, the physical layout of the line (including the station extents) is fixed, because it can usually not be changed in the short-run.

Planning Period. Within a short planning period (e.g., a shift, a day, or a week), certain demands of the different models have to be satisfied by production. These demands often result from individual customer orders with fixed delivery dates or from expected requirements, respectively. Note that *expected* demand proportions of the models are considered when balancing the line (by the average model and possibly the objective of MALBP). As in Section 3.2, we use the term *model mix* for the combination of the model demand values.

Launching Discipline and Station Types. The production rate of the system is determined by the conveyor speed and the launching frequency. Usually, the model units are launched down the line in constant intervals *(fixed rate launching),* because the production control is much easier when using this strategy instead of *variable rate launching* (cf. Section 1.3.7). Therefore, the units are equi-spaced on the conveyor belt. Depending on the type of operations, the stations may be *closed* or *open* (cf. Section 1.3.6). Since the analysis of the system is more difficult for the open station case, we assume that the stations are closed. However, there is no principle difference between both station types (cf. Ziegler (1990, pp. 38) for a discussion of the characteristics in the open station case).

Operators and Floaters. It is assumed that each station is manned by one operator. In the case where more operators are required to work simultaneously, the operation times at the respective station are assumed to be equally split up between these workers (note that this assumption is already contained in MALBP). If the regular work time provided by the operators is not sufficient to complete a workpiece, floaters (utility workers) are additionally employed in order to accelerate the operations. It is assumed that the work rate is doubled by adding utility work and that a sufficient number of floaters is available.

Tolerance Time. Each product unit is accessible to each station for a fixed time interval (tolerance time; see pp. 84) depending on the fixed station length and the constant line speed. During the tolerance time, the work load of the respective model has to be performed by the operator, while he rides downstream on the conveyor. For the sake of simplicity, it is assumed that the operator returns upstream to the left border of the station or to the next unit, whatever is reached first, in zero time after finishing the work load on the current unit.[2] This assumption seems not to be a hard restriction, because the line speed is usually much slower than the walking speed of the workers.

Work Overload and Idle Time. When an operator is not able to complete his operations on a unit before it leaves the station, a utility worker is additionally employed to compensate this work overload situation (see above). If the operator reaches the left border of his station before the next model unit comes into the station area, idle time occurs, i.e., the worker has to wait for the arrival of this unit (see p. 85). Furthermore, we assume that the operators of all stations start their operations as early as possible (late start schedules are considered by Dar-El (1978)).

We briefly summarize the main assumptions:

- given number of closed stations
- given work loads (station times) for all models in every station
- constant conveyor speed
- given station lengths
- fixed rate launching
- given demands for the models in the actual period (production program)
- zero operator return times
- early start of operations
- work overload is compensated by utility workers

Notations. For a precise statement of MSP, we require some notations (cf. Figure 3.7), a part of which is already introduced in Section 3.2. Since a constant conveyor speed is assumed, all distances can be measured in time units (TU for short).

Feasible solutions of MSP are given by sequences of length I which contain d_m units of each model $m = 1,...,M$. The operator start positions s_{ki} describe the distance of the operator from the left border of station k (in terms of time), just before he starts processing the unit at position i of the sequence. The operation times τ_{mk} are assumed not to exceed the time equivalents l_k of the station lengths. This assumption is not restrictive, because a utility work of at least $\tau_{mk} - l_k$ TU is unavoidable if $\tau_{mk} > l_k$. Therefore, the problem can be reduced by eliminating such excess times.

---

2 As a matter of convenience, we assume that the line moves from the left to the right.

M	number of models, index: $m = 1,...,M$
d_m	demand for model m during the planning period
I	length of the sequence, total demand $(= \Sigma_m d_m)$, index: $i = 1,...,I$
c	fixed launch interval (cycle time, in TU)
K	number of stations, index: $k = 1,...,K$
l_k	length of station k (in TU; = physical length / conveyor speed)
τ_{mk}	processing time per unit of model m in station k $(\tau_{mk} \leq l_k)$
$\bar{t}_m$	average operation time per station and unit of model m $(= \Sigma_k \tau_{mk}/K)$
τ'_k	sum of operation times in station k per period $(= \Sigma_m d_m \cdot \tau_{mk})$
Δ_{mk}	displacement of the operator in station k by a unit of model m $(= \tau_{mk} - c)$
x_{mi}	binary variable, $= \begin{cases} 1 & \text{if a unit of model m is assigned to seq. position i} \\ 0 & \text{otherwise} \end{cases}$
s_{ki}	operator start position in station k for the i-th unit (in TU)
e_{ki}	operator end position in station k after processing the i-th unit (in TU)
w_{ki}	utility work in station k to complete the i-th unit of the sequence (in TU)

Figure 3.7. Notations for Mixed-Model Sequencing

Example. We consider a problem with $M = 3$ models and the demands $d_1 = 2$, $d_2 = 2$, and $d_3 = 1$. The launch interval is $c = 9$. Each of $K = 4$ stations has the length $l_k = 10$ TU. All sequences which contain the elements 1, 1, 2, 2, 3 in an arbitrary order are feasible. The total available time per station is $I \cdot c = 45$ TU.

The operation times τ_{mk} (obtained from a line balance), the average station times $\bar{t}_m$ per model unit, and the total station times τ'_k are given in Table 3.7.

Table 3.7. Operation Times

m \ k	1	2	3	4	$\bar{t}_m$
1	10	7	10	6	8.25
2	8	10	9	8	8.75
3	9	8	7	10	8.5
τ'_k	45	42	45	38	42.5

We explain the operation of the sequence <1, 1, 2, 2, 3>. This solution is represented by the values $x_{11} = x_{12} = x_{23} = x_{24} = x_{35} = 1$ and $x_{mi} = 0$, in all other cases. In station 1, the operation time $\tau_{11} = 10$ of model 1 is larger than the cycle time $c = 9$. Therefore, the operator starts at $s_{11} = 0$ (left border of station 1) and rides downstream together with the first unit, until it reaches the right station border after 10 TU ($e_{11} = 10$). At this time, the second unit is inside the station 1 at position 1 TU. Since we assume zero return times, the operator reaches the unit at this position ($s_{12} = 1$). The remaining time for the second unit in station 1 is 9 TU, and a utility work of $\tau_{11} - 9 = 1$ TU occurs. The third unit (model 2) is reached when it stays at position $s_{13} = 1$ inside the station. After performing the tasks with operation time $\tau_{21} = 8$, the operator is at position $1 + 8 = 9$ TU. At this time, the next unit (model 2 again) enters the station and is immediately processed ($s_{14} = 0$). This stops after 8 TU, and the operator returns to $s_{15} = 0$. The last unit (model 3) enters the station 1 TU later, and the operator is idle for 1 TU. After processing for $\tau_{31} = 9$ TU, the operator returns to the left border (denoted by the dummy start point $s_{16} = 0$).

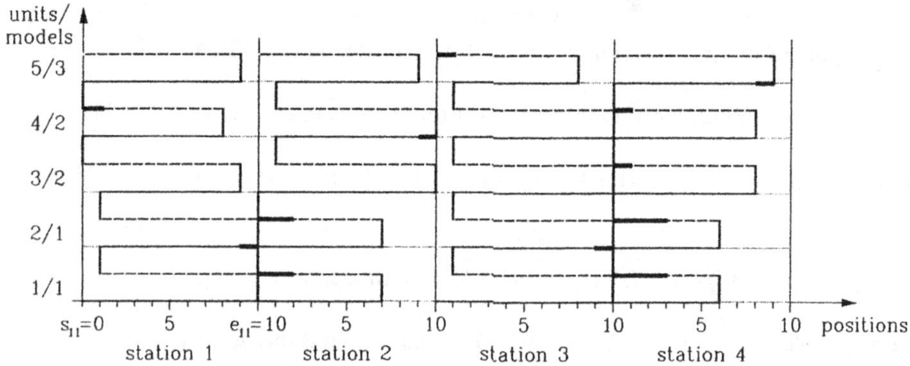

Figure 3.8. Movement Diagram

Movement Diagrams. The movements of all operators necessary to process the units of the sequence can be visualized by so-called movement diagrams. Such diagrams show the changes of the operator positions and the movements of the units. In the case of closed stations, the operators do not interfere with each other. Therefore, the stations $k = 1,\ldots,K$ are completely described by their lengths l_k. Positions inside the stations refer to the left border which is set to zero. On the y-axis, the units $i = 1,\ldots,I$ of the sequence are denoted. Operation times are represented by horizontal lines connecting points s_{ki} and e_{ki}. The upstream return distance of the operator is depicted by dashed lines of length $e_{ki} - s_{k,i+1}$ from the right to the left. Dotted lines connecting positions e_{ki} and $s_{k+1,i}$ represent waiting times during which the respective unit is not being processed by station k or k+1. Bold parts of lines for operation times show utility work, whereas such for return distances represent idle times. For our example with the sequence <1, 1, 2, 2, 3>, the operator movements are shown in Figure 3.8. The total time which has to be performed by utility workers is 4 TU. While the work overload occurring in station 1, 2, and 3, respectively, results from reaching the right station border, the utility work of the last unit in station 4 is necessary in order to return to the left station border after processing the sequence. The sum of idle times is 14 TU.

If the stations (the tolerance times) are remarkably longer than the cycle time, the operators successively depart from the left borders of their stations when several units with operation times exceeding the cycle time are consecutively produced. Contrarily, the consecutive production of models with operation times lower than c leads to idle times, provided that the operators start at the left border.

Displacements. The relative change of the operator's position before and after producing a unit of model m in station k is defined by the displacement:

$$\Delta_{mk} := \tau_{mk} - c \qquad \text{for } k=1,\ldots,K \text{ and } m=1,\ldots,M \qquad (3.8)$$

In the case of $\Delta_{mk} > 0$, the work load of station k causes a cycle time violation for model m (cf. the values Δ^+_{mk}, pp. 85). If possible, the operator drifts to the right by a distance of Δ_{mk} TU. Otherwise, work overload occurs. For $\Delta_{mk} < 0$, the work load of station k for model m results in a slack time $\Delta^-_{mk} = -\Delta_{mk}$, and the operator drifts to the left. If the distance to the left border is smaller than Δ^-_{mk}, idle time appears. For our example, the values of Δ_{mk} are given in Table 3.8.

Table 3.8. Displacements

m \ k	1	2	3	4
Δ_{1k}	1	-2	1	-3
Δ_{2k}	-1	1	0	-1
Δ_{3k}	0	-1	-2	1

Calculation of a Sequence. For a given sequence described by binary variables x_{mi}, the starting points s_{ki} of operating the i-th unit of the sequence in station k can be computed by the following recursive formulae.

$$s_{k1} := 0 \qquad \text{for } k=1,\ldots,K \qquad (3.9)$$

$$s_{ki} := \max\{0, \min\{s_{k,i-1} + \Sigma_m \Delta_{mk} \cdot x_{m,i-1}, l_k - c\}\} \quad \text{for } k=1,\ldots,K,\ i=2,\ldots,I+1 \quad (3.10)$$

The formulae (3.10) can be explained by using additional variables e_{ki} denoting the position after processing the i-th unit and before returning upstream:

$$e_{ki} := \min\{s_{ki} + \Sigma_m \tau_{mk} \cdot x_{mi}, l_k\} \qquad \text{for } k=1,\ldots,K \text{ and } i=1,\ldots,I \quad (3.11)$$

The formulae (3.11) take into consideration that work cannot be continued when the right station border is reached at position l_k. In this case, utility work of duration w_{ki} is required to compensate the work overload.

$$w_{ki} := \max\{0, s_{ki} + \Sigma_m \tau_{mk} \cdot x_{mi} - l_k\} \qquad \text{for } k=1,...,K \text{ and } i=1,...,I \quad (3.12)$$

If the operator has to return to the left border after finishing the last unit I ($s_{k,I+1} := 0$), an additional utility work of $\max\{0, e_{kI} - c\}$ is necessary.

With e_{ki}, the starting point $s_{k,i+1}$ results from comparing the position of the next unit i+1 (note that the upstream distance to unit i is equal to the fixed launch interval c) with the left border of station k (which is set to zero) as shown in (3.13).

$$s_{k,i+1} := \max\{0, e_{ki} - c\} \qquad \text{for } i=1,...,I \text{ and } k=1,...,K \qquad (3.13)$$

If $e_{ki} - c < 0$, the left border is reached before unit i+1 enters the station, and an idle time of $c - e_{ki}$ TU occurs. Otherwise, unit i+1 has to wait for being processed during a time of $e_{ki} - c$ TU after entering the station. The formulae (3.10) directly follow from (3.11) and (3.13).

Definition 3.3. The *actual state* $S_i = (s_{1i}, ..., s_{Ki})$ of an assembly line, immediately before processing the i-th unit of the sequence, is characterized by the positions s_{ki} of the operators relative to the left border of the stations $k = 1,...,K$. The state $(0, 0, ..., 0)$ is termed *initial state.*

3.3.2 Sequencing Objectives

In Section 3.3.1, we describe the operating of mixed-model lines for a given sequence (feasible solution). In this section, we present different objectives considered in the literature which should help to find efficient sequences. Usually, time related goals are used, while only few references propose cost oriented ones. Furthermore, some objectives are utilized to obtain (almost) uniform requirements of the used materials in order to facilitate the realization of the just-in-time concept.

Time Related Objectives. We discuss a number of goals which consider the minimization of either a sum or a maximum of certain time values. While the minimization of the sum of flow times is a frequent objective in many scheduling problems, this objective is inappropriate for MSP due to the constant production rate of the assembly line. Therefore, the sum of flow times is independent of the model sequence (cf. Schneeweiß and Söhner (1991, pp. 25)).

Total Work Overload. A possible objective consists of minimizing the total amount of work overload (cf., e.g., Yano and Rachamadugu (1991)). It is identical with the minimization of the sum of times during which utility workers have to be employed in the planning period. A formal statement of this objective is given by (3.14).

$$\text{Minimize } WO = \sum_{k=1}^{K} \sum_{i=1}^{I} w_{ki} \tag{3.14}$$

Total Idle Time. Partly, idle times can be influenced by the model sequence. Their sequence-dependence increases with the lengths of the stations and the differences in the station times of the models. Since idle times represent unused capacities of the line, a possible objective is to minimize the sum of idle times. A corresponding objective function is expressed as follows (cf. the definition of e_{ki} in (3.11)).

$$\text{Minimize } IT = \sum_{k=1}^{K} \sum_{i=1}^{I} \max\{0, c - e_{ki}\} \tag{3.15}$$

The function (3.15) considers that idle times occur whenever an operator returns to the left border of his station before the next unit arrives. It includes the *intermediate* idle times only, i.e., the idle times between consecutive units of the sequence and the idle times immediately after the last unit. This is based on the assumption that the line is not empty at the beginning of the planning period. In the case of an empty line, all stations but the first are idle until the first unit arrives. However, these initial idle times are independent of the sequence and are not relevant for optimization. Equivalently, the line is assumed to continue its production after the last unit of the sequence.

Remark 3.2. The total idle time of the line may contain times which are unavoidable due to the unequal division of labour among the stations. Each station $k = 1,...,K$ provides a total productive time of $PT = c \cdot I$. The time required for processing all units is given by τ_k^I. In the case of $\tau_k^I < PT$, a total idle time of duration $PT - \tau_k^I$ occurs independently of the chosen model sequence. Note that the inequalities $\tau_k^I \leq PT$ represent the cycle time restrictions of MALBP (cf. Section 3.2.2.1).

Equivalence of Objectives. The total work overload and the total (sequence-dependent) idle time are *identical* if the unavoidable balance-dependent idle times (see Remark 3.2) are eliminated (cf. Yano and Bolat (1989)). This is easily derived from the fact that the requisite total processing time τ_k^I is supplied either by the regular operator or by a utility worker in every station k. Hence, the total work time of the operator is $\tau_k^I - \Sigma_i w_{ki}$ TU, while he is idle for $PT - \tau_k^I + \Sigma_i w_{ki}$ TU. Since the idle time $PT - \tau_k^I$ is unavoidable, the sequence-dependent idle time of station k is equal to the work overload of the station. Hence, the two objectives "Minimize the total work overload" and "Minimize the total idle time" are equivalent. As a consequence, it is not necessary to consider both goals simultaneously as proposed by Wester and Kilbridge (1964).

Related Objectives. Some references describe objectives which are very similar to the minimization of the total work overload and the total idle time, respectively.

Schneeweiß and Söhner (1991, pp. 36) propose the objective of minimizing the sum of the total excess time and the total idle time for a restricted problem with open stations of equal length. This station type allows that operators complete operations in an overlap area which is used in common with the subsequent station. The *excess time* is defined as the time interval during which the overlap area is needed. Usually, this area is rather small such that work overload is likely to occur when excess times are present. Hence, the excess times provide upper bounds on the real work overload. Furthermore, they may result in idle times of subsequent stations, because the respective operators are usually not allowed to interfere with each other.

Decker (1993 a, p. 62) considers positive and negative excess times. While positive excess times are defined as above, the negative excess times measure the (positive) distance to the right station border when processing of a unit is finished. As objective, the minimization of the difference between the sum of positive excess times and that of negative excess times is used. This objective considers that negative excess times are able to equalize positive excess times (cf. the discussion on cycle time violations and slack times in Section 3.2.2.2). Nevertheless, it has the major disadvantage that the sum-criterion does not reflect whether or not such compensations really occur. Even sequences which have only large negative excess times at the first positions and only large positive excess times at the late positions may be preferred if the totals of both are almost equal.

Another objective concerns the minimization of the maximal distance any operator drifts to the right in his station (cf. Mitsumori and Takada (1972) as well as Okamura and Yamashina (1979)). This objective is used as a surrogate for the goal of minimizing the risk of stopping the conveyor. It has the drawback that neither utility work nor off-line repair are considered as actions for compensating work overload (instead of stopping the line).

Line Length. In contrast to the assumptions used so far, some references assume that the station extents and the line length are variable even in the short-run (cf. Dar-El and Cother (1975), Dar-El (1978), Bard et al. (1992)). Accordingly, one may aim at minimizing the line length subject to the restriction that neither idle time nor work overload occurs. In order to avoid idle times, it may be necessary that the operators start their work on the first unit of the sequence later than it arrives at the station *(late start schedule)*. The station lengths have to be chosen such that no work overload occurs. A further objective is the minimization of the (average) flow time of the models which only slightly differs from the time equivalent of the line length (cf. Bard et al. (1994)).

However, the assumption of a short-term variability of station lengths seems to be reasonable only in the special case where a sufficiently long conveyor belt is available and no fixed equipment is required.

Cost Objectives. Time related objectives are sometimes not appropriate, because they may give identical attention to different types of inefficiencies. For example, work overload at a bottleneck station is more critical than at another station. Therefore, it may be necessary to consider cost oriented objectives (cf., e.g., Felbecker (1980, pp. 65), Hackstein and Felbecker (1981)).

Total Labour Costs. A possible objective is to minimize the sum of labour costs, which are obtained by using wage rates for the regular work of operators and the utility work. These wage rates may vary depending on the models and the work performed at the stations, respectively (cf. Ziegler (1990, pp. 73)). In the case of open stations, additional costs may be considered for working in the overlap areas (cf. Thomopoulos (1967)). Moreover, costs for working at off-line repair stations may be included (cf. Vrat and Virani (1976)).

Total Operating Costs. Some type of station (e.g., paint stations) may require considerable sequence-dependent set-up activities when changing from one model to another model. Bolat (1994) as well as Bolat et al. (1994) consider the objective of minimizing the sum of set-up costs and costs of utility work. Chakravarty and Shtub (1985, 1986 b) propose the minimization of the sum of labour costs, set-up costs, and in-process-inventory costs for a stochastic mixed-model line.

Equivalence of Cost and Time Related Objectives. If the wage rates are equal for all stations and models, the objective of minimizing the total labour cost is equivalent to the minimization of the total work overload and the total idle time, respectively (cf. Ziegler (1990, p. 84)).

Just-in-Time Objectives. The objectives discussed above only concern the operating conditions of the line but do not consider the material supply at the stations. While material stocks are always rather expensive, they become critical in the case of a mixed-model production, because a large number of different parts has to be stored. In order to reduce the inventory levels of all parts as much as possible, a *just-in-time system* should be implemented. That is, the parts should not be provided before they are required, and the output should be matched with the demand (cf., e.g., Hall (1983)). The application of the just-in-time concept is facilitated if the usage rates of all parts are levelled (cf., e.g., Miltenburg (1989), Kubiak and Sethi (1991), Monden (1998)).

Similar Part Requirements. In the case where the models have similar product structures, i.e., the models require (almost) the same number and mix of parts, it is sufficient to consider the final products (models). Almost constant part usage rates are obtained by levelling the rate of production for each model. Several objectives are proposed to achieve this goal. In order to describe a respective objective function, we define the following values:

$y_{mi} := \sum_{q=1}^{i} x_{mq}$ cumulated production volume of model m at sequence position i

$r_m := d_m / I$ proportion of the demand of model m

In the case of $y_{mi} = i \cdot r_m$ for all m = 1,...,M and i = 1,...,I, the cumulated productions and demands of all models are perfectly balanced at each sequence position. Usually, this ideal situation is not achievable. Therefore, objectives are required which consider and weight the differences between cumulated production volumes and demands. Similar to objectives for balancing work load variations in mixed-model balancing problems, a large variety of such objective functions is possible (cf. Miltenburg (1989), Ding and Cheng (1993), Ng and Mak (1994)). A simple function which minimizes the sum of squared deviations of cumulated demands and output volumes is the following:

$$\text{Minimize} \quad \sum_{m=1}^{M} \sum_{i=1}^{I} (y_{mi} - i \cdot r_m)^2 \tag{3.16}$$

With a similar objective, MSP corresponds to a single-machine scheduling problem with symmetric earliness and tardiness penalties (cf. Inman and Bulfin (1991) as well as Domschke et al. (1997, chapter 5.3.5)).

Different Part Requirements. In a multi-level production system, different processes manufacture parts (components and sub-assemblies) which are required by a process at the next level. The final items are often assembled on a mixed-model line. Following the just-in-time concept, each process produces parts such that they are completed just when a process at a higher level requires them. Miltenburg and Sinnamon (1989, 1992) as well as Inman and Bulfin (1992) propose generalized objectives which consider the usage rates of the parts at all levels.

Combination with Time Related Goals. The exclusive levelling of part usage rates may introduce considerable operational imbalances into the system. Therefore, it makes sense to combine work load and part usage aspects. Miltenburg and Goldstein (1991) as well as Bard et al. (1994) propose objective functions which include weighted sums of usage and loading criteria. Sumichrast et al. (1992) combine levelling criteria with work overload and idle time criteria. While the above approaches try to include the different goals in the objective function, Parrello et al. (1986) as well as Drexl and Jordan (1995) formulate constraints which cover work load and part usage considerations.

3.3.3 A Mathematical Formulation for MSP-W

The discussion in Section 3.3.2 shows that the minimization of the total work overload is an important objective in mixed-model sequencing. The corresponding problem (MSP-W for short) is formulated as a *mixed-integer linear program* with binary variables x_{mi} and continuous variables s_{ki} and w_{ki}, which are defined as in Section 3.3.1 (cf. Figure 3.7, p. 97).

In order to facilitate the presentation, we abbreviate the operation time of the i-th unit of the sequence in station k by:

$$\rho_{ki} := \Sigma_m \tau_{mk} \cdot x_{mi} \qquad \text{for } k=1,...,K \text{ and } i=1,...,I \qquad (3.17)$$

MSP-W is defined by the objective function (3.18), which is equal to (3.14), and by the constraints (3.19) - (3.25).

$$\text{Minimize } WO(x,s,w) = \sum_{k=1}^{K} \sum_{i=1}^{I} w_{ki} \qquad (3.18)$$

subject to

$$\sum_{m=1}^{M} x_{mi} = 1 \qquad \text{for } i=1,...,I \qquad (3.19)$$

$$\sum_{i=1}^{I} x_{mi} = d_m \qquad \text{for } m=1,...,M \qquad (3.20)$$

$$s_{k,i+1} \geq s_{ki} + \rho_{ki} - w_{ki} - c \qquad \text{for } k=1,...,K \text{ and } i=1,...,I \qquad (3.21)$$

$$s_{ki} + \rho_{ki} - w_{ki} \leq l_k \qquad \text{for } k=1,...,K \text{ and } i=1,...,I \qquad (3.22)$$

$$s_{ki} \geq 0, \; w_{ki} \geq 0 \qquad \text{for } k=1,...,K \text{ and } i=1,...,I \qquad (3.23)$$

$$s_{k1} = 0, \; s_{k,I+1} = 0 \qquad \text{for } k=1,...,K \qquad (3.24)$$

$$x_{mi} \in \{0,1\} \qquad \text{for } m=1,...,M \text{ and } i=1,...,I \qquad (3.25)$$

The objective function (3.18) minimizes the total work overload. The constraints (3.19), (3.20), and (3.25) ensure that each requisite unit is assigned to exactly one position of the sequence. The inequalities (3.21) express that the processing of a unit cannot start before the preceding unit has been completed by using regular or utility work. The work is restricted to the station area by the constraints (3.22) and the non-negativity constraints for s_{ki} in (3.23). The amount of utility work is bounded by (3.23) and the objective function (3.18). The equations (3.24) ensure that the line is in the initial state before and after producing all units.

Remark 3.3. In the above form, the model does not guarantee that all operations start as early as possible. Let s^e_{ki} denote the earliest starting point for processing the i-th unit in station k. In the case of $s^e_{ki} + \rho_{ki} < c$, the operation may start at any time within the interval $[s^e_{ki}, c - \rho_{ki}]$ without influencing the remaining units, because there is an idle time of $c - s^e_{ki} - \rho_{ki}$ TU. Earliest starting of operations is achieved by adding a term $\epsilon \cdot \Sigma_k \Sigma_i s_{ki}$ with $\epsilon < 1/(I \cdot \Sigma_k l_k)$ to the objective function. However, it should not be a major drawback that idle times occur after reaching a unit instead of before. Furthermore, solutions are easily transformed by moving operations towards the left border as near as possible.

In the case where $l_k - s_{ki} < w_{ki}$ holds, more floaters than one have to be employed simultaneously. Such solutions may be excluded by adding the constraints:

$$w_{ki} \leq l_k - s_{ki} \qquad \text{for } k = 1,...,K \text{ and } i = 1,...,I \qquad (3.26)$$

In a rolling planning environment (see below), the restrictions (3.20) should be modified to '≥'-conditions, because it might be advantageous to anticipate future demands with respect to the performance of the production sequence. This is only possible if more than $\Sigma_m d_m$ units can be produced in the considered period, i.e., the sequence length I can be exceeded. In order to avoid that only such additional models are produced which have small work loads, upper bounds on the production quantities of each model should be introduced.

The given formulation is similar to former ones of Ziegler (1990, pp. 145), Yano and Rachamadugu (1991), Bard et al. (1992), and such for permutation flow shop problems (cf., e.g., Domschke et al. (1997, pp. 372)).

Non-linear formulations of problems with objectives for levelling the part usage rates are given by, e.g., Miltenburg (1989), Miltenburg and Sinnamon (1989, 1992), Miltenburg and Goldstein (1991) as well as Inman and Bulfin (1991, 1992). Kubiak and Sethi (1991) propose a transformation to an assignment problem. A survey of mixed-model problems arising in just-in-time systems is given by Kubiak (1993).

3.4 A Hierarchical Planning Approach

In this section, we outline a hierarchical planning approach which represents a framework for integrating balancing and sequencing decisions with respect to their different planning horizons and their interdependencies.

The *hierarchical production planning* (HPP) is a methodology which decomposes a complex production planning problem into a hierarchy of subproblems which are less complex than the entire problem (cf. Buffa (1983, pp. 359)). This decomposition usually reflects different levels of aggregation as well as the organizational structure of a firm. The problems at high (aggregate) levels are based on aggregated input data and include long-term or medium-term decisions. The lower levels have to take detailed short-term decisions subject to constraints or parameters which are imposed by the higher levels. Unlike a simultaneous optimization of the entire problem, a hierarchical decomposition allows postponing detailed decisions until more accurate data are available.

Due to interdependencies between the decisions to be made, the different levels influence each other. In order to distinguish several influence types, we assume that there are only two levels, the top (aggregate) level and the bottom (detailed) level (a more detailed discussion can be found in Schneeweiß (1992, pp. 76, and 1994) and Stadtler (1996)). Decisions at the top level restrict possible decisions at the bottom level by imposing constraints or prescribing objectives to be considered *(top down influence)*. The bottom level may influence the top level in several ways *(bottom up influence)*. At first, currently available informations about the bottom level are taken into consideration at the top level *(feedforward influence)*. Secondly, possible reactions of the bottom level on top level decisions are estimated by so-called *anticipation functions* in order to reduce the loss of optimality caused by the decomposition of the problem. Thirdly, decisions at the bottom level may be used as a *feedback* for (partly) revising top level decisions if the bottom level decisions differ from the anticipated ones.

In order to reduce the uncertainty of data the top level is faced with, planning is often performed in a rolling manner as follows. A number of periods of the planning horizon (beginning with the current period) defines a so-called *rolling horizon*. For this horizon, an aggregate plan is developed at the top level and is executed for the first period, while the decisions for the remaining periods are only temporary. At the bottom level, a detailed plan is developed for this first period based on the realized aggregate decisions. Then the rolling horizon is moved to the next period, and a new aggregate plan is prepared utilizing updated (more accurate) data as well as feedback informations of the bottom level which concern consequences of the former decisions at the top level (cf., e.g., Nahmias (1989, pp. 271) and Steven (1994, pp. 55)). The rolling horizon approach takes into consideration that the data show a decreasing degree of certainty with an increasing planning horizon. That is, the input data for aggregate decisions may be subject to considerable forecasting errors such that changes in the aggregate plan are necessary when these errors become obvious.

Starting with the fundamental work of Hax and Meal (1975), a variety of HPP approaches has been described for several types of production systems in the literature. Surveys and classifications of such planning systems are given by Stadtler (1988, chapter 3) as well as by Steven (1994, chapter 3). Furthermore, see Dempster et al. (1981). A capacity oriented production planning and control system (PPC), which is based on a hierarchical concept, is proposed by Drexl et al. (1994).

As explained in Section 3.1, the balancing and sequencing problems which arise in mixed-model assembly line production systems cover different planning horizons and show strong interdependencies. Therefore, they define two natural levels of a planning hierarchy. Three (sub-) levels which represent different levels of aggregation with respect to the certainty of data available and planning horizons covered may be defined for the balancing level. At the sequencing level, the usual short-term sequencing is complemented by a so-called master sequencing which is introduced in order to couple the two basic levels. The resultant five-level hierarchy is illustrated in Figure 3.9.

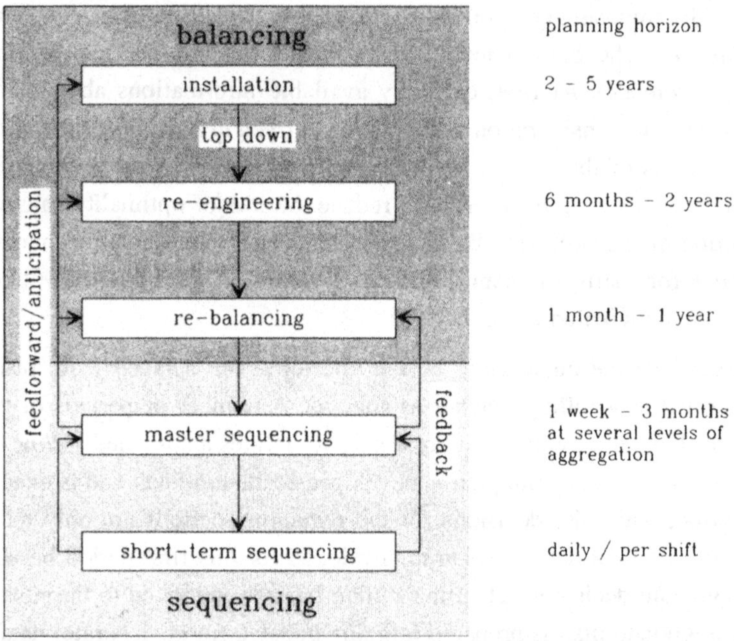

Figure 3.9. Planning Hierarchy

The different levels are explained with respect to the type of input data available, the planning horizons considered, and the decision problems to be solved. Furthermore, the interdependencies between the levels are discussed. Since the presentation mainly concentrates on balancing and sequencing aspects, it considers only a part of the decisions required for installing and operating mixed-model assembly lines. The planning horizons are specific for actual production systems and can be specified only roughly.

Installation. This level is concerned with long-term decisions connected with the installation of a mixed-model assembly line. When it has been decided to use such a production system, the line layout has to be determined. This includes, among others, decisions on the number of (parallel) lines, the material handling equipment, the types and layouts of stations, the machinery as well as the degree of the division of labour (cf. Section 1.3). These decisions are based on data concerning the costs of equipment, the models to be produced, the production processes as well as the expected demand proportions of the models (relative model mix). The planning horizon corresponds to the expected operating life of the system which depends on the lifecycles of the products. Hence, it may vary between 2 and 5 or even more years. The usual long-term objective consists of maximizing the total (expected) profit.

The long-term decisions mentioned above require that profit oriented line balancing problems with layout considerations are solved. These problems are based on an aggregate model which is defined by the production processes of the different models to be manufactured on the line(s) and the expected relative model mix (cf. Section 3.2). In analogy to the single-model case (cf. Section 2.3.3), such problems may be solved by modified procedures for MALBP-E (cf. Section 3.2.2.1). The profit is mainly determined by the production rate (cycle time) and the number of stations, because these parameters influence the fixed and variable costs as well as the marketable output (cf. Section 1.3.10). Additionally, it depends on the costs related with operating the line. These costs cannot be predetermined accurately in the long-term, because they depend on actual sequences to be processed in the short-term.

Therefore, it is important to *anticipate* operational effects of line balances using the *feedforward informations* available. Since it is known that line balances with smoothed station loads tend to produce efficient short-term sequences (provided that capacity oriented sequencing objectives are used), objectives for equalizing station utilizations (cf. the discussion in Section 3.2.2.3) should be added to the balancing problem. Furthermore, line balances may be evaluated with respect to their adequacy for typical sequencing problems which are expected to arise in the short-term. Basically, the same bottom up influences are relevant for the next two levels, too, which are also concerned with balancing decisions.

Re-Engineering. During the operating life of an assembly line, major changes may take place with respect to the models to be produced, the production process or the demand structure (model mix), respectively. In such a case which may occur about once or twice a year, a new balance has to be computed for the existing line based on the modified data.[3] Since the basic line layout is already defined *(top down influence* of the installation level*)*, the number of stations is fixed or can be modified only slightly. Therefore, the main attention is given to equalizing the work

---

[3] Note that no clear distinction is possible for the re-engineering and the re-balancing levels with respect to the planning horizon.

loads of the stations and maximizing the production rate. That is, the problems MALBP-2 or MALBP-E are relevant together with objectives for smoothed work loads (see above). In order to realize the new balance, movements of equipment may be necessary.

Re-Balancing. Operational experiences may reveal forecast errors and inefficiencies in the current line balance (*feed back* from sequencing levels). Furthermore, actual demand trends or even given demands may lead to modifications of the model mix. Then it may be necessary to perform re-arrangements in the actual line balance. Opposite to the re-engineering described above, the re-balancing usually covers a shorter planning horizon (1 month to 1 year) and is restricted to minor modifications which do not require expensive movements of equipment. That is, only a part of the operations is allowed to be moved to other stations. This is achieved by applying MALBP-2 with assignment restrictions. Sometimes, it is possible to re-arrange task assignments even in the short-run, especially, when highly qualified workers are employed and no fixed equipment is needed. As in the higher balancing levels, it is important to use additional objectives for smoothed work loads and to anticipate possible short-term sequences.

Master Sequencing. This level is introduced to connect the long-term and medium-term balancing problems with the short-term sequencing problem. Decisions are based on deterministic demands which are given by individual customer orders. Therefore, the planning horizon covered corresponds to a number of periods for which orders are already known (1 week to 3 months).

For the demands of such a planning horizon, a *master sequencing problem* is considered which is to build a sequence of all requisite units. That is, an instance of the mixed-model sequencing problem (MSP, see Section 3.3) is defined for this horizon, covering all corresponding customer demands. Since the master sequencing problem tends to be very large, the models are aggregated to *model families* in order to reduce their variety. These families are built by grouping together models being quite similar with respect to their production processes and, especially, their operation times. This may be done by some clustering method (cf., e.g., Gaul and Schader (1986)). Basically, the master sequencing problem is solvable by any method of the corresponding version of MSP which is considered in the short-run. However, the length of the master sequence calls for specialized solution methods. Such a procedure is given for the problem version MSP-W in Section 6.2.4.3. Master sequencing problems can be defined for different planning periods at several levels of aggregation.

The master sequencing problem may be utilized to facilitate planning and to link the decision levels as follows.

Bottleneck Identification. The master sequencing problem may help to find bottleneck stations, i.e., stations which are critical with respect to the performance of the line. This knowledge may be utilized for evaluating and re-arranging line balances. That is, results of the master sequencing can be incorporated in the re-balancing or even higher levels as a *feedback.* If re-balancing is possible in the short-run, the master sequencing and re-balancing levels may be coupled in a *rolling* manner. That is, the line balance can be modified successively according to results of the master sequencing problem depending on the demand data currently available.

Furthermore, it is sometimes helpful to restrict the short-term sequencing problem to such bottleneck stations (cf. Lorenz (1982), Rachamadugu and Yano (1994)).

Short-Term Production Programs. An important task of production planning for customer-oriented mixed-model assembly systems consists of combining single orders to short-term production programs. Usually, the program generation process involves several levels of aggregation (programs for weeks, days, and shifts) and is oriented by desired or promised delivery dates (cf., e.g., Decker (1993 a, pp. 18)). Furthermore, it is attempted to achieve almost homogeneous programs (e.g., the daily programs of a week should be very similar) in order to obtain a smoothed capacity utilization. The daily programs or even the shift programs define instances of MSP by fixing the short-term model mix.

Even though the above mentioned way of generating short-term programs includes capacity considerations, it does not really reflect operational aspects. Therefore, it may be advantageous to utilize the master sequencing problem by subdividing the master sequence into subsequences which can be processed during a day or a shift. This *anticipatory* approach tends to produce MSP instances which have efficient sequencing solutions. Nevertheless, a sequencing problem remains, because the master sequence is based on model families instead of single models. In the case where only minor differences are present within those families, the respective part of the master sequence provides a good solution for the short-term problem.

The master sequencing approach should be applied in a *rolling planning* environment. For example, a master sequence may be determined for a week. From this sequence, an actual daily production program is derived, and the rolling horizon is moved by one day defining a modified master sequencing problem. This rolling planning approach enables the planner to determine appropriate *delivery dates* instead of fixing these dates independently of operational aspects. Furthermore, it is possible to react on short-term demand changes and to incorporate *feedback* informations of the short-term sequencing by defining modified model families or excluding inefficient partial sequences.

Supply Restrictions. Besides considering the (time) capacity provided by the line, the master sequencing problem may include restrictions with respect to material supply. If only a small number of particular parts required by one or more special models is available per day or week, the number of respective model units is restricted to that maximal value in the corresponding part of the master sequence. Such restrictions are formulated by Drexl and Jordan (1995).

Material Requirements Planning. Master sequencing can help to determine the quantities of components and raw materials needed for assembling the different models and the dates by which they are required. In this sense, a master sequence may serve as a *master production schedule* from which detailed production/procurement schedules are derived for the parts (cf., e.g., Silver and Peterson (1985, pp. 510)).

Short-Term Sequencing. The operating sequence is built for a day or even a shift based on the actual production program provided by the master sequencing (top down influence). That is, a version of MSP has to be solved. It is assumed here that the short-term objective is to minimize the total work overload (MSP-W).

To summarize, the hierarchical approach outlined above attempts to integrate sequencing aspects into balancing decisions as a look-ahead mechanism. Furthermore, it is intended to support the generation of short-term production programs and the determination of appropriate delivery dates. This connection between medium-term and short-term problems is achieved by defining a new decision level, the master sequencing problem, which is related to the balancing levels and the short-term sequencing by planning in a rolling manner.

Another hierarchical planning concept is proposed by Decker (1993 a, pp. 198). Additionally to balancing and sequencing problems, this approach includes the long-range decision problems of determining floater and buffer capacities. It differs from the planning hierarchy presented here mainly with respect to the definition of the medium-term balancing level, because no informations about the expected model mix and the expected (desired) production rate are used. Furthermore, no equivalent for the master sequencing level is contained. The medium-term balancing is performed only with respect to equalizing station loads for each model. Since neither cycle time constraints nor expected demands are observed, it is not sure that the capacity is sufficient to produce certain production programs. Therefore, it may be necessary to reduce the desired production rate or to enlarge the work force in the short-term (cf. Decker (1993 a, pp. 44)). Thus, the determination of the cycle time is considered as a short-term decision. Besides adapting the cycle time and the work force, respectively, a local re-balancing is considered for equalizing differences in the work loads of the stations for a given production program.

Part II: Solution Methods

4 Exact Procedures for Single-Model Assembly Line Balancing Problems

With this chapter, we introduce the second part of the work which addresses solution methods for assembly line balancing and sequencing problems. Chapter 4 deals with exact procedures and Chapter 5 with heuristic procedures for single-model assembly line balancing problems. Chapter 6 considers exact and heuristic methods for solving mixed-model balancing and sequencing problems.

Chapter 4 is organized as follows. In Section 4.1, we describe and classify existing and new exact algorithms for SALBP-1 which are mostly based on the well-known branch and bound method. Due to the relationship between SALBP-1 and SALBP-2, procedures for the first problem type can be applied to obtain solutions for the latter. Respective search methods are described in Section 4.2 together with two new branch and bound approaches directly solving SALBP-2. In Section 4.3, we propose SALBP-1 based and SALBP-2 based solution procedures for SALBP-E. Finally, Section 4.4 describes how to transfer the solution concepts to the different versions of UALBP.

4.1 Exact Solution Procedures for SALBP-1

This section is intended to present different solution concepts for exactly solving SALBP-1. All techniques fall into two categories which are well-known in combinatorial optimization: dynamic programming and branch and bound methods. Thus, we do not give lengthy introductions to these basic concepts but provide references and focus on their application to SALBP-1. Among respective approaches described in the literature, the branch and bound methods are clearly superior with respect to computation time and storage space requirements. As a consequence, we concentrate on such procedures and perform a conceptual comparison of two recent efficient approaches, namely FABLE and EUREKA. By analyzing their individual pros and cons, we develop a new branch and bound method which clearly outperforms the latter two (cf. Section 7.2.2.2). This procedure is complemented by a number of additional components including a bidirectional branching strategy.

Comprehensive surveys of exact procedures for SALBP-1 are provided by Baybars (1986 a) as well as Domschke et al. (1994 a and 1997, chapter 4). Earlier surveys, partly including heuristic procedures, are given by Kilbridge and Wester (1962), Ignall (1965), Bussmann et al. (1968) and Klenke (1977, pp. 60).

4.1.1 Notations and Definitions

We use the notations and definitions given in Section 2.1 (cf. Figure 2.1, p. 24). Some of the notations which need be specified for SALBP-1 are repeated in Figure 4.1. Furthermore, it contains some notations used for describing procedures.

c	given cycle time
LB	lower bound on the number of stations
UB, $\bar{m}$	upper bound on the number of stations, maximal number of stations
LLB_k	local lower bound at level k of an enumeration tree
BD(m)	total idle time (balance delay time) for m stations ($= m \cdot c - t_{sum}$)
E_j	earliest station of task j; cf. (2.1) and (2.37)
$L_j(m)$	latest station of task j if the line has m stations; cf. (2.2) and (2.38)
$SI_j(m)$	station interval of task j for m stations ($= [E_j, L_j(m)]$)
$B_k(m)$	set of tasks potentially assignable to station k ($= \{j \mid k \in SI_j(m)\}$)

Figure 4.1. Notations for SALBP-1

Additionally, we require some definitions which facilitate the presentation of solution procedures. Most algorithms are based on enumerating feasible solutions by successively assigning tasks or subsets of tasks to stations. Therefore, these algorithms consider partial solutions containing a number of already assigned tasks and partial station loads S_k ($k=1,2,...$), while the remaining tasks and station capacities constitute a residual problem (cf. Section 4.1.2.1).

Definition 4.1. A not assigned task j is *(forward) available* if all preceding tasks $h \in P_j^*$ have already been assigned to a station. A not assigned task j is *backward available* if all succeeding tasks $h \in F_j^*$ have already been assigned to a station. That means, a task is backward available if it is forward available in the reverse precedence graph, and vice versa.

Definition 4.2. A (forward) available task is *(forward) assignable* to a station k if all preceding tasks have already been assigned to one of the stations 1,...,k, i.e.,

$P_j^* \subseteq S_1 \cup S_2 \cup ... \cup S_k$, and the resulting station time $t(S_k \cup \{j\})$ does not exceed the cycle time c. Correspondingly, a backward available task is *backward assignable* to a station k if all succeeding tasks have already been assigned to one of the stations $k, k+1, ..., \bar{m}$ and the cycle time in station k does not get violated. That is, a task is backward assignable if it is forward assignable in the reverse precedence graph.

Definition 4.3. The load S_k of a station is *maximal* if no further task can be added without violating the precedence and cycle time constraints. The following sufficient condition can be used to check the maximality of a station load currently considered in a solution procedure: A station load S_k is maximal if the loads $S_1, S_2, ...,$ S_{k-1} ($S_{k+1}, S_{k+2}, ...$) are maximal and no task is forward (backward) assignable to station k.

Definition 4.4. A subset $S \subseteq \{1,...,n\}$ of tasks is called *feasible* if S contains the predecessors of each included task as well, i.e., $P_j^* \subset S$ for all $j \in S$.

4.1.2 Existing Branch and Bound Procedures - Survey and Discussion

In this section, we describe and compare different existing branch and bound procedures for SALBP-1. First, we discuss components of such procedures and classify different concepts. Subsequently, we comprehensively investigate two algorithms, namely FABLE and EUREKA, which seem to be most effective. Finally, we survey further branch and bound procedures proposed in the literature.

4.1.2.1 Components of Branch and Bound Procedures

The branch and bound method is a well-known general solution concept in combinatorial optimization. Therefore, we only give a brief statement of its components and show possibilities to arrange them for SALBP-1, but we renounce detailed descriptions which can be found in the literature in a great variety (cf., e.g., Mitten (1970), Geoffrion and Marsten (1972), Minoux (1986, pp. 248), Voß (1990, pp. 44), Schildt (1994, pp. 119)), Domschke et al. (1997, pp. 41), Scholl et al. (1997b)). The function of this section is to give a framework for further discussion.

Branch and bound algorithms consist of the two main components *branching* (enumeration) and *bounding*. Additionally, *dominance* and *reduction rules* should be included to reduce the solution effort.

Branching (Enumeration). During the branching process, the *initial problem* is subdivided *(branched, developed)* into several *subproblems*. By continuously developing such subproblems, a multi-level enumeration tree (with subproblems as nodes) is constructed. The first level of this tree consists of the *root node* (initial problem), whereas further levels contain *descending nodes* which are built by developing an *ancestor node* at the preceding level, in each case. Subproblems for which the optimal solution is already known need not be branched and are referred to as *leaf nodes*. The same term is used for nodes which are fathomed (see below). A path from the root to any other node of the tree is called a *branch*.

Branch and bound algorithms are usually subdivided with respect to the sequence in which nodes of the enumeration tree are generated and branched.

- In a **depth-first-search** (DFS), a single branch of the tree is developed until a leaf node is reached. On its way back to the root, the search follows the first possible alternative branch, i.e., each node is completely developed before its ancestor nodes are revisited. Most commonly, DFS is organized as *laser search*, i.e., in each node of the current branch, only one descending node is built and developed at a time. Alternatively, all descending nodes may be generated and sorted by a priority rule (e.g., according to increasing lower bound values). The one with the highest priority is branched first. The remaining nodes are stored in a *candidate list* and are developed according to the priority order at each revisit of the node.
- A **minimal-lower-bound-strategy** (MLB) always chooses a not yet developed node which has the minimum value of a lower bound (see the paragraph on bounding below) from a candidate list. This node is completely branched by constructing all descending nodes. The latter are stored in the list, while the current node is dropped. At the beginning, the list only contains the root node.

While DFS only considers one branch from the root to a leaf node at a time, MLB has to manage a large number of not yet developed nodes considerably restricting its applicability to SALBP-1 and other combinatorial optimization problems (cf. Liao (1994) for strategies to deal with this difficulty).

In case of SALBP-1, two basic branching strategies may be distinguished.

- **Taskoriented procedures** iteratively choose a single available task and add it to the earliest station to which it is assignable. Alternatively, only one station is considered at a time, and a single task which is assignable to that station is chosen for branching.
- **Stationoriented procedures** build a complete load for one station in each step.

These strategies lead to two kinds of subproblems both consisting of a *partial solution* and a *reduced problem* (residual problem). In either case, the reduced problem is based on the *reduced precedence graph* which is created by deleting all tasks already assigned. Stationoriented branching results in subproblems of the same structure as the initial problem, whereas in case of taskoriented branching one or more stations may only partially be loaded with one or several tasks.

Bounding. Generally, bounding is applied to reduce the size of enumeration trees. This is achieved by computing *lower bounds* (LB) on the number of stations, at least necessary for a feasible solution, in each node. Lower bounds are obtained by solving *relaxations* which are derived from the problem considered by omitting or relaxing constraints (cf. Section 2.2.2.1). The bound of the root node is referred to as *initial* or *global lower bound.* Those of all other nodes which correspond to residual problems are called *local lower bounds.*

The minimum number of stations for which a feasible solution is currently known represents an *upper bound* (UB) for the initial problem (cf. Section 2.2.2.2). As soon as an upper bound equal to the global lower bound is found, the procedure stops, because the corresponding solution is optimal. All nodes whose local lower bound is not smaller than UB are *fathomed,* i.e., they are excluded from further consideration, because they cannot lead to a solution with less than UB stations.

Dominance Rules and Reduction Rules. Besides bounding procedures, *logical tests* like dominance and reduction rules may be used to reduce the size of the enumeration tree. Such tests utilize the special problem structure in order to simplify the problem or to identify branches which can be fathomed without losing optimality (cf. Voß (1990, pp. 49)).

Dominance Rules. By bounding, a node is only fathomed when an already known feasible solution provides an upper bound UB which is not larger than the local lower bound of the node. By means of dominance rules, it is even possible to exclude branches for solutions which are not better than some other one before a corresponding (dominating) solution is really found or when the values of local lower bounds are too small for fathoming, respectively. This is achieved by identifying dominance relationships between subproblems based on properties of partial solutions and reduced problems, which allow the conclusion that the solution of a subproblem cannot be better than that of another one.

An example is the *maximum load rule* which excludes partial solutions containing at least one completed but non-maximal station load (cf. our argumentation for UB_2 in Section 2.2.2.2 as well as Jackson (1956)).

Reduction Rules. Such rules try to modify problem data, i.e., task times or precedence relations, such that the number of solution alternatives can be reduced or bounds may get improved values. However, the reduced problem must have at least one optimal solution in common with the original problem.

For example, the task times t_j of all tasks j which cannot share a station with another task may be set to c. This rule is referred to as *task time incrementing rule* (cf. Remark 2.4, p. 58). A sufficient condition for incrementing the task time of a task j is $t_j + t_{min} > c$.

4.1.2.2 The Algorithms FABLE and EUREKA

Recently, a number of branch and bound procedures for SALBP-1 has been proposed in the literature. Among these, especially the algorithms FABLE (Johnson (1988)) and EUREKA (Hoffmann (1992)) have been shown to be effective. We comprehensively investigate and compare these procedures for the following reasons. On the one hand, they extremely differ concerning the enumeration process as well as the extent of applying fathoming rules. On the other hand, they contain most of the solution concepts presented in the literature so far. Hence, they are well suited for a comparison and may be seen as prototypes for further developments. Table 4.1 gives a comparative survey of the different components of FABLE and EUREKA.

Table 4.1. Comparison of FABLE and EUREKA

	FABLE	EUREKA
Enumeration	• taskoriented • forward planning • depth first search • laser search • no heuristic needed • task renumbering using priority rules	• stationoriented • forward and backward planning • depth first search • lower bound method with laser search • additional heuristic • original task numbering
Bounding	• includes several bounds (cf. Section 2.2.2.1): - $LB_1 := \lceil t_{sum}/c \rceil$ - LB_2, LB_3: bounds for bin packing problem - LB_4: bound for one-machine problem	• simple bound computation: total available idle time for the current lower bound
Logical Tests	• includes various logical tests: - maximum load rule - Jackson dominance rule - labeling dominance rule - first station dominance rule - task time incrementing rule	• no logical tests

The following discussion can be well illustrated by means of an example. Figure 4.2 repeats the precedence graph of Figure 2.5 (see p. 43), which contains 10 tasks with $t_{sum} = 48$ and $t_{max} = 9$. For a cycle time $c = 10$, the simple lower bound is $LB_1 := \lceil 48/10 \rceil = 5$. The balance delay time is $BD(5) = 5 \cdot 10 - 48 = 2$.

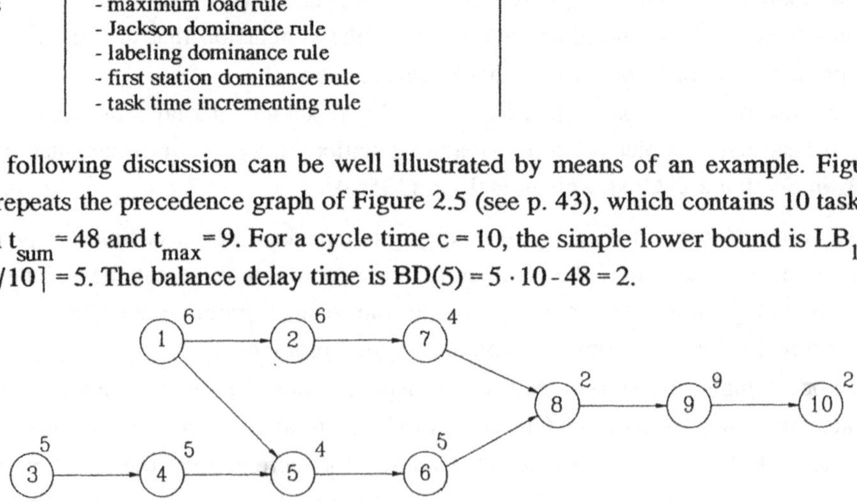

Figure 4.2. Precedence Graph

Branching (Enumeration)

FABLE. It uses a taskoriented DFS, organized as laser search. Subproblems are constructed by adding an assignable task to the currently considered station k (starting with station 1). If no such task exists, the current station load is maximal, and the consecutive station k+1 is opened. In order to avoid unnecessary permutations of tasks inside stations, loads are constructed as follows. Whenever a station is opened, the task with the smallest number among the assignable tasks is added. Any further tasks in the station must have a larger number than the task assigned in the ancestor node. After assigning all tasks, a feasible solution is obtained. Now, the current branch is traced back by removing task assignments until an alternative branch can be followed, i.e., the latest removed task can be replaced by another task which is assignable to the current station and has a larger number. A detailed description of the enumeration technique is given by Domschke et al. (1994 b).

Since the search is based on the task numbers, FABLE uses the following procedure for *renumbering tasks* which preserves a topological ordering. Initially, all tasks are not marked. In each of n iterations i = 1,...,n, one non-marked task with largest operation time which has no predecessor or only marked predecessors gets the number i and is marked. Ties are broken in favour of tasks with the largest number of immediate successors; second-level ties, in favour of the smallest original task number. In our example, the renumbering procedure leads to the original task numbers.

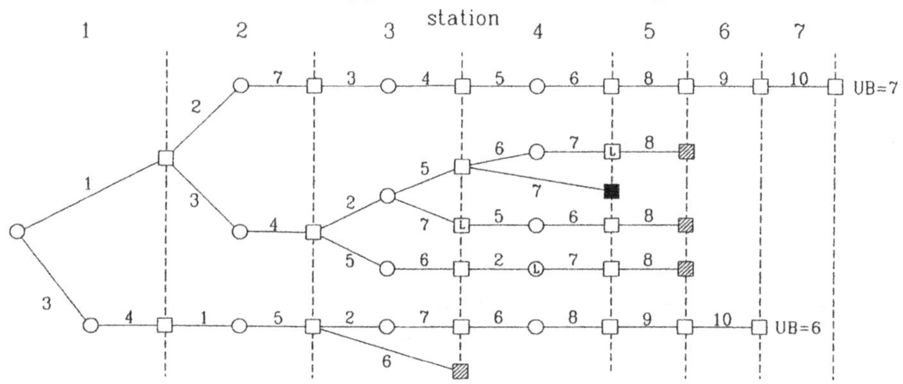

Figure 4.3. Enumeration Tree of FABLE

Example. Figure 4.3 shows the enumeration tree which is built for the example by FABLE. The tree is constructed from left to right and from up to down. Assigned tasks are denoted as arc weights. Quadratic nodes symbolize the completion of a station, whereas circular nodes indicate that another task can be added to a station. Hatched nodes are fathomed, because the local application of the lower bound LB_1 shows that no improved solution can be found. Tasks 9 and 10, which require two stations, form the reduced problems of the first three hatched nodes. In each case, no solution with less than 7 stations can result, because 5 stations are already loaded maximally. The black node is fathomed by the maximum load rule. The character 'L' inside some nodes is explained in the context of dominance rules.

EUREKA. It applies a so-called *Lower Bound Method* which involves the repeated application of a *stationoriented* laser search procedure. Station loads are built by considering tasks in order of increasing numbers. Starting with the global lower bound $LB = LB_1$, the procedure tries to find a feasible solution which does not contain more than LB stations. Hence, nodes can be fathomed if the cumulated idle time of the respective partial solution (i.e., the already built station loads) exceeds the total amount of available idle time $BD(LB) := LB \cdot c - t_{sum}$ (balance delay time). If the search is not successful, the global lower bound is increased by 1, and the enumeration procedure is started once more. This process is repeated until a feasible (= optimal) solution is determined.

The algorithm is subdivided into a *forward* and a *backward procedure.* The forward procedure builds station loads in the order 1,2,..., whereas stations are considered in the reverse order by the backward procedure. This is achieved by applying the same enumeration method to both the original and the reverse precedence graph. When the forward procedure fails to determine a feasible solution within a prespecified time limit, the backward search is initiated for the same time interval, starting with the current global lower bound obtained by the forward procedure. As soon as a feasible solution is found in either direction, the algorithm stops, because this solution is optimal. In order to determine a feasible solution when both the forward and the backward procedure exceed the time limit (note that no feasible solution is known before the procedure stops), the heuristic proposed by Hoffmann (1963) is applied in both directions without time limit (cf. Section 5.1.2 for a description of this procedure).

Example. In forward direction, EUREKA has to check two enumeration trees which are shown in Figure 4.4. It starts with the lower bound $LB_1 = 5$ and constructs only a small tree to verify that no feasible solution with 5 stations exists. Since the balance delay time is $BD(5) = 2$, only loads with a station time of at least 8 are to be considered. The hatched node is fathomed, because no load not exceeding the available idle time can be found. In the next iteration, the algorithm searches for a solution with $LB = 6$ stations and a balance delay time of $BD(6) = 12$. Again, the hatched nodes are fathomed, because all respective loads for station 5 would exceed the *remaining idle time* (BD(6) - cumulated idle time of stations 1 to 4). For example, consider the first branch consisting of the loads $S_1 = \{1\}$ with idle time 4, $S_2 = \{2,7\}$ and $S_3 = \{3,4\}$ without idle time as well as $S_4 = \{5,6\}$ with idle time 1. The remaining idle time of 7 would be exceeded by the only load $S_5 = \{8\}$ of station 5. In the last branch, a feasible solution is found, and the algorithm terminates. Note that the boldfaced part of the tree is enumerated twice. Furthermore, Figure 4.4 shows that large parts of the enumeration tree are built due to non-maximal station loads, e.g., the loads $S_2 = \{2\}$ and $S_2 = \{3\}$.

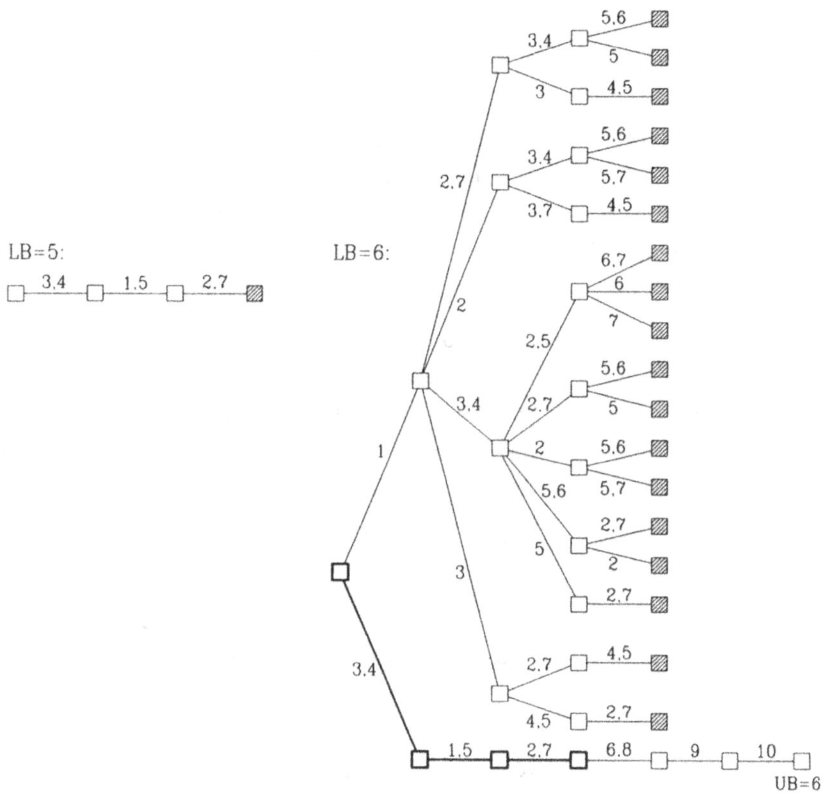

Figure 4.4. Enumeration Tree of EUREKA

Comparison. Empirical results have shown that optimal solutions of SALBP-1 instances tend to be close to the lower bound LB_1 (cf. Section 7.2.2). Therefore, FABLE may unnecessarily explore branches if they lead to an upper bound larger than the optimal number of stations. Note that this may be true even if an improvement of the current upper bound is achieved. This problem is avoided by the enumeration technique of EUREKA accepting the disadvantage that parts of an (expanding) enumeration tree may be considered repeatedly. Whenever the global lower bound is increased, the same enumeration tree is built again. Furthermore, additional branches due to the enlarged amount of available idle time have to be evaluated. However, if a solution with LB_1 stations exists, EUREKA has to examine a rather small tree.

As remarked above, the performance of FABLE strongly depends on finding a good upper bound in the first branches of the tree. This is supported by its renumbering procedure which uses priorities shown to be effective by Talbot et al. (1986) and our computational experiments (cf. Section 7.2.2.3). EUREKA does without renumbering, because the enumeration tree cannot be reduced if there is no feasible solution with a certain trial number of stations. Only the tree which is constructed for the optimal number of stations may get smaller by a proper renumbering scheme.

It is reported in the literature that solving problems may be easier either in forward or backward direction (cf. Saltzman and Baybars (1987)). This observation is exploited by EUREKA in a stationary manner, i.e., each direction is examined but no effort is made to recognize the superior direction. FABLE only performs a forward enumeration.

The backward application of EUREKA to our above example results in very small enumeration trees. Immediately after assigning task 10 in the root node, it becomes obvious that no solution with 5 stations exists. The tree for LB = 6 stations (see Figure 4.5) contains only one complete branch leading from the root node to a feasible solution.

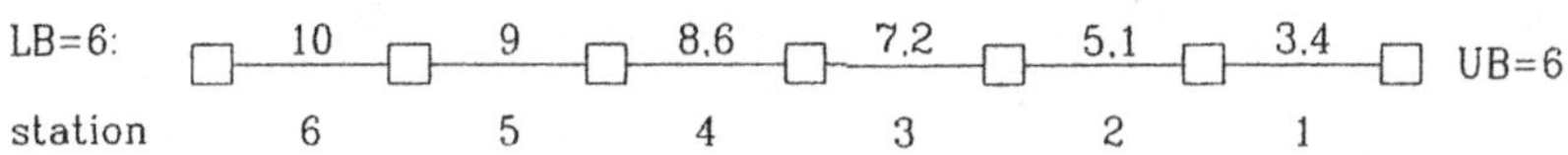

Figure 4.5. Enumeration Tree of EUREKA in Backward Direction

Bounding

FABLE. It utilizes the simple lower bound LB_1, the bin packing bounds LB_2 and LB_3 as well as the one-machine bound LB_4 (cf. Section 2.2.2.1). The first three bound arguments are applied to each node. Due to its computational complexity, the bound LB_4 is only computed in the root node to obtain an initial lower bound. Besides, LB_4 is determined in a restricted version, because the one-machine relaxations are only defined on direct successors (i.e., n_{j4} is based on F_j instead of F_j^*) for computing the tails. During the enumeration, the tails are used as follows. Just prior to assigning a particular task j to a station k, its tail n_j is added to the current station requirement including task j in order to check whether or not a solution with less than UB stations can be found. This requirement is given by $k - 1 + t(S_k \cup \{j\})/c$.

EUREKA. Only the simple bound LB_1, which is implicitly contained in the enumeration process, is applied. For each trial station number LB, only such loads not exceeding the remaining idle time are considered. This is equivalent to the assumption that a solution with UB = LB + 1 stations is already known.

Comparison. The taskoriented branching strategy of FABLE allows to use bounding rules after the assignment of each single task so that building a complete station may be avoided. However, bounds are most effective after completing a station load, because the idle time of the station is not known earlier. Therefore, the successive generation of a station load may enforce FABLE to assign a number of tasks and to evaluate bound arguments for the respective nodes even though fathoming may not be possible before the load is completed. On the contrary, a station-oriented procedure like EUREKA may not build a new node at all if the station load is inferior, and it calculates bounds less often.

However, good lower bounds are frequently able to reduce the size of the enumeration tree on a large scale. The initial computational effort for determining the bound LB_4 seems to be justified, because our experimental results indicate that it often results in sharper bounds than the other bound arguments (see p. 242). Furthermore, the local lower bounds can be computed in linear time, and their benefits should compensate this effort. As a consequence, it seems not to be justified that EUREKA avoids any bounding rules but LB_1.

Dominance and Reduction Rules

Whereas EUREKA excludes even the most common *maximum load rule*, FABLE incorporates this rule in its enumeration process and uses a number of additional logical tests. The simplest one is the *task time incrementing rule* (see p. 117). For our example, this simple rule (with a linear time complexity) increases the operation time of task 9 to $c = 10$, because task 9 cannot be combined with any other task due to $t_{min} = 2$.

Labeling Dominance Rule. This rule identifies partial solutions which contain the same set of tasks than another, previously enumerated one. A node is fathomed if the set of assigned tasks does not require less stations than an assignment of the same set considered earlier. This rule necessitates to store the current minimum number of stations (not necessarily integral) for each feasible subset of tasks (cf. Definition 4.4). An efficient method is based on a labeling scheme.

Labeling Scheme of Schrage and Baker (1978). In order to store the station requirements of all feasible subsets, each subset needs a unique address in a computer memory. The address of any feasible subset S is obtained from adding up labels L(j) for all tasks $j \in S$. The task labels L(j) for the tasks $j = 1,...,n$ are computed by the recursive formulae (4.1) which use the sets

$$P'_j := \{h \mid h < j \text{ and } h \notin P^*_j\} \qquad \text{for } j = 1,...,n$$

of tasks which have smaller numbers than task j and are not predecessors of j.

$$L(j) := \sum_{h \in P'_j} L(h) + 1 \qquad \text{for } j = 1,...,n \tag{4.1}$$

The uniqueness of addresses is explained by the following argument. The task labeling ensures that the label of each task j without predecessors (i.e., a source node which builds a feasible subset itself) is larger than the address of any feasible subset of tasks with numbers $h < j$. The same is true for an arbitrary feasible subset S', because it may be defined as an aggregate source node, which gets the largest number j' of a contained task. All feasible subsets of tasks with a smaller number than j', which are not contained in this aggregate node, have an address smaller than the label $L(j')$ and, thus, a smaller address than S'. Note that a task may get the same label as one of its immediate predecessors.

Complexity. The task labels are computed in $O(n^2)$ time. In each node, a constant time is required to update the address of the current feasible subset as well as to check and possibly modify the respective entry in the memory. While the time complexity is moderate, the storage space requirements quickly get intractable when the problem size increases. For example, a problem with n tasks and no precedence relations needs 2^n memory positions. In this case, the labels are $L(j) = 2^{j-1}$ for all j. Even if precedence relations are existent, the memory requirements are huge so that only small problems can be treated by the labeling dominance rule completely.

Example. The task labels for our example problem are given in Table 4.2. For instance, the label $L(7) = 11$ is obtained from adding up the labels of the tasks 3,...,6, which are not predecessors of task 7, and increasing this sum by 1. The addresses for the subsets {1,2} and {1,2,7} are 2 and 13, respectively. All feasible subsets of {3,...,6} have addresses between those two values. The sum of all labels, i.e., the maximal address, is 26. Therefore, a memory with 26 positions is needed to store the station requirements of all feasible subsets.

Table 4.2. Task Labels

j	1	2	3	4	5	6	7	8	9	10
L_j	1	1	3	3	2	2	11	1	1	1

In the branch and bound tree of FABLE, the labeling dominance rule fathoms three nodes which are marked by the character 'L' in Figure 4.3. In the case of the first marked node, the feasible subset {1,...,7} has already been assigned to four stations in the first branch of the tree.

Jackson Dominance Rule (cf. Jackson (1956)). This rule is applied to reduce the number of alternative station loads which have to be considered for a certain station. It is based on the following definition.

Definition 4.5. A task h *potentially dominates* another task j if $F_j \subseteq F_h$ and $t_j \leq t_h$ hold. If $t_j = t_h$ and $F_j = F_h$, the lower-numbered task is defined to be dominating.

The Jackson dominance rule excludes a maximal station load S_k from consideration if the following conditions are fulfilled:

- At least one task $j \in S_k$ can be replaced by an available not yet assigned task h which potentially dominates j, and
- the resulting station time $t(S_k) - t_j + t_h$ does not exceed the cycle time c.

This rule utilizes the fact that all successors of task j are successors of task h as well and cannot start before task h is performed (first condition of Definition 4.5). Hence, the sequence of j and h is not important for the successors of j. The condition $t_j \le t_h$ of Definition 4.5 guarantees that the station time will not decrease if h replaces j in S_k.

Complexity. The Jackson dominance rule causes a computation time of order $O(n^3)$ to determine the potential dominances before the enumeration starts. In each node which represents a maximal station load, a list of potentially dominating tasks has to be examined for all tasks of this load. The time requirement is small, because only a few tasks are usually assigned to one station. The potential dominances may be stored in the same data structure as the precedence graph, because they are only required for pairs of tasks which are not related by precedence.

Example. Since only two potential dominances occur (task 1 potentially dominates task 4, 6 potentially dominates 7), the Jackson dominance rule does not help to reduce the size of the tree in Figure 4.3.

First Station Dominance Rule. It is applied to nodes corresponding to maximal station loads. It excludes all maximal station loads S_k' (with $k > 1$) which are a subset of any station load S_1 that has already been constructed for the first station.

The rule is derived from the reflection that solutions are *equivalent*, if they contain the same station loads but in different orders. In this sense, all solutions containing a load S_k' equal to some load S_1 for the first station, are equivalent to some solution in the subtree following S_1. If a load S_k' is a proper subset of some S_1, it is equivalent to a corresponding solution with a non-maximal first station load.

Complexity. The first station dominance rule requires that all (or a subset of all) loads for the first station are stored. Usually, their number is not very large and a moderate amount of storage space is needed. In each node, which corresponds to a maximal station load, this load has to be compared with the first station loads stored in the memory.

Example. Assume that in the root node of the tree in Figure 4.3 the second branch would be chosen prior to the first. Then the first station dominance rule would fathom those quadratic nodes which represent the loads {3,4} in station 3 and 2, respectively.

Comparison. FABLE uses a variety of dominance and reduction rules, while EUREKA does not contain any respective rule. On the one hand, these rules require some additional computation time, on the other hand, they frequently reduce the size of the enumeration tree considerably (cf. Section 7.2.2.2, pp. 243). Especially in case of such simple rules like the maximum load rule and the task time incrementing rule, which do not require much time, it is not justified to accept a large number of unnecessary enumeration steps.

4.1.2.3 Further Branch and Bound Procedures

In the last three decades, a considerable number of branch and bound approaches has been proposed in the literature. We only outline the essential features of those algorithms, because most of their components are already included in the discussion of FABLE and EUREKA above. Table 4.3 classifies the approaches with respect to the used branching strategies (cf. Section 4.1.2.1). The descriptions are given in a chronological order.

Table 4.3. Classification of Branch and Bound Algorithms in the Literature

	stationoriented	taskoriented
DFS	• Mertens (1967) • Johnson (1981) • Betts and Mahmoud (1989) • Berger et al. (1992) • Hoffmann (1992): EUREKA	• Talbot and Patterson (1984) • Saltzman and Baybars (1987) • Johnson (1988): FABLE • Nourie and Venta (1991)
MLB	• Jaeschke (1964) • Van Assche and Herroelen (1979) • Hackman et al. (1989)	

Jaeschke (1964) proposes the first simple approach which performs a stationoriented branching and uses the bound LB_1. It utilizes an MLB strategy, i.e., nodes with a minimal cumulated idle time are chosen for branching. Ties are broken in favour of nodes at the highest level of the tree, i.e., nodes with the largest number of loaded stations are preferred.

Mertens (1967) describes a stationoriented DFS procedure based on the same renumbering of tasks which is used by FABLE. For fathoming, the simple bound LB_1 is used.

The procedure of *Van Assche and Herroelen* (1979) combines a stationoriented branching and MLB. Several penalty and tie breaking rules are proposed for choosing the node to be developed in the case of identical lower bounds. The algorithm contains the maximum load rule and an extension of this rule to feasible subsets (cf., furthermore, Jackson (1956)). The latter rule fathoms a node if it corresponds to a subset of the tasks which are contained in the partial solution of another node at the same level of the enumeration tree (cf. the labeling dominance rule of FABLE). In their implementation, Van Assche and Herroelen restrict the number of nodes stored in the candidate list to a value of 3,500. It is not reported whether the procedure performs a DFS or reduces to a heuristic when this limit is reached.

The algorithm of *Johnson* (1981) performs a stationoriented branching with the maximum load rule and uses a combined DFS and MLB strategy. Among all nodes descending from a current node, one with minimal cumulated idle time is chosen and developed. Besides the bound LB_1, simple counting criteria similar to LB_2 and LB_3 are applied. They count operation times exceeding a certain part of the cycle time and add up corresponding weights. For example, all task times exceeding a quarter of the cycle time are given the weight 1/3, because at most three such tasks can share a station.

The procedure of *Talbot and Patterson* (1984) is a generalization of the additive algorithm of Balas (1965), which uses integer variables instead of binary variables. It is discussed in more detail, because it differs from all other approaches significantly. The procedure can be interpreted as a branch and bound algorithm with a taskoriented branching and DFS. At each level j = 1,...,n of the enumeration tree, the task with number j is assigned to one of the stations $E_j, E_j+1, ..., L_j(UB-1)$, at a time. This results in (at most) $L_j(UB-1) - E_j + 1$ descending nodes. E_j denotes the earliest station and $L_j(UB-1)$ the latest station to which a task has to be allocated in order to improve the current upper bound UB (see p. 28 and p. 48). Since the tasks are considered in the *fixed order* of the original numbering 1,...,n, some stations may not be loaded maximally, before one or more subsequent stations are opened by assigning tasks to them.

The effort of the enumeration is reduced by two strategies based on chains and cuts. A *chain* is an ordered set of consecutively numbered tasks which are related by arcs (j,j+1) in the precedence graph. In our example of Figure 4.2, tasks 7, 8, 9, and 10 form a chain. Such subgraphs are exploited by immediately backtracking to the highest-numbered task outside the chain.

A station s is called *cut* if there is a task j_s with $E_{j_s} = s+1$, and $E_h \geq s+1$ holds for all tasks with numbers $h > j_s$. Obviously, the loads of the stations 1,...,s cannot be extended by a further task when such a task j_s is to be assigned, because tasks are considered in the order of increasing numbers. This knowledge is exploited by several fathoming and backtracking rules.

For example, a node can be fathomed if the cumulated idle time of the stations 1,...,s exceeds the balance delay time BD(UB-1) of an improved solution searched for. The test can be strengthened by computing a lower bound on the idle times occurring in the stations beyond the cut s. This is achieved by solving a *knapsack problem* for each of the stations $k=s+1,...,UB-1$. The cycle time is interpreted as capacity of the knapsack, the tasks j with $k \in [E_j, L_j(UB-1)]$ are interpreted as items and the task times as item weights. The unused capacity of an optimal knapsack solution provides a lower bound on the idle time of station k (cf. the bound LC_5, p. 58).

The algorithm uses an extension of the *task time incrementing rule* (see p. 117) which works as follows. The operation time of a task j is set to the value of the cycle time c if no other task can share a station with j. This is true if

$$t_j + t_h > c \quad \text{or} \quad [E_j, L_j(UB-1)] \cap [E_h, L_h(UB-1)] = \phi$$

holds for all other tasks h which are immediate predecessors or successors or are not related to j by precedence.

The procedure of *Saltzman and Baybars* (1987) performs the enumeration as a task-oriented DFS. Tasks are renumbered according to non-decreasing operation times. An initial value of the upper bound UB is determined by a priority based heuristic which assigns tasks in order of non-increasing task times (cf. the rule MaxTime in Section 5.1.1.1).

The extended task time incrementing rule of Talbot and Patterson is applied before the initial lower bound is computed by using LB_1, the counting criteria of Johnson (1981), and the bound LB_5 (see p. 48).

After finding a maximal station load, the current partial solution is completed to a trial heuristic solution using the priority based heuristic with rule MaxTime for the reduced problem. This solution leads to an improvement of the incumbent solution if it needs less than UB stations. Only if the resulting number of stations is equal to the local lower bound of the corresponding node, the node can be fathomed. The extended task time incrementing rule of Talbot and Patterson is applied to each reduced problem in order to strengthen the bound if possible.

In order to take advantage of the fact that solving a problem may be easier in forward or backward direction, respectively, the branch and bound procedure outlined above is simultaneously applied to the original problem and the reverse one. This is done by simulating a two-processor computer on a one-processor machine. Each of both procedures is allowed to develop one node in turn, before the other procedure is activated. Global lower and upper bounds are kept in a shared memory and are updated and utilized by both procedures.

The algorithm of *Betts and Mahmoud* (1989) is almost identical with that of Johnson (1981) but differs with respect to the method of building station loads. While Johnson's algorithm is based on the enumeration scheme of Held et al. (1963), the algorithm of Betts and Mahmoud applies the method contained in the heuristic of Hoffmann (1963) (cf. Section 4.1.3.2 and Section 5.1.2).

The procedure proposed by *Hackman et al.* (1989) performs a stationoriented branching with MLB similar to the algorithm of Jaeschke and uses some dominance rules. Additionally, it applies priority based heuristics to determine an initial upper bound and to complete partial solutions (cf. Wee and Magazine (1981 a)). Furthermore, heuristics are used to choose nodes for fathoming. While doing so, the algorithm itself becomes a heuristic.

Nourie and Venta (1991) present the branch and bound procedure OPTPACK which applies the branching scheme of FABLE and uses heuristics to find an initial upper bound. Like EUREKA, it computes only the simple lower bound LB_1. Besides the maximal load rule, OptPack applies the **tree dominance rule** which is an extension of FABLE's labeling dominance rule. Instead of the memory wasting Schrage and Baker labeling scheme a more compact and thus more effective data structure is used for storing and retrieving information on precedence-feasible subsets already enumerated. Within a binary (out-)tree each subset is represented by a path through this tree, where nodes correspond to tasks and the minimum number of stations required is stored in the terminal node of the path. The arcs indicate whether a task is contained in the subset (1-arc) or not (0-arc).

Berger et al. (1992) describe an extension of the algorithm of Hackman et al. for a restricted version of SALBP-1 which allows at most one predecessor for each task, i.e., the precedence graph is an *out-tree*. In contrast to the algorithm of Hackman et al., it uses a DFS strategy. A heuristic version of the algorithm builds only a subset of promising station loads in every node. Besides the simple bound LB_1, the lower bound LB_6 is used (see p. 49).

4.1.3 Dynamic Programming Procedures

Like branch and bound, dynamic programming is a general approach to many types of problems including most combinatorial optimization problems. However, specialized algorithms have to be developed for particular problems. In what follows, we give only a short introduction to the basic elements of dynamic programming procedures, because detailed descriptions are available in a large variety (cf., e.g., Schneeweiß (1974), Cooper and Cooper (1981), Denardo (1982), Hillier and Lieberman (1988, chapter 11) as well as Domschke and Drexl (1998, chapter 7)).

4.1.3.1 Basic Elements of Dynamic Programming

Dynamic programming involves the optimization of *multi-stage decision processes* (Denardo and Mitten (1967)). That is, a given problem is divided into subproblems *(stages)* which are sequentially solved until the initial problem is finally solved.

At each stage, a number of *states* describing all possible conditions of the decision process at this stage (i.e., partial solutions of the original problem) have to be considered. The set of all states included in the decision process is called *state space*.

States at a particular stage k are *transformed* to states at the subsequent stage k+1 by a *decision*. The generation of states is described by *transformation (transition) functions* which depend on the current state and the decision taken. Usually, each decision specifies values for particular problem variables and is related to a cost value (in the case of a minimization problem). A sequence of decisions, which transforms a state at a stage h to a state at a stage k > h, is called *policy*.

Dynamic programming procedures are fundamentally based on the following *principle of optimality* (cf. Bellman (1957, p. 83), Bellman and Dreyfus (1962, p. 15)):

> "An optimal policy has the property that whatever the initial state and initial decision are, the remaining decisions must constitute an optimal policy with regard to the state resulting from the first decision."

This principle states that an optimal policy of the initial problem, connecting the initial state and the terminal state of the decision process, must consist of optimal policies for all included subproblems. This is based on the fact that the current condition of the system is completey described by a state whatever decisions have been made at former stages.

The application of dynamic programming as a principle solution technique requires the formulation of a problem as a multi-stage decision process, i.e., it has to be divided into subproblems which can be solved separately. Two different approaches for solving the decision process are distinguished:

Forward Recursion. Starting in the initial state, the stages are considered in the order 1,2,... until the terminal state is reached. At each stage, all possible states are generated before the next stage is examined. The optimal policy which transforms the initial state to the terminal state constitutes an optimal solution of the original problem.

Backward Recursion. The procedure works stage-by-stage from the terminal state towards the initial state. Each stage k contains all states which can be transformed to the terminal state by an optimal policy. At a stage k - 1, all states which result in at least one state at stage k by a decision are generated. The optimal policy connecting the terminal and the initial state represents an optimal solution of the original problem.

4.1.3.2 Dynamic Programming Procedures for SALBP-1

We survey solution methods for SALBP-1 which are based on dynamic programming. By analogy with branch and bound algorithms, two types of procedures may be distinguished depending on the used decisions.

Stationoriented Procedures. The stages correspond to the stations $k=1,2,...,$ and the states at stage k are described by feasible subsets of tasks which can be assigned to stations 1,...,k. The loads which can be built for station k based on a state at stage $k-1$ constitute the decisions transforming this state to one at stage k. Policies describe partial solutions, i.e., sequences of station loads. The initial state at stage 0 is given by the empty set, the terminal state by the set $\{1,...,n\}$ of all tasks. While the terminal state is known, the corresponding stage is unknown, because it corresponds to the number of stations required.

Procedure of Jackson. Jackson (1956) proposes a stationoriented procedure with the above specifications except the definition of states which originally describe sequences of station loads. This definition has several drawbacks which are eliminated by the one given above (proposed by Held et al. (1963)). First, it ignores that states should be independent of corresponding policies. Secondly, a large amount of memory space is needed to store this type of states.

The procedure performs a forward recursion by means of two lists of states for the stages $k-1$ and k. Initially ($k=1$), the first list contains the initial state, and the second list is empty. The elements (states) of the first list are successively considered and deleted from the list. For each state, all possible decisions (station loads) are enumerated. This results in a number of states at stage k which are appended to the second list. Note that duplicates are eliminated. When the first list is empty, all states of the current stage k have been enumerated. In order to consider the next stage, the second list becomes the first one and the second list is emptied. The process stops when a state equal to the terminal state is found. The corresponding optimal policy gives an optimal SALBP-1 solution. For a more detailed description see Domschke et al. (1997, pp. 205).

The enumeration effort is reduced by a number of dominance rules. The first rule is the maximum load rule, i.e., only maximal station loads are considered as possible decisions. The second one is based on potential dominances between tasks. This rule is already described in Section 4.1.2.2 as Jackson dominance rule (see p. 124), because it is applied by FABLE, too. Like the maximum load rule, it avoids storing those station loads in the list, which are dominated by some other possible station load. Furthermore, Jackson proposes generalizations of this rule based on an extended definition of the potential dominance. However, these rules may require more computation time than it is saved by the reduced state space.

Discussion. In its original form, the procedure of Jackson considers the same enumeration tree as stationoriented branch and bound procedures like EUREKA. Unlike the latter algorithms, it performs a breadth-first-search, i.e., the complete tree has to be built until the final stage is reached. Even though it reduces the number of states to be considered by dominance rules, the computation time and storage space requirements are immense even for small problems. The modification of Held et al. achieves that only one node is considered for each feasible subset at a stage but cannot eliminate the major shortcomings.

Remark 4.1. If states (feasible subsets) are represented by nodes and decisions (station loads), by arcs which are weighted with the corresponding station idle times, SALBP-1 is transformed to a *shortest path problem*. Each path in this graph corresponds to a feasible solution and each shortest path to an optimal solution of SALBP-1 (cf. Gutjahr and Nemhauser (1964) as well as Mansoor (1967), furthermore see Klein (1963)). Since the number of nodes grows exponentially with the number of tasks, it is generally not possible to construct the complete graph. Therefore, Easton et al. (1989) incorporate lower and upper bounds as well as dominance rules to reduce the size of the graph. Furthermore, their procedure is designed to be applied in forward and backward direction, respectively.

Taskoriented Procedures. The decision process is subdivided into stages $k=0,...,n$ which correspond to the numbers of tasks already assigned. Policies are built by sequences of tasks. Similar to the stationoriented approach, the states are described by feasible subsets of tasks. In each state at a stage k, k tasks are already assigned and the decisions are to assign one of the currently available tasks to position k+1 of the sequence. The initial state at stage 0 is given by the empty set; the terminal state at stage n, by the set $\{1,...,n\}$ of all tasks. Similar to FABLE, the task sequences are related to stations by using the maximum load rule. Therefore, each state corresponds to a current station requirement and each decision to the assignment of a task to the current station. The optimal policy, which represents the optimal SALBP-1 solution, contains the minimum number of stations.

Like stationoriented procedures, the taskoriented ones enumerate all feasible subsets. In the literature, two basic strategies are used (cf. Denardo and Fox (1979)).

Pulling. The optimal station requirement of a state z_k at a stage k is obtained by examining all *(predecessor)* states at the stage k-1 which are transformed to z_k by a decision.

Reaching. For a current state z_k at stage k, all possible decisions which result in some *(successor)* state z_{k+1} at stage k+1 are evaluated. The best station requirements currently known for the respective states z_{k+1} are modified if they can be improved.

In the following, we outline several taskoriented procedures. They differ from each other with respect to the above strategies, the order in which states are considered, and the way of storing and addressing the states.

Procedure of Held et al. A taskoriented forward recursion with pulling strategy is proposed by Held et al. (1963). All generated states are stored in a memory together with their station requirement (resulting from the corresponding optimal policy). This is done by means of an addressing function which produces a *compact addressing*, i.e., each memory position corresponds to exactly one state. The addressing function is based on a counting procedure which determines the cardinality of the state space (denoted by C). This function ensures that the addressing is *monotonous*, i.e., each state z_k gets a larger address than all states at lower stages which are connected with z_k by a policy. The addressing function is very complex, because it is based on recursively partitioning the precedence graph. As a consequence, it is not described here.

The procedure successively considers the memory positions in the order 1,...,C (due to the monotonous addressing). At each position, the corresponding state is determined by computing the inverse of the addressing function based on the counting procedure mentioned above. The pulling strategy requires that the addresses of all predecessor states are computed. The resultant minimal station requirement is stored at the memory position currently considered. Note that this enumeration technique does not strictly work stage-by-stage.

Held et al. propose a heuristic version of their procedure in order to reduce the computation time and storage space requirements. It is based on building fictitious tasks by combining feasible task sequences.

Procedure of Schrage and Baker. The addressing function proposed by Held et al. is computationally very expensive. Therefore, a simpler addressing method is developed by Schrage and Baker (1978). It is based on the task labels which are utilized by the labeling dominance rule of FABLE, too (see p. 123). The address of a feasible subset is simply computed by adding up the labels of the included tasks. Unfortunately, the achieved addressing is usually not compact, i.e., a part of the memory cannot be used, because no feasible subsets correspond to particular addresses. Therefore, the procedure tries to obtain a more compact addressing by different labeling techniques (cf. Baker and Schrage (1978)).

The states are enumerated according to a *lexicographic ordering*, i.e., all states z_{k-1} which may be transformed to a state z_k are considered prior to z_k in order to perform a pulling strategy. The required addresses are easily computed by means of the task labels.

Procedure of Kao and Queyranne. The enumeration procedure of Schrage and Baker does not work stage-by-stage. Therefore, states of all stages have to be stored simultaneously, resulting in immense storage space requirements. Kao and Queyranne (1982) propose a modification of the algorithm which reduces these requirements. It contains another enumeration technique which generates the states according to increasing addresses. Since the addresses of states at neighbouring stages differ by at most the largest task label L_{max}, a (cyclical) memory with L_{max} positions is sufficient to store the required states in each step of the procedure.

Procedure of Lawler. In contrast to all taskoriented procedures mentioned above, Lawler (1979) presents an enumeration technique which generates the states stage-by-stage. Therefore, the procedure needs only two lists like the procedure of Jackson. The first list contains the states at the current stage k; the second list, the states at stage k+1. Furthermore, a reaching strategy is chosen, whereas all other procedures use pulling strategies.

Discussion. These procedures consider multi-stage decision processes which are similar to the enumeration trees of taskoriented branch and bound procedures like FABLE. As a difference, dynamic programming procedures consider only one node for each subset of tasks. Usually, they have to enumerate the state space (almost) completely, whereas branch and bound procedures may reduce the size of their trees by bounding rules and logical tests on a large scale. Furthermore, a large amount of storage space is needed which restricts the applicability of those procedures to small problems.

Comparison. Stationoriented as well as taskoriented dynamic programming procedures have the same major drawbacks concerning their excessive computation time and storage space requirements. Therefore, these algorithms are not competitive to versatile branch and bound procedures like FABLE and EUREKA. This judgement is supported by some computational comparisons with older (less efficient) branch and bound approaches (cf. Johnson (1981), Baybars (1986 a)). Among the dynamic programming algorithms, the procedure of Jackson seems to be superior (cf. Johnson (1981)). This finding is supported by some computational experiences of the author. The performance of the various taskoriented procedures seems to be quite similar (cf. Kao and Queyranne (1982)).

4.1.4 The Branch and Bound Procedure SALOME-1

In this section, we propose a new algorithm based on the main ideas of FABLE and EUREKA. It is called SALOME-1 (Simple Assembly Line Balancing Optimization Method for Type 1; cf. Scholl and Klein (1997 a, 1999)). On the one hand, this approach integrates and extends the most powerful components of both former algorithms. On the other hand, it contains a number of new and successful ideas.

Promising features of FABLE are the bounding and dominance rules as well as its early determination of a feasible solution. In the case of EUREKA, the advantages of the Lower Bound Method and the use of both planning directions seem to be worth considering. Its disadvantage of repeatedly enumerating parts of the branch and bound tree is overcome by the new enumeration technique of SALOME-1 which is called *Local Lower Bound Method.* This general technique can also successfully be applied to bin packing and resource-constrained project scheduling problems (cf. Scholl et al. (1997 a), Klein and Scholl (1998 a,b)).

New concepts like *prefixing* and *bidirectional branching* are included to exploit the problem structure of SALBP-1 more intensively than former algorithms do.

4.1.4.1 Combining FABLE and EUREKA - Basic Approach

In Section 4.1.2.2, it is pointed out that EUREKA produces a small enumeration tree if the minimal number of stations is equal to the global lower bound, because many loads can be excluded from consideration. On the contrary, it has to enumerate expanding parts of the tree repeatedly if the lower bound must be increased. Furthermore, its enumeration procedure may fail to find a feasible solution within a given time limit. While avoiding a repeated enumeration, FABLE may construct unnecessary parts of the enumeration tree whenever the first feasible solutions differ from the optimal one significantly.

Branching (Local Lower Bound Method). In order to obtain feasible solutions as soon as possible, the enumeration is organized as a *laser search.* Furthermore, a *stationoriented* branching is used, because it enables to control the resulting station idle times. Basically, the enumeration tree is built similar to that of EUREKA, but the disadvantages outlined above are removed. This is achieved by modifying the Lower Bound Method to a *Local Lower Bound Method* which works as follows.

For each node, an individual *local lower bound* on the number of stations required is considered, whereas EUREKA uses the same global lower bound in all nodes. The local lower bound of a subproblem at a level k of the tree (LLB_k for short) initially gets the value of the local bound in the ancestor node at level k - 1. The value LLB_k corresponds to a *remaining idle time* RI_k which is equal to the differ ence between the balance delay time $BD(LLB_k) = LLB_k \cdot c - t_{sum}$ (i.e., the total available idle time) and the sum of idle times in the stations 1,...,k already loaded.

$$RI_k := LLB_k \cdot c - t_{sum} - \sum_{h=1}^{k} (c - t(S_h)) \tag{4.2}$$

Through the remaining idle time RI_k, the possible loads of station k+1 can be subdivided into two classes. The first class contains all station loads with an idle time of at most RI_k time units; the second one contains all other possible station loads.

In order to utilize the advantages of the Lower Bound Method, the process of branching a node at a level k is subdivided into two steps. In the first step, all loads of the first class are branched, because these loads are in accordance with the current local lower bound LLB_k. When in the corresponding subtrees a feasible solution with LLB_k stations is found, the node is fathomed. Otherwise, the local lower bound LLB_k is increased by 1. This corresponds to an increase of RI_k by the cycle time c. Hence $RI_k \geq c$, and all loads of the second class become feasible with respect to RI_k, i.e., they are in accordance with the increased local lower bound. After branching all nodes of the second class, the node is completely branched.

This procedure is equivalent to applying a Lower Bound Method *locally* in each node to the corresponding reduced problem, instead of restarting the procedure for the initial problem after fathoming all nodes of the current tree. Therefore, it has been given the name Local Lower Bound Method. Note that the station loads of the first class need not be stored, because those of the second class can easily be identified through their station times.

In order to avoid generating inferior loads, the *maximum load rule* is used. By applying the Local Lower Bound Method, the algorithm finds a feasible solution in the first complete branch of the tree like FABLE.

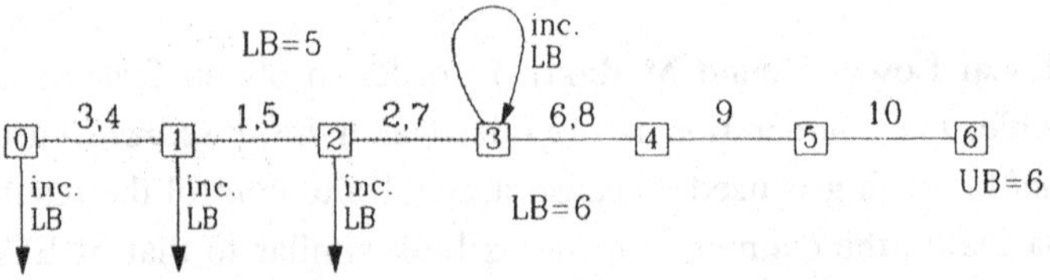

Figure 4.6. Enumeration Tree of the Local Lower Bound Method

Example. The Local Lower Bound Method is best explained by means of our example. The initial lower bound is LB = 5, the corresponding balance delay time is $BD(5) = 5 \cdot 10 - 48 = 2$. The resulting enumeration tree is shown in Figure 4.6. It parallels the tree of EUREKA for LB = 5 (cf. Figure 4.4, p. 121) until the node 3 is reached. In this node, no loads with their idle times not exceeding $RI_3 = 2$ are available. Unlike EUREKA, which traces back in such a situation, our approach applies a Lower Bound Method to the reduced problem. This problem consists of the tasks 6, 8, 9, and 10 as well as the respective precedence constraints. For the increased local lower bound $LLB_3 = 6$ with $RI_3 = 12$, the branch with station loads $S_4 = \{6,8\}$, $S_5 = \{9\}$, and $S_6 = \{10\}$ leads to a first feasible solution with UB = 6. After finding the new incumbent solution, the procedure traces back to node 2, because the local bounds of the nodes 3 to 6 are equal to 6. In node 2, the value of the current local lower bound LLB_2 is still 5, and other loads feasible for this number of stations are examined. The only remaining maximal load for station 3 is $S_3 = \{6\}$, the idle time of which exceeds $RI_2 = 2$. Thus, the current local lower bound is increased to 6, and the node is fathomed. Since no further maximal station load is available for the nodes 1 and 0 (root node), these nodes are fathomed, too, and the algorithm terminates.

Discussion. The example indicates that our approach improves the concept of EUREKA with respect to the following aspects. No repeated enumeration of subtrees takes place, because the Local Lower Bound Method continues extending the tree which is built by EUREKA for LB = 5 (underlayed in Figure 4.6) instead of fathoming node 3. Loads with large amounts of idle time are not considered before it is necessary. While EUREKA considers the trial load $S_1 = \{1\}$ of station 1, which leads to a subtree of significant size, when it enumerates for LB = 6, our approach takes the load $S_1 = \{3,4\}$ without idle time at first. The load $S_1 = \{1\}$ would not have been built before increasing the global lower bound. Even if the Local Lower Bound Method starts with an increased global lower bound of 6, the tree is much smaller due to the maximum load rule.

The concept of FABLE is improved, because the Local Lower Bound Method generates loads in order of increasing lower bounds instead of performing trial task assignments which may lead to station loads their idle time being too large to hold the current lower bound. Since our approach increases the bound as late as possible, it finds a better first feasible solution for the example than FABLE (cf. Figure 4.3).

Nevertheless, the sequence in which tasks are combined to trial station loads is still important, because the remaining tasks of a reduced problem decide on the possibilities to form loads for later stations. On the one hand, stations at the beginning of the line may be loaded with a small amount of idle time by combining tasks with small task times first. On the other hand, remaining large tasks possibly not able to share a station with each other may lead to very bad loads in later stations. In order to find good feasible solutions in the first branches of the tree, it seems to be promising to prefer tasks with large task times and such which make available many successors. This is achieved by using the *renumbering scheme* of FABLE.

Bounding. Besides the simple bound LB_1 which is implicitly contained in the logic of the Local Lower Bound Method, the bound arguments LB_2, LB_3, and LB_4' are used to determine a global lower bound. Since LB_2 and LB_3 are computed in linear time, they are additionally applied in each node to strengthen the local lower bounds.

Instead of the restricted version of the bound LB_4 used by FABLE, the new procedure SALOME-1 utilizes the generalized bound LB_4' (see pp. 47). Since it is computationally more expensive than LB_2 and LB_3, this bound is only determined in the root node. Note that the bound LB_4 used by FABLE is restricted to computing tails for immediate successors, while LB_4' considers heads and tails as well as transitive precedence relations.

Example. While $LB_1 = LB_2 = LB_3 = 5$, LB_4' provides a global lower bound of 6 (see p. 48 for a detailed computation).

Discussion. Computational experiments show that the chosen bounds are frequently very successful in reducing the size of the tree (see pp. 242). Therefore, it is reasonable to utilize a part of the computation time for bound computations. However, only simple bounds should be applied in all nodes, while more complex bounds are advantageous in the root node, because an increased initial bound affects all nodes of the tree.

Dominance and Reduction Rules. Among the logical tests of FABLE, we choose the maximum load rule, the task time incrementing rule, and the Jackson dominance rule. The labeling dominance rule and the first station dominance rule are excluded.

Maximum Load Rule. As stated above, this rule is contained in the enumeration process and can easily be applied in linear time. A station load is identified as being non-maximal if an available unassigned task can be added without violating the cycle time restriction (cf. Definition 4.3, p. 115).

Task Time Incrementing Rule. It is used prior to enumeration in linear time by checking the condition $t_j + t_{min} > c$ for each task $j = 1,...,n$.

Jackson Dominance Rule. This rule is used in an extended manner which is based on a modified definition of potential dominances including transitive successors (cf. Definition 4.5, p. 124):

Definition 4.5'. A task h *potentially dominates* another task j if $F_j \subseteq F_h^*$ and $t_j \leq t_h$ hold. In the case of $t_j = t_h$ and $F_j = F_h$, the lower-numbered task is defined to be the dominating one.

It is easy to verify that the foundation given for the Jackson dominance rule (see p. 125) holds for the modified definition of potential dominances, too.

Example (see p. 118). While the original definition identifies only two potential dominances, the new one finds six pairs of tasks with a potential dominance relation. For example, task 7 is potentially dominated by task 5 ($t_7 = t_5$; $F_7 \subseteq F_5^*$) or task 6 by task 2. In the case of the enumeration tree of FABLE (see Figure 4.3, p. 119), two nodes would be fathomed by the extended Jackson dominance rule. They correspond to the partial solutions with $S_1 = \{1\}$, $S_2 = \{3,4\}$, and $S_3 = \{2,7\}$ as well as $S_3 = \{5,6\}$ (both loads of station 3 are dominated by the alternative load $\{2,5\}$).

Modification of the Task Renumbering. In order to find feasible solutions as fast as possible, the task renumbering of FABLE is modified according to potential dominances. By giving each task j which is potentially dominated by a task h a larger number than h, dominating station loads are built prior to dominated ones.

Labeling Dominance Rule. In the basic version of SALOME-1, this dominance rule is not included, because its applicability is very restricted due to the extensive memory requirements of the Schrage and Baker labeling scheme. As computational experiments (see p. 245) show, the *tree dominance rule* (Nourie and Venta (1991), see p. 129) overcomes this disadvantage.

First Station Dominance Rule. The rule is excluded, because it seems to be arbitrary and needs defining and managing an additional data structure for storing trial loads of the first station. Furthermore, our computational experiences show that the rule does not fit frequently enough.

4.1.4.2 Additional Components

We extend our basic approach of SALOME-1 by some additional components which exploit the problem structure by using existing as well as new ideas.

Dynamic Prefixing. Prefixing is a reduction rule directly exploiting the concept of earliest and latest stations (cf. Section 2.2.1.1) and can be viewed as avoiding predictable bounding. Before finding station loads of a current station k by enumeration, prefixing is used to assign tasks which cannot be assigned to a later station than k. In the framework of our Local Lower Bound Method, prefixing is performed with respect to the current local lower bound LLB_k, i.e., a task j is prefixed to the currently considered station k+1 if $L_j^!(LLB_k) = k+1$ (cf. the definition of $L_j^!(m)$ by (2.38), p. 49). The prefixed tasks remain in station k+1 until the local lower bound LLB_k is increased. Due to the equation $L_j^!(m+1) = L_j^!(m) + 1$, no new prefixing can occur in station k for the increased local bound.

Example. The earliest and latest stations for our example are given in Table 2.6 (see p. 49). Imagine that the tasks 3 and 4 have already been assigned to station 1. Then in the next node at level 2, task 1 is prefixed to its latest station 2. In the stations 4, 5, and 6, the tasks 8, 9, and 10 are prefixed, respectively.

Simple Permutation Rule. Since the labeling or tree dominance rule may require a large amount of memory space for large-sized problem instances, we describe a very simple dominance rule for detecting certain partial solutions which contain the same station loads as another already considered one. In order to obtain a rule which is inexpensive with respect to computation time and storage space requirements, it is restricted to two consecutive levels of the tree:

> In any node at level k, a load S_{k+1} need not be considered if the condition $\max\{j \in S_{k+1}\} < \min\{j \in S_k\}$ holds for S_{k+1} and the preceding load S_k.

Due to the topological numbering of the tasks, the condition of the rule ensures that no task in S_{k+1} is a successor of a task in S_k. Hence, the loads S_k and S_{k+1} of the stations k and k+1 can be exchanged without violating the precedence constraints. Since loads are enumerated in the order of increasing task numbers, the pair S_k and S_{k+1} of station loads has already been considered in the reverse order.

Example. Imagine that a partial solution with $S_1 = \{1,2\}$ and $S_2 = \{3,4\}$ has already been built (impossible in our example). If the tasks 3 and 4 have none of the tasks 1 and 2 as predecessors, the reverse sequence $S_1 = \{3,4\}$, $S_2 = \{1,2\}$ of these loads will be obtained in a later step of the procedure. The load S_2 of this reverse sequence need not be considered, because the same reduced problem as in the first case would result.

Further Bound Arguments. In addition to LB_1, LB_2, LB_3, and LB_4', we use the bounds LB_5, LB_6, and LB_7 (cf. Section 2.2.2.1) in order to improve the initial as well as the local lower bounds.

Bound LB_5. In the root node, LB_5 is computed by utilizing the improved earliest and latest stations E_j' and $L_j'(m)$ (cf. (2.37) and (2.38), p. 49), because they can easily be derived from the heads and tails computed for LB_4' anyway. In all other nodes, LB_5 is based on E_j and $L_j(m)$ which are computed by the formulae (2.1) and (2.2) (see p. 28).

Bound LB_6. The bound LB_6 takes linear time in all nodes but the root, because the tasks have to be sorted according to non-increasing operation times only initially.

Bound LB_7. It uses the same task ordering as LB_6 and performs a few linear time iterations in each node. Even though the number of iterations may be large in the worst case, no more than two iterations occurred in our tests.

Discussion. The large variety of bounds used in our procedure enlarges the probability that at least one of the bound arguments provides a sharp bound on the number of stations. The required computation time is very small in all nodes but the root node.

4.1.4.3 Bidirectional Branching

As indicated by the results of Saltzman and Baybars (1987) as well as Hoffmann (1992), the planning direction may considerably influence the size of an enumeration tree. Though some existing algorithms use both planning directions, none incorporates criteria to decide on the favourable direction or to change the planning direction within the enumeration. A successive search in both directions, as it is performed by EUREKA, may cause the enumeration of a large tree in the forward phase, while the optimal solution may be found in the backward phase soon. Then by reversing the sequence of search directions, a significant acceleration of the algorithm could be obtained. Such a case occurs in our example (cf. Figures 4.4 and 4.5).

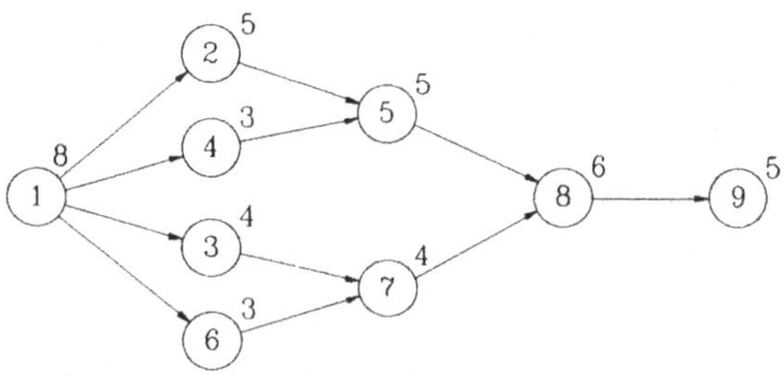

Figure 4.7. A Further Example Problem

Example. By means of another example, we demonstrate the importance of having data-dependent criteria for deciding on the planning direction. Figure 4.7 shows the precedence graph and the operation times of a problem with 9 tasks. The cycle time is assumed to be $c = 10$. With $t_{sum} = 43$, we obtain the initial lower bound $LB_1 = 5$ and the available total idle time $BD(5) = 7$.

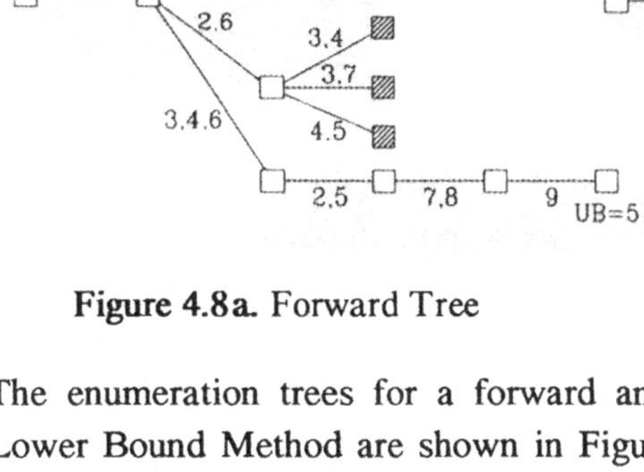

Figure 4.8a. Forward Tree

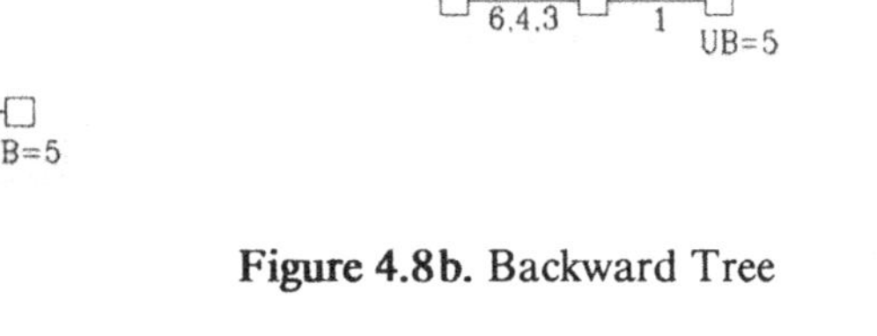

Figure 4.8b. Backward Tree

The enumeration trees for a forward and a backward application of our Local Lower Bound Method are shown in Figure 4.8 a and 4.8 b, respectively. Hatched nodes indicate that all station loads exceed the remaining idle time (with respect to the current lower bound of 5 stations). So, the local lower bounds of these nodes are increased to 6 and the nodes are fathomed, because a solution with 6 stations is already known in all cases. The forward tree contains much more nodes than the backward one. This is due to the fact that after assigning task 1 to the first station in the forward tree more tasks become available than after assigning task 9 to the first station in the backward tree (= last station in forward direction). Consequently, more station loads have to be checked for succeeding levels in the forward tree. Beyond that, the first backward station shows a greater amount of unavoidable idle time than the first forward one. Therefore, the remaining idle time is smaller in the backward case resulting in a reduced number of trial loads for LB = 5.

The above observations can be summarized as follows. At the first levels of the enumeration tree, only few station loads should exist and as much unavoidable idle time as possible should occur. As stated earlier, it seems to be promising to assign tasks with large operation times first, because they may be connected with unavoidable idle time. Furthermore, their assignment gets more difficult in later stations when the available idle time decreases. Additionally, it makes sense to prefer tasks for which the probability of choosing a disadvantageous station is low. This probability depends on the number of stations $E_j^!,...,L_j^!$ (LLB) to which a task j is *potentially assignable* for a current local lower bound LLB. That is, it depends on the length of the station interval SI_j(LLB). Furthermore, the probability of building inferior station loads is expected to be low if the set of tasks potentially assignable to a currently considered station k is small.

Example. For the problem of Figure 4.7 with c = 10 and LB = 5 (lower bound of the root node), the earliest and latest stations are as follows.

Table 4.4. Earliest and Latest Stations

j	1	2	3	4	5	6	7	8	9
$E_j^!$	1	2	2	2	3	2	3	4	5
$L_j^!(5)$	1	3	3	3	3	3	4	4	5

In this case, it is difficult to decide prior to enumeration whether the forward or the backward direction is the better choice. On the one hand, the larger operation time of task 1 compared to task 9, each of which is the single assignable task in the first and the last station, respectively, might argue for a forward planning. On the other hand, only two tasks are assignable to the station next to the last and four tasks to the second station. Hence, the probability for wrong decisions is much larger in the forward direction.

Our example shows that some difficulties arise from a single (static) decision on the planning direction. First, it is not obvious which parts of the precedence graph or which stations should influence the decision. Secondly, the properties of the reduced problems to be considered may extremely change during the enumeration. Therefore, it can be advantageous to change the direction *dynamically*, i.e., to decide in each node on the further planning direction. Such a *bidirectional* approach is possible, because a stationoriented branching results in reduced problems of the same type at each level of the tree. For example, the procedure may start in forward direction with station 1 and proceed with the last station in backward direction. The first choice is referred to as *forward step*, the latter one as *backward step*. In general, a forward (backward) step consists of choosing the earliest (latest) not yet considered station k_f (k_b) for branching.

Remark 4.2. A backward step represents a forward step in the reverse precedence graph. In accordance with the Local Lower Bound Method, we assume the last station (= the first one in backward direction) always to be the station LLB (current local lower bound). Hence, the number of the last station has to be increased whenever LLB is changed. However, this assumption is only important with respect to the following presentation of a bidirectional strategy. Generally, the last station may get a sufficiently large number.

We propose a decision rule which enables SALOME-1 to choose the branching direction dynamically. It is based on the following definition.

Definition 4.6. Let B_k(LLB) denote the set of tasks potentially assignable to a station k (cf. Figure 4.1, p. 114) in some node of the enumeration tree which has the current local lower bound LLB (for short, we write B_k). The *average task time* T_k of tasks which may be assigned to station k is defined by (4.3).

$$T_k := [\sum_{j \in B_k} (t_j / (L_j^!(\text{LLB}) - E_j^! + 1))] \;/\; |B_k| \qquad k = k_f, \ldots, k_b \tag{4.3}$$

A large value of T_k implies that one or more of the following conditions are true: Large tasks are assignable to k, only a few assignable tasks exist (small cardinality $|B_k|$ of the set B_k), or tasks assignable to station k are assignable to a few further stations only (small difference $L_j^!(\text{LLB}) - E_j^!$). Following our above discussion and that on stage-dependent complexity measures in Section 2.2.1.5 (see pp. 40), a large value indicates that it is promising to consider the respective station soon. These findings are utilized by a dynamic decision rule.

Bidirectional Branching Rule. In each node, the following rule is applied to decide on the planning direction for the next branching step:

> Choose a forward step if $T_{k_f} > T_{k_b}$ or $T_{k_f} = T_{k_b}$ and $|B_{k_f}| \le |B_{k_b}|$, otherwise, perform a backward step.

Example. For our example, the first decision is whether to start with station 1 or with station LB = 5. By means of the values in Table 4.4, we obtain $T_1 = 8$ and $T_5 = 5$. Hence, the procedure starts with a forward step and assigns task 1 to station 1 without affecting the value of T_5. Due to $T_2 = (5/2+4/2+3/2+3/2)/4 = 1.875$, a backward step assigns task 9 to station 5. A further backward step follows, because $T_4 = (4/2+6)/2 = 4$ is larger than T_2. Station 4 is chosen, and the load $S_4 = \{8,7\}$ is assigned. Due to $T_3 = 12.5/5 = 2.5$, the procedure continues with planning in the backward direction and forms the load {5,2} for station 3. The remaining tasks 6, 4, and 3 are assigned to station 2. The enumeration stops, because a solution with LB = 5 stations has been found. The corresponding enumeration tree is a single branch consisting of arcs for the station loads $S_1 = \{1\}$, $S_5 = \{9\}$, $S_4 = \{8,7\}$, $S_3 = \{5,2\}$, and $S_2 = \{6,4,3\}$.

Remark 4.3. According to the bound LB_5, the earliest and latest stations needed for computing the values T_k are based on heads and tails in the root node. In all other nodes, the weaker formulae (2.1) and (2.2) are used.

4.2 Exact Solution Procedures for SALBP-2

In this section, we consider exact solution procedures for SALBP-2, the version of SALBP which is to minimize the cycle time for a given number of stations. In the literature, corresponding approaches are usually based on repeatedly solving related SALBP-1 instances in the framework of a search procedure. Up to now, only one general branch and bound procedure which is able to solve SALBP-2 directly has been described (cf. Charlton and Death (1969)). Therefore, we propose two different branch and bound procedures which are based on the specific properties of SALBP-2. The first approach is a flexible taskoriented procedure which contains a number of interesting components. The second procedure is a non-trivial modification of our efficient procedure SALOME-1 for SALBP-1.

4.2.1 Notations and Definitions

We use the notations and basic definitions of Sections 2.1 and 4.1.1. In Figure 4.9, we repeat some SALBP-2 specific notations and introduce additional ones which are required for the description of procedures. The definitions of Section 4.1.1 which concern the cycle time c are valid for SALBP-2 if a maximal cycle time $\bar{c}$ is used (see pp. 114). That is, a task is (forward or backward) assignable if the conditions of Definition 4.2 are fulfilled and the cycle time $\bar{c}$ is not violated. Definition 4.3 is modified by the condition that no task can be added such that the station time does not exceed the cycle time $\bar{c}$.

m	given number of stations
c	(realized) cycle time $(= \max\{t(S_k) \mid k = 1,...,m\})$
LC	lower bound on the cycle time
UC, $\bar{c}$	upper bound on the cycle time, maximal cycle time
$I_k(c)$	idle time in station k for cycle time c $(= c - t(S_k))$
$LI_k(c)$	lower bound on the idle time in station k for cycle time c
BD(c)	total idle time (balance delay time) for cycle time c $(= m \cdot c - t_{sum})$
SL(c)	lower bound on the station time to realize c $(\geq \max\{0, c - BD(c)\})$
$E_j(c)$	earliest station of task j for cycle time c; cf. (2.54) and (2.57), p. 56
$L_j(c)$	latest station of task j for cycle time c; cf. (2.55) and (2.58), p. 56
$SI_j(c)$	station interval of task j for cycle time c $(= [E_j(c), L_j(c)])$
$B_k(c)$	set of tasks potentially assignable to station k $(= \{j \mid k \in SI_j(c)\})$
P_μ	node of a branch and bound tree
LLC_μ	local lower bound on the cycle time in node P_μ

Figure 4.9. Notations for SALBP-2

4.2.2 SALBP-1 Based Search Methods

Following our argumentation in Section 2.2.1.3, SALBP-2 (minimize c for given m) can be formulated as the problem of finding the smallest cycle time c for which a corresponding SALBP-F instance (m,c) has a feasible solution. Hence, SALBP-2 can be solved by successively solving SALBP-F instances with m stations and various *trial cycle times*. The latter are restricted to an interval [LC,UC] which is bounded by a lower and an upper bound on the cycle time (cf. Section 2.2.3). Usually, the simple bounds LC_1 and UC_1 are used (cf. Hackman et al. (1989)).

SALBP-1 Procedures for SALBP-F. Any procedure for SALBP-1 (cf. Section 4.1) can easily be modified for SALBP-F instances (m,c) by considering the given cycle time c and by excluding all solutions which have a larger number of stations than m.

For example, EUREKA is adapted to SALBP-F by starting it with the given number of stations as the initial lower bound. If the modified procedure finds a feasible solution with that number of stations, the current SALBP-F instance is solved. Otherwise, the procedure stops with the result that no feasible solution exists when the initial lower bound has to be increased.

The other branch and bound algorithms, including FABLE and SALOME-1, fathom every node of the tree when its local lower bound gets larger than the given m.

Search Methods. The following general search procedures are distinguished with respect to the sequence in which trial cycle times of the interval [LC,UC] are considered. Such procedures are applied by, e.g., Mansoor (1964), Mansoor and Yadin (1971), Hardeck (1974), Dar-El and Rubinovitch (1979), Wee and Magazine (1981 b) as well as Hackman et al. (1989).

Lower Bound Search. Starting with the lower bound LC, the trial cycle time c is successively increased by 1 until the respective SALBP-F instance (m,c) is feasible. Note that this approach is analogous to the Lower Bound Method for SALBP-1 used by EUREKA.

Upper Bound Search. It starts with an upper bound UC. The cycle time c is successively decreased by 1, until SALBP-F is infeasible for c-1 or c is equal to LC. In each iteration, the maximal station time of the feasible solution found is used as the new value of c.

Binary Search. The search interval [LC,UC] is successively subdivided into two subintervals by choosing the mean element $c = \lfloor (LC+UC)/2 \rfloor$. If SALBP-F is feasible for c, the upper bound UC is set to the maximal station time in the corresponding solution. Otherwise, LC is set to c+1. The search stops with an optimal cycle time UC when UC=LC.

Fibonacci Binary Search. The method contains two steps. First, a *Fibonacci Search* is performed following the basic idea of the Lower Bound Search. The trial cycle times are LC, LC+F(1), LC+F(1)+F(2),... using the *Fibonacci numbers* which are defined as follows:

$$F(1) := 1, \quad F(2) := 2, \quad F(i) := F(i-1) + F(i-2) \quad \text{for } i \geq 3 \tag{4.4}$$

If a SALBP-F instance with a trial cycle time c is not feasible, LC is set to c+1. When a feasible cycle time c is obtained, the Binary Search is applied to the remaining interval [LC,c]. Note that the Fibonacci Search is usually defined somewhat different (cf., e.g., Minoux (1986, pp. 66)).

Binary Search with Prespecified Entry Point. The method differs from the Binary Search only in determining the first trial cycle time. It is based on the observation that for many problem instances the cycle times close to LC are more likely to be optimal than larger values. Therefore, the first trial cycle time is determined by the maximal value of c for which $m = \lceil t_{sum}/c \rceil$ holds. The Binary Search is applied to the remaining search interval.

Example. We consider the problem with its precedence graph given in Figure 4.10 (cf. Figure 2.8, p. 54) and m = 5 stations. The simple lower bound LC_1 is 10.

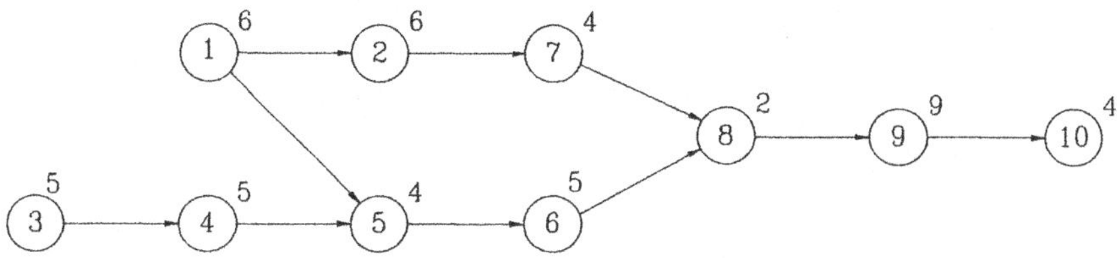

Figure 4.10. Precedence Graph

Figure 4.11 illustrates the application of the *Lower Bound Search* to the example. The SALBP-F instances are solved by the Local Lower Bound Method (the enumeration procedure of SALOME-1; cf. Section 4.1.4.1). The search starts with $LC_1 = 10$. Due to BD(10) = 0, only station loads without idle time are feasible for the SALBP-F instance (m,c) = (5,10). One station load is possible in each of the first three stations. The hatched node is fathomed, because no load without idle time is available for station 4. For c = 11, the balance delay time is 5, and the station times must be in the interval [6,11]. In the first branch of the corresponding tree, no idle time occurs, until the station load $S_3 = \{5,6\}$ with an idle time $I_3(11) = 2$ is built. Now, only loads with station times not smaller than 8 avoid that the total idle time is exceeded. Since only the maximal load $S_4 = \{6,8\}$ with station time 7 exists for station 4, the corresponding hatched node is fathomed. The other hatched nodes are fathomed accordingly. After examining an enlarged tree for c = 12 without success, a feasible (= optimal) solution is found in the first branch of the tree for c = 13. The bold parts of the trees are considered repeatedly.

The *Upper Bound Search* requires (up to) 8 iterations if it is started with $UC = UC_1 = 20$ (cf. p. 60).

If the *Binary Search* is applied to the interval $[LC_1, UC_1] = [10,20]$, it has to consider SALBP-F instances for c = 15, 12, 14, and 13.

The *Fibonacci Binary Search* solves SALBP-F instances for c = 10, 11, 13, and 12.

The first trial cycle time for the *Binary Search with Prespecified Entry Point* is c = 12. Since this cycle time is not feasible, the Binary Search has to be applied to the remaining interval [13,20] and considers the trial cycle times 16, 14, and 13.

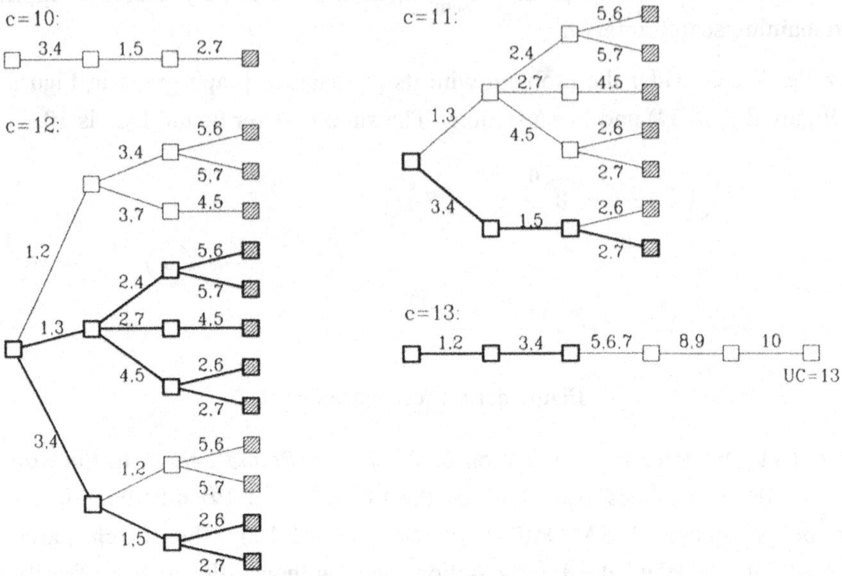

Figure 4.11. Lower Bound Search

4.2.3 A Taskoriented Branch and Bound Procedure (TBB-2)

We propose a branch and bound procedure for SALBP-2 which branches in a task-oriented manner, i.e., a single task is assigned to one station in each step. This branching scheme is similar to that of the branch and bound procedure of Charlton and Death (1969), which is able to solve flow shop and job shop problems with minor changes, too. We do not describe this algorithm, because it does not sufficiently utilize the particular structure of assembly line balancing problems (cf. Domschke et al. (1997), pp. 227).

In the following sections, we describe the components of a taskoriented branch and bound approach which is referred to as TBB-2 (Taskoriented Branch and Bound Procedure for SALBP-2; cf. Scholl (1994)). It represents a toolbox for constructing different branch and bound procedures, because it contains a number of optional components.

4.2.3.1 Branching

After describing the representation of subproblems together with some definitions, we distinguish two taskoriented branching strategies.

Representation of Subproblems. A subproblem (node of the enumeration tree) is defined by an *assignment vector* $z = (z_1,...,z_n)$, where z_j denotes the number of the station to which a task j is assigned (cf. Formulation 3 in Section 2.2.1.2, p. 31).

Values $z_j = 0$ indicate that task j is currently unassigned. Subproblems are denoted by P_μ, where the index μ is used to distinguish the problems. The initial problem (root node) P_0 is represented by the assignment vector $z = (0, ..., 0)$.

By the assignment vector z, each subproblem is connected with *station loads*

$$S_k = \{j \mid z_j = k\} \qquad \text{for } k = 1, ..., m$$

and corresponding *station times* $t(S_k)$.

Maximal Cycle Time. We assume that an initial upper bound UC on the cycle time is known (cf. Section 2.2.3.2 for methods to obtain such a bound). Therefore, the branch and bound procedure can be restricted to solutions with the cycle time not exceeding the maximal cycle time $\bar{c} = UC - 1$.

Earliest and Latest Stations. Based on the maximal cycle time $\bar{c}$, we define the earliest stations $E_j(\bar{c})$ and the latest stations $L_j(\bar{c})$ for each task j by the formulae (2.54) and (2.55) or (2.57) and (2.58), respectively (see p. 56). The earliest and latest stations define station intervals $SI_j(\bar{c}) = [E_j(\bar{c}), L_j(\bar{c})]$.

As a matter of convenience, we use E_j, L_j, and SI_j instead of $E_j(\bar{c})$, $L_j(\bar{c})$, and $SI_j(\bar{c})$, whenever it is clear that the values refer to the maximal cycle time $\bar{c}$.

Definition 4.7. The station intervals restrict the assignment of tasks to stations. In each node of the tree, a set of tasks *potentially assignable* to a station k is defined as follows (cf. Figure 4.9 for the definition in the root node):

$$B_k(\bar{c}) := \{j \mid k \in SI_j(\bar{c}) \ \wedge \ t(S_k) + t_j \le \bar{c} \ \wedge \ z_j = 0\} \tag{4.5}$$

Definition 4.8. In a subproblem P_μ, a station k is said to be *closed* if the set $B_k(\bar{c})$ is empty. Otherwise, the station is called *open*.

Branching Strategies. Branching is organized as a *taskoriented laser search,* i.e., subproblems are generated by assigning one task to one station at a time. Two strategies are distinguished:

Task Selection. In each node (subproblem), a not yet assigned task j ($z_j = 0$) is selected by one of the following rules or by a combination of those rules. That is, one rule is used as a primary rule, another one as a tie breaker.

- Task Time Rule: Select a task j with largest operation time t_j.
- Interval Rule: Select a task j with smallest station interval SI_j.
- Precedence Rule: Select a forward or backward available task j, respectively.

By assigning the chosen task j to the stations $E_j, E_j + 1, ..., L_j$ at a time, we obtain $L_j - E_j + 1$ subproblems of the current node.

The subproblems are generated in *alternating order*, i.e., task j is assigned to station $M_j := \lfloor (E_j + L_j)/2 \rfloor$ at first. At the subsequent re-visits of the node, the other subproblems are built in the order M_j+1, M_j-1, $M_j+2, \ldots, E_j, L_j$. Another possibility is to take the reverse order.

Station Selection. In each node, an open station is selected. As in case of task selection, several orders in which tasks are considered are possible:

- Unidirectional Selection: The stations are considered in increasing (forward)/ decreasing (backward) order of their numbers. That is, in each node the earliest/ latest open station is selected. The forward selection is similar to the approach of FABLE (cf. Section 4.1.2.2).
- Strict Bidirectional Selection: The stations are considered in the order 1, m, 2, $m-1, \ldots, \lceil (m+1)/2 \rceil$. Each currently considered station is repeatedly selected until it is closed. Alternatively, one may start with station m.
- Flexible Bidirectional Selection: Either the earliest open station k_f *(forward step)* or the latest open station k_b *(backward step)* is selected. The decision between k_f and k_b is based on the mean operation time T_k of tasks which are already assigned to station k or are potentially assignable to k (cf. a similar definition in (4.3)):

$$T_k := \Big[t(S_k) + \sum_{j \in B_k} \big(t_j / (L_j - E_j + 1)\big) \Big] \, / \, |S_k \cup B_k| \tag{4.6}$$

Furthermore, a non-negative parameter α is utilized. It controls the frequency of changing the direction (from a forward step to a backward step, and vice versa). In the root node P_0, α is set to 1. In all other nodes of the tree, its value depends on the last branching step. If the last step was a forward one, α is set to a value (α_1) which is smaller than 1. Analogously, it is set to a value (α_2) which is larger than 1, if the last step was a backward one. The more the values α_1 and α_2 differ from 1, the less often the branching direction is changed. Furthermore, the direction is not changed anymore, whenever the "mean station" m/2 has been crossed in either direction. That means, the procedure continues branching in a unidirectional manner.

Using the above parameters, the stations are selected according to the following rule:

> If $k_f > m/2$, then select k_f. Otherwise, select k_b, if $k_b < m/2$. If neither condition is true, select k_f in case of $T_{k_f} \geq \alpha \cdot T_{k_b}$. Otherwise, select k_b.

When a station k is selected, the respective node P_μ is branched into several subproblems by alternatively assigning all forward or backward available tasks depending on whether a forward or backward step is performed (cf. the precedence rule above). The corresponding tasks are subdivided into two groups. The first group contains all tasks which do not increase the current local lower bound LLC_μ of the node (cf. Section 4.2.3.2). The second one includes the remaining tasks which increase the local lower bound, because their assignment results in a station time exceeding LLC_μ, i.e., $t(S_k) + t_j > LLC_\mu$.

In order to obtain solutions with small cycle times first (cf. the Lower Bound Search in Section 4.2.2), the first group is taken for branching prior to the second group. Within each group the tasks can be ordered with respect to one of the above task selection rules (e.g., the task time rule as primary rule and the interval rule as tie breaker). If the first group is empty, i.e., no task is assignable to station k without increasing LLC_μ, an additional subproblem is generated prior to all subproblems resulting from the second group of tasks. This subproblem is obtained by *closing* the station k, i.e., no further tasks can be added in later nodes of the subtree which follows P_μ.

4.2.3.2 Bounding and Fathoming

We describe methods to obtain upper and lower bounds on the cycle time for every node of the branch and bound tree. The usage of these bounds for fathoming is considered.

Initial Upper Bound. At the beginning of the branch and bound procedure, an initial upper bound UC is computed. Basically, UC may be obtained from a theoretical upper bound (cf. Section 2.2.3.2) or any heuristic algorithm for SALBP-2. Since theoretical bounds are usually rather weak, we propose to utilize a heuristic combination of a priority rule based procedure and a tabu search procedure such as described in Sections 5.1.1 and 5.2.3. In order to achieve a fast computation of the initial upper bound, the tabu search procedure should be given a short time limit.

Local Upper Bounds. A subproblem is solved (it is represented by a leaf node of the tree) if all tasks are assigned to a station, i.e., $z_j > 0$ for $j = 1,...,n$. The realized cycle time c (local upper bound) is given by the maximal station time of this solution. If $c < UC$, the global upper bound is set to c and the solution serves as new incumbent solution. The maximal cycle time $\bar{c}$ is set to $UC - 1$, and the earliest and latest stations are updated for this new value in all nodes which are currently not completely branched.

In each non-leaf node, the current partial assignment of tasks may be completed by some modified heuristic procedure (e.g., a priority rule based procedure). The realized cycle time of such a heuristic solution provides a local upper bound of the node. Possibly, the local upper bound improves the global upper bound, but the node can only be fathomed if the local upper bound and the local lower bound are equal.

Global Lower Bound. In the root node, some or all of the bound arguments of Section 2.2.3.1 may be used for computing an initial lower bound LC. The branch and bound procedure stops when all nodes are completely branched or a feasible solution with a cycle time being equal to LC is found.

Local Lower Bounds. For each node P_μ, the value of the local lower bound LLC_μ is inherited from the local lower bound LLC_σ of the ancestor node P_σ. Furthermore, it may be increased by the branching operation, i.e., the assignment of a task to a station:

$$LLC_\mu := \max\{LLC_\sigma, \max\{t(S_k) \mid k=1,...,m\}\} \tag{4.7}$$

If $LLC_\mu \geq UC$, then the problem P_μ is *fathomed.* In order to strengthen the local lower bound, further bound arguments may be used. However, only computationally efficient bounds like LC_2 or LC_6 (with the SALBP-1 bounds LB_2 and LB_3) should be applied in nodes for subproblems (cf. Section 2.2.3.1). Eventually, the application of the bounds may be varied depending on the problem structure and the effectiveness of the bound arguments in P_0.

4.2.3.3 Logical Tests

We describe a number of logical tests which are used in every node of the tree in order to fathom the node or to reduce the data of the subproblem, respectively. Especially, lower bounds $LI_k(c)$ on the idle times of the stations are recognized.

Prefixing. In case of $E_j = L_j$ for a task j, it is assigned to the station E_j before branching the respective node. The prefixing test is always applied when the maximal cycle time $\bar{c}$ is updated, and, hence, the earliest and latest stations are modified.

Task Test. A node P_μ can be fathomed if at least one unassigned task cannot be assigned to any station. This case occurs for a task j either if the condition $E_j > L_j$ is true or none of the stations in SI_j shows enough idle time with respect to the maximal cycle time $\bar{c}$ to include task j. Note that the first case corresponds to the lower bound LC_3 (see p. 56).

Station Test 1. A station k is *closed* if no unassigned task can be added without exceeding the maximal cycle time $\bar{c}$. This is true if B_k is empty or $t(S_k) + \Delta_{min} > \bar{c}$ holds, where Δ_{min} denotes the smallest operation time of a task in B_k. In this case, a lower bound on the idle time depending on the current local lower bound LLC_μ of the node P_μ is obtained by

$$LI_k(LLC_\mu) = \max\{0, LLC_\mu - t(S_k)\}. \tag{4.8}$$

Station Test 2. All tasks potentially assignable to a station k can be assigned to this station if the following condition is true:

$$t(S_k) + \sum_{j \in B_k} t_j \leq LLC_\mu \tag{4.9}$$

The station gets closed and a lower bound on the idle time of the station is given by

$$LI_k(LLC_\mu) = \max\{0, LLC_\mu - t(S_k) - \sum_{j \in B_k} t_j\}. \tag{4.10}$$

Station Test 3. A node P_μ with local lower bound LLC_μ is fathomed if the sum of unavoidable idle times of the stations exceeds the maximal amount of idle time in a solution which improves UC (= balance delay time for the maximal cycle time $\bar{c} = UC - 1$). That is, a node P_μ is fathomed if the following condition is true:

$$\sum_{k=1}^{m} LI_k(LLC_\mu) > BD(\bar{c}) \tag{4.11}$$

Otherwise, a lower bound $SL(LLC_\mu)$ on the station time of all *open* stations h (with $LI_h(LLC_\mu) = 0$) is given by subtracting the remaining maximal amount of idle time from the current local lower bound on the cycle time:

$$SL(LLC_\mu) := LLC_\mu - (BD(\bar{c}) - \Sigma_k\, LI_k(LLC_\mu)) \tag{4.12}$$

If in a station this lower bound on the station time can only be achieved by assigning all potentially assignable tasks, these tasks are assigned and an unavoidable idle time may occur. If the resultant station time exceeds the maximal cycle time $\bar{c}$, the node P_μ is fathomed.

Remark 4.4. Basically, lower bounds $LI_k(LLC_\mu)$ on the idle time can be obtained by solving a subset sum problem with "capacity" $\bar{c}$ for each station (cf. the lower bound LC_5, p. 58). Nevertheless, this problem is NP-hard and its solution usually takes too much computation time to be applied for several times in each node.

Modification of the Precedence Graph. By modifying the precedence graph, i.e., introducing additional arcs or nodes, one may obtain improved values E_j and L_j or improved lower bounds, respectively.

Additional Precedence Relations. For a pair h and j of not related tasks an additional arc is introduced if $L_h \leq E_j$. This modification is possible, because the sequence of tasks is optional within a station.

Station Nodes. The precedence graph is modified by introducing nodes for the stations which are denoted by $N_1,...,N_m$. In the initial problem, the station nodes are connected by arcs (N_i, N_{i+1}) for $i = 1,...,m-1$, i.e., they build a chain. Additionally, node N_1 serves as fictitious source and node N_m, as fictitious sink of the modified graph. All station nodes get the operation time 0, i.e., the stations are unloaded.

When a task j is assigned to a station k by branching or logical tests, the node for j and N_k are merged and their operation times are added up. Furthermore, the sets of predecessors and successors of both nodes are united, respectively. The modified node is called N_k, too, because it represents the station load of station k. If both nodes are connected by a path in the precedence graph, all intermediate nodes have to be merged into the node N_k in order to avoid cycles.

With this concept, a subproblem P_μ can be represented by a modified precedence graph $G_\mu = (V_\mu, A_\mu, t_\mu)$, where V_μ, A_μ, and t_μ denote the modified node set, arc set, and operation times, respectively.

Each branching operation eliminates at least one task node by merging it into a station node. Feasible solutions are obtained when the modified graph is serial and contains station nodes only.

By analogy with task nodes, we may define heads and tails as well as earliest and latest stations for station nodes, too. Since N_k represents the station k, the following conditions are true: $a_{N_k} \geq k-1$, $n_{N_k} \geq m-k$, $E_{N_k} = L_{N_k} = k$

The fictitious station nodes allow successive strengthening of precedence relations, because predecessor and successor sets are united when two nodes are merged. Thus, heads and tails as well as earliest and latest stations may be improved. Besides, the following simple conditions may help to improve the value of the lower bound LC_4 if it is applied to a modified precedence graph G_μ (see p. 57):

$$L_j \leq k \text{ for } j \in P^*_{N_k} \quad \text{and} \quad E_j \geq k \text{ for } j \in F^*_{N_k} \qquad \text{for } k = 1,...,m \tag{4.13}$$

Remark 4.5. This representation of subproblems is well suited to treating station related assignment restrictions (cf. Section 1.3.4). If a task must be performed in a certain station, the respective nodes are simply merged for the initial problem resulting in a reduction of the problem. Further assignment restrictions can easily be incorporated into the taskoriented branch and bound procedure, because they only reduce the number of elements in the station intervals used for branching.

4.2.3.4 An Example for TBB-2

In the preceding sections, we present several possible components of a taskoriented branch and bound approach. The scope of this section is to consider an explicit version of this (general) procedure more precisely. In order to concentrate on the basic approach, the chosen procedure is only characterized by the following simple components.

- Branching: task selection with task time rule, alternating station order
- Bounding: application of the bounds LC_1 and LC_6 (with LB_2, LB_3) in all nodes

• Logical Tests: prefixing, station nodes

To be more specific, the procedure selects an unassigned task j with maximal task time in each branching step and generates subproblems according to the alternating order of stations in the interval SI_j. In each node, the bound LC_1 is used together with (4.7). Furthermore, the bounds LB_2 and LB_3 for SALBP-1 (see p. 44) are utilized to strengthen the bounds. This type of bound is referred to as LC_6 in Section 2.2.3.1 (see p. 59) and is computed by (2.62). The earliest and latest stations are determined by (2.57) as well as (2.58) and are utilized by prefixing (see p. 56). The subproblems are represented by modified precedence graphs with station nodes.

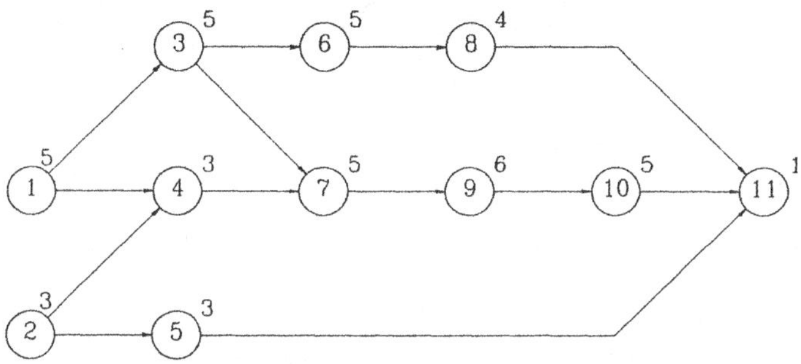

Figure 4.12. Precedence Graph

We consider an example problem with 11 tasks and the precedence graph of Figure 4.12 (cf. Figure 2.2, p. 26). The operation times are characterized by $t_{sum}=45$, $t_{max}=6$, and $t_{min}=1$. The given number of stations is $m=5$.

Initial Problem (P_0). The computation of initial upper and lower bounds as well as branching of the initial problem P_0 are considered.

Table 4.5. Computation of Earliest and Latest Stations in P_0

j	1	2	3	4	5	6	7	8	9	10	11
p_j	0.5	0.3	0.5	0.3	0.3	0.5	0.5	0.4	0.6	0.5	0.1
a_j	0	0	0.5	1	0.3	1	2	1.5	3	4	4.5
n_j	3.5	3	3	2.5	0.1	0.5	2	0.1	1	0.1	0
E_j	1	1	1	2	1	2	3	2	4	5	5
L_j	2	2	2	3	5	5	3	5	4	5	5

Initial Upper Bound. We assume that a solution with UC = 11 is determined by a heuristic procedure. Since such procedures are presented in Section 5.1.3, we do not describe here how the solution is obtained. The maximal cycle time of an improved solution is $\bar{c}=10$. Earliest stations $E_j(\bar{c})$ and latest stations $L_j(\bar{c})$ are given in Table 4.5 (abbreviated by E_j and L_j; the station requirements of the nodes are denoted by p_j; the computation of heads a_j and tails n_j is explained in Section 2.2.2.1, pp. 44).

Prefixing. Due to $E_j = L_j$, the tasks 7, 9, 10, and 11 are assigned to their single station in each case. The resultant modified precedence graph with station nodes N_k being symbolized by rectangles is given in Figure 4.13. In Table 4.6, the modified earliest and latest stations are presented.

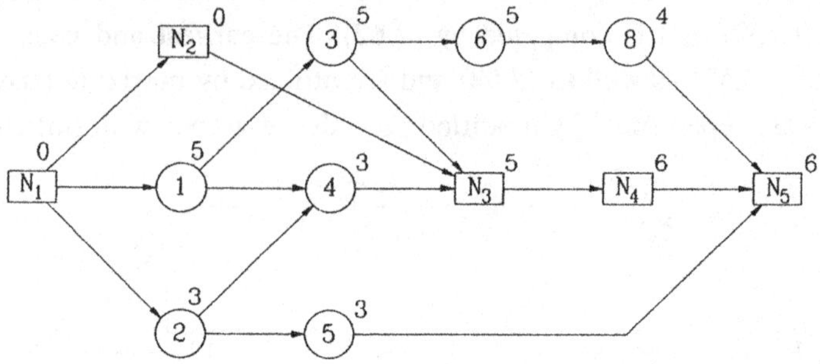

Figure 4.13. Modified Precedence Graph after Prefixing in P_0

Initial Lower Bound. The simple bound is $LC_1 = \lceil 45/5 \rceil = 9$. The bound LC_6 requires the computation of LB_2 and LB_3 for the modified problem with cycle time 9:

$$LB_2(9) := 6 + \lceil 1/2 \cdot 0 \rceil = 6 \quad \text{and} \quad LB_3(9) := \lceil 0 + 2/3 \cdot 2 + 1/2 \cdot 5 + 1/3 \cdot 3 \rceil = 5$$

Due to $LB_2(9) = 6$, no solution with 5 stations can exist for cycle time 9, and the initial bound LC can be increased to 10. For this cycle time, the SALBP-1 bounds are given by $LB_2(10) = LB_3(10) = 4$. Hence, the initial lower bound is $LC = 10$.

Table 4.6. Modified Earliest and Latest Stations in P_0

j	N_1	N_2	1	2	3	4	5	6	N_3	8	N_4	N_5
p_j	0	0	0.5	0.3	0.5	0.3	0.3	0.5	0.5	0.4	0.6	0.6
a_j	0	1	0	0	0.5	1	0.3	1	2	1.5	3	4
n_j	4.5	2	3.5	3	3	2.5	0.6	1	2	0.6	1	0
E_j	1	2	1	1	1	2	1	2	3	2	4	5
L_j	1	2	2	2	2	3	5	4	3	5	4	5

Branching. Following the task selection rule, the tasks 1, 3, and 6, each of which has the operation time 5, may be chosen for branching. Among these, the task 6 is chosen arbitrarily. According to the alternating station order, task 6 is first assigned to station $M_6 = \lfloor (2+4)/2 \rfloor = 3$ (and later to station 2 and 4, respectively). The resultant problem P_1 contains the modified precedence graph of Figure 4.14. The modified earliest and latest stations (for $\bar{c} = 10$) are given in Table 4.7.

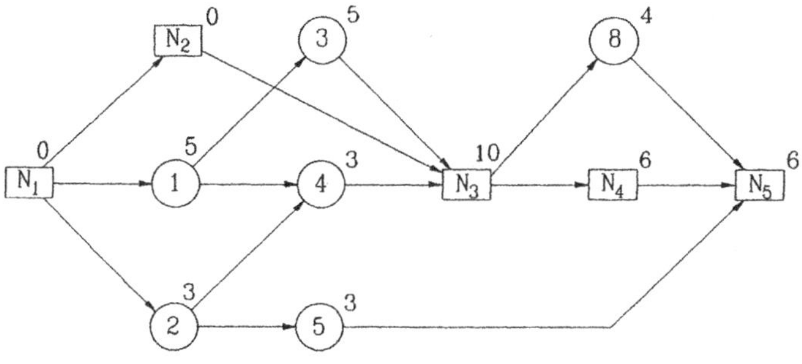

Figure 4.14. Modified Precedence Graph of Problem P_1

Table 4.7. Earliest and Latest Stations in P_1

j	N_1	N_2	1	2	3	4	5	N_3	8	N_4	N_5
p_j	0	0	0.5	0.3	0.5	0.3	0.3	1	0.4	0.6	0.6
a_j	0	1	0	0	0.5	1	0.3	2	3	3	4
n_j	4.6	3	4	3.3	3	3	0.6	2	0.6	1	0
E_j	1	2	1	1	1	2	1	3	4	4	5
L_j	1	2	1	2	2	2	5	3	5	4	5

Problem P_1. P_1 is derived from P_0 by assigning task 6 to station 3.

Prefixing. Task 1 is prefixed to station 1 and task 4 to station 2. The modified precedence graph, which includes an additional arc $(3,N_2)$ due to $L_3 \leq E_{N_2}$, is shown in Figure 4.15. The earliest and latest stations are given in Table 4.8.

Local Lower Bound. $LLC_1 = \max\{10, \max\{5,3,10,6,6\}\} = 10$ (cf. (4.7), p. 152).

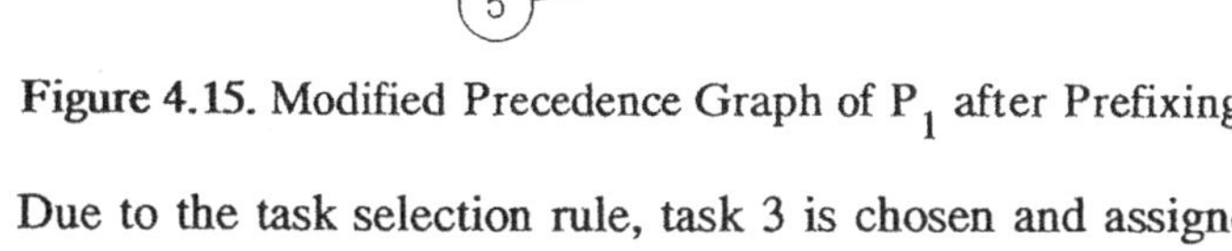

Figure 4.15. Modified Precedence Graph of P_1 after Prefixing

Branching. Due to the task selection rule, task 3 is chosen and assigned to station 1. The resultant problem P_2 is characterized by the precedence graph in Figure 4.16 and the modified values in Table 4.9.

Table 4.8. Modified Earliest and Latest Stations in P_1

j	N_1	2	3	N_2	5	N_3	8	N_4	N_5
p_j	0.5	0.3	0.5	0.3	0.3	1	0.4	0.6	0.6
a_j	0	0.5	0.5	1	1	2	3	3	4
n_j	4.1	3.3	3.3	3	0.6	2	0.6	1	0
E_j	1	1	1	2	2	3	4	4	5
L_j	1	2	2	2	5	3	5	4	5

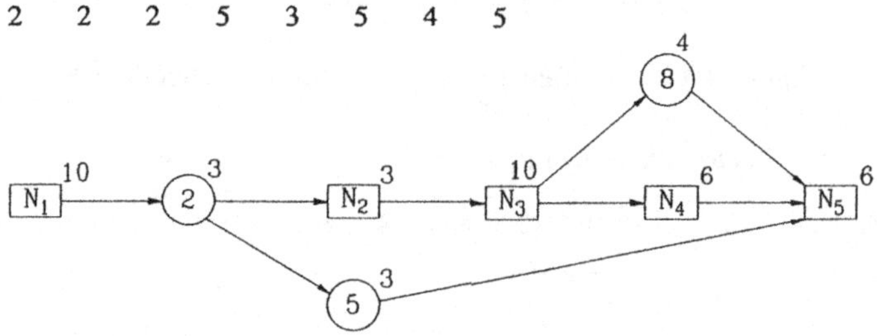

Figure 4.16. Modified Precedence Graph of Problem P_2

Table 4.9. Modified Earliest and Latest Stations in P_1

j	N_1	2	N_2	5	N_3	8	N_4	N_5
p_j	1	0.3	0.3	0.3	1	0.4	0.6	0.6
a_j	0	1	1.3	1.3	2	3	3	4
n_j	4	3.3	3	0.6	2	0.6	1	0
E_j	1	2	2	2	3	4	4	5
L_j	1	2	2	5	3	5	4	5

Problem P_2. It is derived from P_1 by assigning task 3 to station 1.

Prefixing. Task 2 is assigned to station $E_2 = L_2 = 2$. The earliest and latest stations of the remaining tasks 5 and 8 are not changed.

Local Lower Bound. $LLC_2 = \max\{10, \max\{10, 6, 10, 6, 6\}\} = 10$.

Branching. According to the task selection rule, task 8 is chosen and assigned to station 4. The only remaining task 5 can be added to station 2 (or station 5). A feasible solution with UC = 10 and the following station loads is obtained:

$$S_1 = \{1,3\},\ S_2 = \{2,4,5\},\ S_3 = \{6,7\},\ S_4 = \{8,9\},\ S_5 = \{10,11\}$$

Termination of the Procedure. Due to UC = LC = 10, the procedure is terminated and the determined solution is optimal.

4.2.4 The Branch and Bound Procedure SALOME-2

We present a stationoriented branch and bound approach for SALBP-2. It is called SALOME-2 (**S**imple **A**ssembly **L**ine Balancing **O**ptimization **M**ethod for Type **2**; cf. Klein and Scholl (1996)) and represents a modification of the procedure SALOME-1 (see Section 4.1.4).

4.2.4.1 Local Lower Bound Method

The Local Lower Bound Method for SALBP-2 is a branching concept which includes an intelligent bounding component. Therefore, bounding aspects are considered together with the enumeration technique in this section.

Introduction. Following properties of existing search methods and possible problem instances can be identified. Our computational experiments (cf. Section 7.2.3.2) confirm these findings.

- All search methods of Section 4.2.2, which are based on SALBP-1 algorithms, have the major drawback not to use results of computations performed for previously considered trial cycle times (i.e., SALBP-F instances). Hence, large portions of enumeration trees may be constructed repeatedly. In our example, four trees containing common parts (boldfaced in Figure 4.11, p. 148) have to be built by the Lower Bound Search.
- These methods are based on the assumption that all values of the search interval are potential cycle times. This may lead to unnecessary iterations if no combination of tasks exists which has a station time equal to a trial cycle time. For example, imagine the simple case of a problem with operation times which are multiples of 100. Then only such multiples are candidates for the optimal cycle time.
- The practical application of solution procedures is often restricted by limited computation times. In such a case, search methods (like the Lower Bound Search), which examine infeasible SALBP-F instances at first, may not provide a feasible solution at all.
- For a large number of problem instances, the relative difference between the minimal cycle time and an initial lower bound is very small.

These observations lead to following properties a reasonable SALBP-2 procedure should have. It should obtain feasible solutions soon, avoid to consider different SALBP-F instances separately, orientate on the lower bound, and exclude inefficient trial cycle times from consideration. Since the first three aspects are observed by the Local Lower Bound Method used in the procedure SALOME-1, it is obvious to apply a similar procedure to SALBP-2.

Outline of the Local Lower Bound Method for SALBP-2. Branching is performed as a *stationoriented laser search* by successively building loads for the stations 1,...,m. Resultant subproblems (nodes of the enumeration tree) are reduced problems of the same type with less stations and a reduced precedence graph.

The method starts with a lower bound LC on the cycle time and proceeds like a *Lower Bound Search* as long as possible in order to find a feasible solution with cycle time LC. Hence, in each node only such maximal station loads whose idle times do not exceed the remaining idle time (with respect to LC) are branched.

After examining all loads feasible for the trial cycle time LC without success (no feasible completion of the partial solution with cycle time LC exists), the lower bound of the node is *locally* increased. Note that this node is fathomed by a Lower Bound Search. In order to avoid trying cycle times impossible for the reduced problem, the new value LC' of the *local lower bound* is determined as the smallest one, for which at least one not yet considered station load is feasible. For LC' only those maximal loads which have not been feasible for former bound values are examined.

Whenever loads for all stations but station m are constructed, a feasible solution is obtained by assigning the remaining tasks to station m. The cycle time of the best known feasible solution serves as an upper bound UC. A node is fathomed when its local lower bound is equal to or exceeds the value of UC.

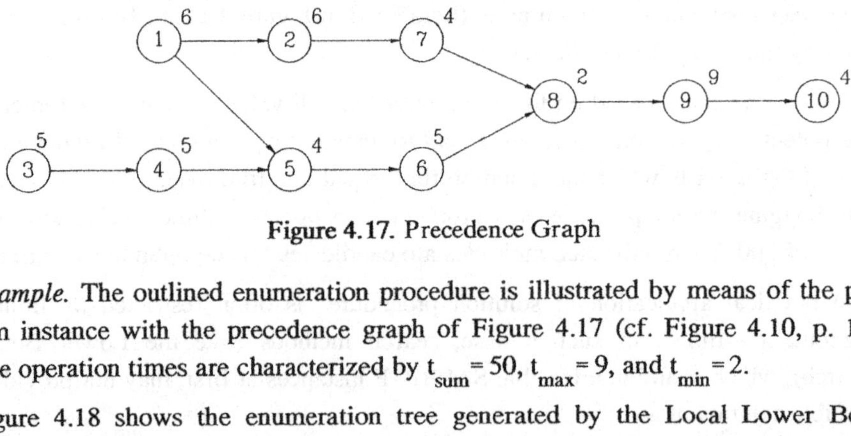

Figure 4.17. Precedence Graph

Example. The outlined enumeration procedure is illustrated by means of the problem instance with the precedence graph of Figure 4.17 (cf. Figure 4.10, p. 147). The operation times are characterized by $t_{sum} = 50$, $t_{max} = 9$, and $t_{min} = 2$.

Figure 4.18 shows the enumeration tree generated by the Local Lower Bound Method. The nodes are numbered in order of their generation. The current local lower bounds are given as node weights together with the respective remaining idle times. Loops mark the increase of a local bound. Parts of the tree which correspond to the same value of the local lower bound are underlayed with the same pattern. In our explanation below, we use the symbol LLC_k for the local lower bound of a node at a *level* k which represents a partial solution with the stations 1,...,k being loaded.

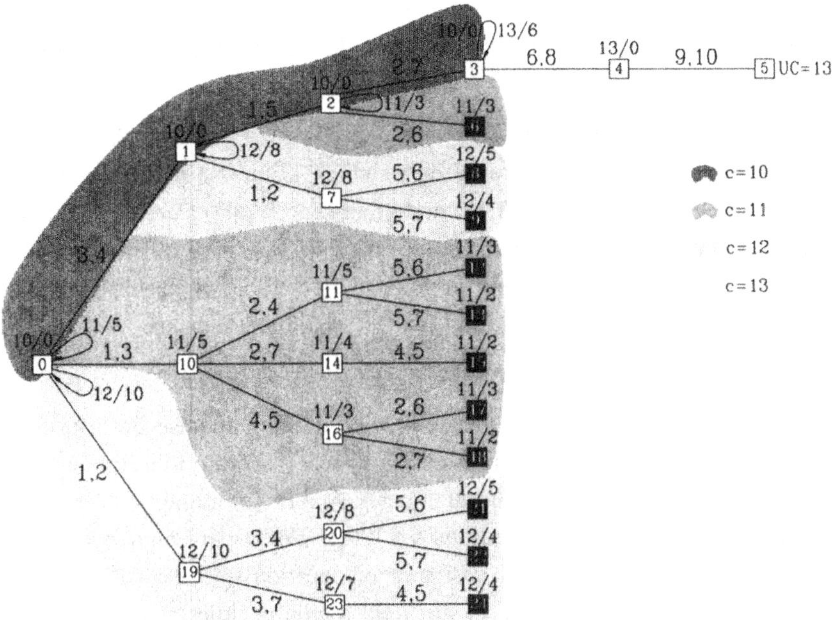

Figure 4.18. Enumeration Tree of the Local Lower Bound Method

The procedure starts with the lower bound $LLC_0 = 10$ as well as the total balance delay time $BD(10) = 5 \cdot 10 - 50 = 0$ and builds the same station loads as the Lower Bound Search for c = 10 (cf. Figure 4.11, p. 148) at first.

In node 3, the only maximal load $S_4 = \{6,8\}$ for station 4 has an idle time $I_4(10) = 3$. Hence, the partial solution $S_1 = \{3,4\}$, $S_2 = \{1,5\}$, $S_3 = \{2,7\}$ cannot be completed to a feasible solution with cycle time 10. In contrast to solving a SALBP-F instance only, the Local Lower Bound Method does not fathom the node, but it increases the local lower bound LLC_3 to 13. Due to the available station loads, the cycle times 11 and 12 are not possible in this node. For the cycle time 11, the first three loads of the partial solution show idle times $I_1(11) = I_2(11) = I_3(11) = 1$. Since the only maximal load $S_4 = \{6,8\}$ of station 4 would result in the idle time $I_4(11) = 4$, the cumulated idle time of 7 would exceed the total balance delay time $BD(11) = 5$. Due to idle times $I_1(12) = I_2(12) = I_3(12) = 2$ and $I_4(12) = 5$, the cycle time 12 with $BD(12) = 10$ cannot be realized, too.

With the local lower bound $LLC_3 = 13$, a feasible solution is found in node 5 (UC = 13). Since the nodes 4 and 3 both have the local lower bound value 13, the enumeration traces back to node 2 with the current local lower bound $LLC_2 = 10$. This local bound must be increased, because no further maximal load of station 3 exists for cycle time 10. Due to the station load $S_3 = \{2,6\}$, the next trial cycle time is $LLC_2 = 11$ with remaining idle time 3. The resultant node 6 is fathomed after having increased its local lower bound to 13. The same is true for all other black coloured nodes.

Discussion. The example shows that the Local Lower Bound Method avoids enumerating same parts of the tree repeatedly, which is the main disadvantage of the Lower Bound Search (cf. Figure 4.11) and other search procedures of Section 4.2.2. Furthermore, it provides a feasible solution in the first branch of the tree. Note that the new enumeration technique builds almost the same tree as the Lower Bound Search for $c = 12$, but generates station loads in another order according to increasing bounds.

Idle Times and Station Loads. In order to give a more precise description of the enumeration technique, formalizations of the terms remaining idle time and feasible load are required. Consider a node at level k of the enumeration tree, i.e., loads have already been built for the stations $h = 1,...,k$. For a particular cycle time c, a respective load S_h causes a *station idle time* of duration $I_h(c) = c - t(S_h)$. In order to obtain a solution with cycle time c, the total available idle time is given by the *balance delay time* $BD(c) = m \cdot c - t_{sum}$ (cf. Definition 2.5, p. 25). Thus, the *remaining idle time* $RI_k(c)$ available for stations $k+1,...,m$ is defined by (cf. (4.2), p. 136):

$$RI_k(c) := BD(c) - \sum_{h=1}^{k} I_h(c) \qquad \text{for } k = 0,...,m-1 \tag{4.14}$$

The cycle time c can only be realized if the station times of the stations $k+1,...,m$ do not fall below the lower bound $SL_k(c)$.

$$SL_k(c) := \max\{0, c - RI_k(c)\} \qquad \text{for } k = 0,...,m-1 \tag{4.15}$$

Definition 4.9. A station load S_{k+1} of station k+1 is called *feasible for cycle time c* if its station time $t(S_{k+1})$ falls into the interval $[SL_k(c), c]$.

Definition 4.10. A station load is *maximal with respect to cycle time c* if no task not yet assigned can be added to the respective station neither exceeding the cycle time nor violating the precedence constraints (cf. Definition 4.3, p. 115).

The Enumeration Procedure. The algorithm starts with the global lower bound LC obtained by computing the bound arguments LC_1, LC_2, and LC_3, because these are the only linear time bounds (cf. Section 2.2.3.1). The global upper bound UC is initialized with the value computed as proposed in Remark 2.5 (see p. 60).

The enumeration is realized by the procedure *Branch* in Figure 4.19 which represents a recursive version of a laser search branch and bound algorithm. The local lower bound LLC_0 of the root $(k=0)$ is set to the value of the global bound LC before calling *Branch(0)*.

```
Procedure Branch(k);
begin
  if k = m - 1 then
  begin
    UC := LLC_k;
    fix all unassigned tasks to station m; store the solution as current incumbent;
  end else
  while LLC_k < UC do (* outer loop *)
  begin
    SL_k(LLC_k) := max { 0, LLC_k - RI_k(LLC_k) }; (* cf. the formulae (4.15) *)
    find first maximal load S_k+1 for station k+1 with t(S_k+1) ∈ [SL_k(LLC_k), LLC_k];
    while S_k+1 exists do (* inner loop *)
    begin
      LLC_k+1 := LLC_k;
      Branch(k+1); (* recursive call of procedure branch *)
      find next maximal load S_k+1 with t(S_k+1) ∈ [SL_k(LLC_k), LLC_k];
    end;
    increase LLC_k;
  end;
end;
```

Figure 4.19. Procedure Branch

The local lower bounds LLC_{k+1} of nodes at level k+1 are initially set to the value of the bound LLC_k of the respective ancestor node at level k. In each node at level k, all maximal station loads possible for station k+1 are enumerated explicitly or implicitly (cf. the inner loop of procedure *Branch*).

In order to hold the current local lower bound LLC_k, only station loads feasible for the trial cycle time LLC_k are chosen for branching. Equivalently, the station times of those loads must be in the interval $[SL_k(LLC_k), LLC_k]$. The corresponding maximal station loads are systematically constructed by considering tasks in increasing order of their numbers. Note that the precedence graph is assumed to be topologically ordered and that the task order inside a station is not relevant for optimization. The enumeration scheme, which is not formulated in procedure *Branch,* is identical with that of EUREKA. In order to find good feasible solutions in the first branches of the tree, the tasks are renumbered by the procedure of FABLE. For both the enumeration scheme and the renumbering procedure see Section 4.1.2.2.

After examining all station loads feasible for the current local lower bound LLC_k, it must be increased to a value $LLC_k^{\prime}$ (see the paragraph below for details). This leads to an enlarged interval $[SL_k(LLC_k^{\prime}), LLC_k^{\prime}]$ which results in additional feasible station loads. Since the new interval includes the former one $[SL_k(LLC_k), LLC_k]$ as a subinterval, only maximal loads with their station times being in the intervals $[SL_k(LLC_k^{\prime}), SL_k(LLC_k))$ or $(LLC_k, LLC_k^{\prime}]$ are to be considered for the new local lower bound value $LLC_k^{\prime}$. The interval enlargement is repeatedly done until LLC_k is equal to or exceeds the upper bound UC (cf. the outer loop of procedure *Branch*).

The recursion terminates at level $m-1$, where loads for $m-1$ stations are fixed (cf. the first condition in procedure *Branch*). In this case, all remaining tasks are assigned to station m, and a solution with cycle time LLC_k results.

Adjusting Lower Bounds. In the enumeration procedure described above, the incrementing of local lower bounds plays an important role. Due to the integrality of task times, an obvious increment is 1. In order to avoid checking impossible trial cycle times and to enlarge the local lower bound by a minimum value concurrently, station loads not being part of the current interval of station times are considered. Let S^+ symbolize the smallest station time of a load S_{k+1} larger than the cycle time LLC_k, and let S^- denote the largest station time smaller than $SL_k(LLC_k)$ of a load S_{k+1} which is maximal with respect to LLC_k. Using these terms, the next possible value $LLC_k^{\prime}$ of the local lower bound LLC_k is determined by formula (4.16).

$$LLC_k^{\prime} := \begin{cases} S^+ & \text{if } S^- = 0 \\ \min\{S^+, \lceil (t_{sum} - \sum_{h=1}^{k} t(S_h) - S^-)/(m-k-1) \rceil \} & \text{if } S^- > 0 \end{cases} \tag{4.16}$$

In case of $S^- = 0$, no maximal load with a station time smaller than $SL_k(LLC_k)$ exists. Then only loads with station times not smaller than S^+ are left for branching, and $LLC_k^{\prime}$ can be set to S^+. Otherwise, it has to be checked whether the load with station time S^+ or S^- leads to a lower increment of the current local lower bound when it is realized for station k+1. In case of considering $t(S_{k+1}) = S^-$, the new value of LLC_k follows from applying the bound LC_1 to the reduced problem with stations k+2,...,m.

Example (cf. Figure 4.18, p. 161). For the problem instance considered above, the initial lower bound of the root node 0 is $LLC_0 = 10$. Hence, $BD(10) = RI_0(10) = 0$, and $SL_0(10) = 10$. Only the load $S_1 = \{3,4\}$ is feasible for station 1, because it has a station time of 10.

After examining the corresponding subtree, LLC_0 has to be increased. Due to the loads {1} and {1,3}, the values of S^- and S^+ are 6 and 11, respectively. The new local lower bound is $LLC_0^l := \min\{11, \lceil (50-6)/4 \rceil\} = 11$ with $BD(11) = RI_0(11) = 5$ and $SL_0(11) = 11 - 5 = 6$. Now, the only maximal load $S_1 = \{1,3\}$ is used for branching.

At the next re-visit of the root node, the local bound is increased to $S^+ = 12$, the station time of the load $S_1 = \{1,2\}$, because no load exists which is maximal for cycle time 11 and has a station time smaller than 6 ($S^- = 0$).

After checking the subtree with the load {1,2}, the lower bound is increased to $S^+ = 16$, and the root node is fathomed due to $UC = 13$.

In node 3, the new value of the local lower bound results from $S^- = 7$ and $S^+ = 16$. It is computed by $LLC_3^l := \min\{16, \lceil (50-30-7)/1 \rceil\} = 13$.

Note that the values S^+ and S^- for adjusting the bounds can be determined while enumerating the loads which are feasible for the respective current local lower bounds.

4.2.4.2 Logical Tests and Fathoming

The Local Lower Bound Method implicitly contains a bounding mechanism. In each node, branching is performed according to monotonously increasing values of the local lower bound LLC_k. Hence, a node is fathomed when the value of LLC_k reaches or exceeds the upper bound UC. In this section, we describe logical tests (dominance rules and reduction rules) which help to reduce the size of the enumeration tree. Most of these rules are modifications of the rules proposed for SALOME-1 (cf. Section 4.1.4).

Maximum Load Rule. As already mentioned, only maximal loads have to be considered. As a consequence, all loads which can be extended by an available task without exceeding the current trial cycle time LLC_k are to be excluded.

Jackson Dominance Rule. A load S_{k+1} is excluded if there is an available task h which potentially dominates a task $j \in S_{k+1}$ and can replace j without exceeding the current local lower bound, i.e., $t(S_{k+1}) - t_j + t_h \leq LLC_k$.

Example. Task 4 is potentially dominated by task 1, task 6 by task 2, and task 7 by tasks 3, 4, 5, and 6. The Jackson dominance rule avoids building the nodes 14 and 23 of the tree in Figure 4.18 (see p. 161) due to the fact that task 7 can be replaced by the dominating task 4 in the respective loads for station 2. Furthermore, the nodes 9, 13, 18, and 22 would be avoided by the rule.

Extended Maximum Load Rule: Consider the current branch of the enumeration tree which leads to a node at level k. Assume Δ_h to be the smallest operation time of a task which has been available in station h = 1,...,k but has not been assigned to it. Then for station k+1, only loads with station times smaller than

$$S^{max} = \min\{\, t(S_h) + \Delta_h \mid h = 1,...,k \,\}$$

have to be considered, and the respective node can be fathomed when the local lower bound reaches or exceeds this value.

This rule takes into consideration that the load of at least one of the stations h = 1,...,k loses the property of being maximal if in station k+1 a load is realized which causes the local lower bound (cycle time) to be larger than or equal to S^{max}.

On the one hand, the extended maximum load rule results in large reductions of the enumeration tree in most cases. On the other hand, it may prevent finding feasible solutions in the first branches of the tree. The latter problem can be illustrated by the following example. Imagine that a load S_1 for station 1 is feasible for the lower bound LLC_0 of the root node. Furthermore, assume that no solution with load S_1 and cycle time LLC_0 exists and that the load S_1 can be extended by an available task to a load with station time $LLC_0 + 1$ (at the next re-visit of the root). Then the extended maximum load rule fathoms all nodes of the subtree which follows S_1 whenever their local lower bounds are increased. Hence, the procedure acts as a Lower Bound Search in this subtree, and no feasible solution is found there. In case of large problem instances, such subtrees may be very large so that no feasible solution is found within a prespecified time limit.

Applying the Extended Maximum Load Rule. Due to this trade-off between reducing the size of the enumeration tree and finding feasible solutions, the application of the rule is controlled by the following strategy which has been found by a parameter adjusting process:

- The rule is not used for the first m^2 nodes of the tree.
- Furthermore, the rule is not applied to a node at a level k in case of

  $$\ln(UC - LC)/(m-k) \geq 1.5 \cdot \ln(LC)/m.$$

The latter condition includes two aspects. First, it is desirable to improve the upper bound UC if the difference between UC and the (global) lower bound LC is large. Secondly, it is not expensive to complete a solution if only few stations remain (small value of m - k). The bound difference is transformed into a logarithmic value in order to make it comparable to the station difference.

Simple Permutation Rule. In any node at level k, a load S_{k+1} need not be considered if the following condition is true:

$$\max\{\, j \in S_{k+1} \,\} < \min\{\, j \in S_k \,\}$$

The same rule is applied by SALOME-1 (see p. 140). As a consequence of the topological numbering of tasks, the condition of the rule guarantees that no task in S_{k+1} is a successor of a task in S_k. Therefore, the loads S_k and S_{k+1} of stations k and k+1 can be exchanged without violating precedence constraints.

Example. Consider the loads $S_1 = \{3,4\}$ and $S_2 = \{1,2\}$ which lead to node 7 in the tree of Figure 4.18. Node 7 can be fathomed, because the combination $S_1 = \{1,2\}$ and $S_2 = \{3,4\}$ is also examined (node 20).

Prefixing. All tasks j with $E_j(UC-1) = L_j(UC-1) = k+1$ are prefixed to station k+1 in a node at level k.

The prefixed tasks must be assigned to the respective station in order to find a feasible solution with a cycle time smaller than UC (cf. Section 4.2.3.3). Note that prefixing is performed with respect to the current local lower bound by the procedure SALOME-1 with the consequence that the prefixed tasks have to be removed from the station after each increase of the lower bound. Since this occurs at most once per node for SALOME-1 and possibly often for SALOME-2, the restricted prefixing rule stated above is considered for the latter approach.

Application of the Rules. The rules are applied to every subproblem at a level k in the following manner. First of all, prefixing of tasks is performed possibly resulting in a reduction of the subproblem. Now, all possible loads of station k are enumerated. For each trial load, the maximum load rule, the permutation rule, and the Jackson dominance rule are applied in the given order (due to the simplicity of the first two rules). The remaining loads result in new subproblems which are treated in the same way. Whenever the local lower bound of a node is increased, the extended maximum load rule is applied according to the above strategy. The incumbent solution is proven to be optimal when all nodes are completely branched or fathomed.

4.2.4.3 Bidirectional Branching

In the preceding sections, we describe a version of our branch and bound procedure SALOME-2 (hereafter called *UniFor*) which works from station 1 to station m in a *forward oriented unidirectional* manner. By analogy with SALOME-1, we describe a modification of SALOME-2 which uses a flexible bidirectional branching strategy.

A simple modification can be made to use SALOME-2 in backward direction by applying it to the reverse problem. In the sequel, this version will be named *UniBack*.

Example. The enumeration tree of Figure 4.20 is built by UniBack for our example problem (see p. 160). A single branch, representing an optimal solution with loads $S_5 = \{10\}$, $S_4 = \{9,8\}$, $S_3 = \{7,6,5\}$, $S_2 = \{4,3\}$, $S_1 = \{2,1\}$, is constructed, because the global lower bound LC can immediately be increased to 12 in the root node.

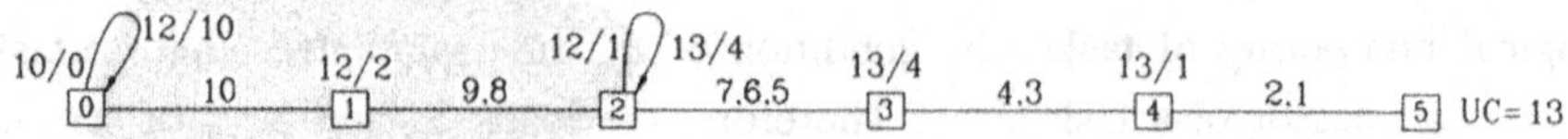

Figure 4.20. Backward Enumeration for the Example Problem

The example shows that the planning direction may have a considerable influence on the computation time of the algorithm. In order to take advantage of this fact, it is desirable to be able to use both planning directions in form of a bidirectional branching simultaneously. Simple bidirectional versions of SALOME-2, *BiFor* and *BiBack*, are obtained by considering the stations in the order $1, m, 2, m-1, \ldots, \lceil (m+1)/2 \rceil$ or $m, 1, m-1, 2, \ldots, \lfloor (m+1)/2 \rfloor$, respectively. These strategies swap between *forward steps* (choose the first not yet considered station k_f) and *backward steps* (choose the last not yet considered station k_b). In backward steps, the reverse precedence graph is considered.

A more flexible approach (*BiFlex*) uses a priority based strategy, which is a modified version of the bidirectional branching rule of SALOME-1 (see p. 143), to decide on the branching direction. Within this strategy, a mean value T_k of operation times per station serves as priority criterion. For its definition, we introduce $\bar{c} := UC - 1$, the maximal cycle time of an improved solution (cf. (4.6), p. 150).

$$T_k := \Big[\sum_{j \in B_k(\bar{c})} \big(t_j / (L_j(\bar{c}) - E_j(\bar{c}) + 1)\big) \Big] / |B_k(\bar{c})| \quad \text{for } k = k_f, \ldots, k_b \tag{4.17}$$

In (4.17), the operation times of all tasks j which are potentially assignable to a station k are proportionately distributed among the stations of the respective station intervals $SI_j(\bar{c})$. The sum of these parts is divided by the number of potentially assignable tasks resulting in a mean value T_k of operation times of tasks for station k.

In each node of the enumeration tree, the following rule decides on the branching direction for the next step, depending on the data of the respective reduced problem:

> Perform a forward step if $T_{k_f} > T_{k_b}$ or $T_{k_f} = T_{k_b}$ and $|B_{k_f}(\bar{c})| \le |B_{k_b}(\bar{c})|$.
> In all other cases, make a backward step.

Example. We illustrate the proceeding of BiFlex by our example of Figure 4.17 (see p. 160) and start with $UC = 15$ ($\bar{c} = UC - 1 = 14$) obtained by some heuristic procedure (cf. Chapter 5). Table 4.10 contains the earliest and latest stations in the root node. Figure 4.21 shows the resultant enumeration tree.

The first decision in the root node is to select among the stations $k_f = 1$ and $k_b = m = 5$. Due to $T_1 = (6/3 + 6/4 + 5/3 + 5/3)/4 = 1.71$ and $T_5 = (4/2 + 9/2)/2 = 3.25$, a backward step is preferred. After increasing LLC_0 from 10 to 12 and building the load $S_5 = \{10\}$, the values of $L_j(\bar{c})$ have to be modified as given in Table 4.11 (those of $E_j(\bar{c})$ do not change).

Table 4.10. Earliest and Latest Stations in the Root Node

j	1	2	3	4	5	6	7	8	9	10
t_j	6	6	5	5	4	5	4	2	9	4
$E_j(\bar{c})$	1	1	1	1	2	2	2	3	4	4
$L_j(\bar{c})$	3	4	3	3	4	4	4	4	5	5

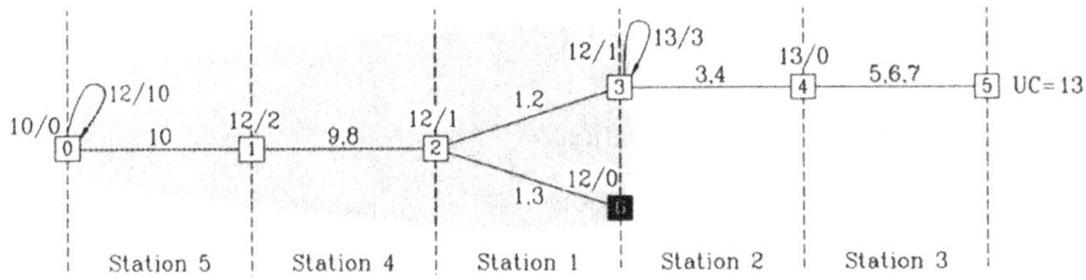

Figure 4.21. Bidirectional Enumeration Tree for the Example Problem

Because of $T_1 = 2.29 < T_4 = 5$, the station 4 is considered in a backward step. The load $S_4 = \{9,8\}$ is branched. Since the latest stations remain valid for the not yet assigned tasks 1,...,7, T_1 keeps its value.

Table 4.11. Modified Latest Stations in Node 1

j	1	2	3	4	5	6	7	8	9
$L_j(\bar{c})$	2	3	2	3	3	3	3	4	4

Due to $T_3 = 2.03$, the next step is a forward one. After assigning the tasks 1 and 2 to station 1, the stations 2 and 3 as well as the tasks 3,...,7 build the reduced problem with the earliest stations $E_j(\bar{c}) = 2$ for $j = 3,4,5,7$, $E_6(\bar{c}) = 3$, and the latest stations of Table 4.11.

Because of $T_2 = T_3 = 2.875$ and $|B_2(\bar{c})| = |B_3(\bar{c})| = 4$, the tie breaker of the rule chooses station 2 in a forward step. Since no loads feasible for the local lower bound of 12 are available, it is increased to 13. For this cycle time, the current partial solution can be completed by $S_2 = \{3,4\}$ and $S_3 = \{5,6,7\}$ to a feasible one. Only in node 2, a further load (of station 1) feasible for cycle time 12 is available, but does not lead to an improved solution.

4.3 Exact Solution Procedures for SALBP-E

This section considers the problem type SALBP-E which is to find a combination of a number m of stations and a cycle time c as well as a corresponding line balance such that the line efficiency is maximized (cf. Section 2.2.4). As stated in Section 2.2.1.3, SALBP-E can be formulated in terms of SALBP-F, because any feasible SALBP-F instance (m,c) with minimal value of $T = m \cdot c$ is optimal for SALBP-E (cf. the equivalence of objectives in Section 2.2.4.1).

4.3.1 Notations and Definitions

c_{min}	minimal cycle time
c_{max}	maximal cycle time
m_{min}	minimal number of stations
m_{max}	maximal number of stations
T	line capacity $(= m \cdot c)$
E	line efficiency $(= 100\,\% \cdot t_{sum}/T)$
BR	balance delay ratio $(= 100\,\% - E)$
BD	balance delay time $(= T - t_{sum})$
LE	lower bound on the line efficiency
UE	upper bound on the line effiency

Figure 4.22. Notations for SALBP-E

We use the notations and definitions given in the previous sections and Section 2.2.4. Some of the notations which are frequently required to specify solution procedures are repeated in Figure 4.22. As a matter of convenience, the line efficiency and the balance delay ratio are expressed in per cent.

4.3.2 Search Methods for SALBP-E

An obvious solution procedure for SALBP-E consists of considering all possible (m,c)-combinations (with $m \in [m_{min}, m_{max}]$ and $c \in [c_{min}, c_{max}]$) and examining their feasibility. A feasible combination with maximal value of E is optimal. This may be done with modified procedures for SALBP-1 or SALBP-2, respectively. However, usually a lot of (m,c)-combinations exist. Furthermore, the SALBP-F instances to be considered are NP-complete. Thus, the outlined approach is expected to be very inefficient, because it usually examines too many instances.

By analogy with search methods for SALBP-2 (Section 4.2.2), we propose a general search procedure for solving SALBP-E (Scholl and Voß (1995)). We describe an Upper Bound Search which corresponds to the SALBP-2 Lower Bound Search. (The number of stations is minimized in SALBP-2, while the line efficiency is maximized in SALBP-E.) Basically, other search methods may be applied to SALBP-E, too, but they seem to be somewhat artificial in this context. This is due to the fact that values of the line efficiency E are generally not integral and that certain values which may be proposed by a search method are not possible at all.

Upper Bound Search. The procedure starts with an interval [LE,UE] of values for the line efficiency E which is bounded by a lower bound LE and an upper bound UE. These values may be determined as proposed in Section 2.2.4.3 (see p. 65).

Basic Idea. We illustrate the basic idea of the search method by our example of Section 2.2.4. The precedence graph with 11 tasks and $t_{sum} = 45$ is repeated in Figure 4.23, and the balance delay function, in Figure 4.24 (cf. Figures 2.11 and 2.14).

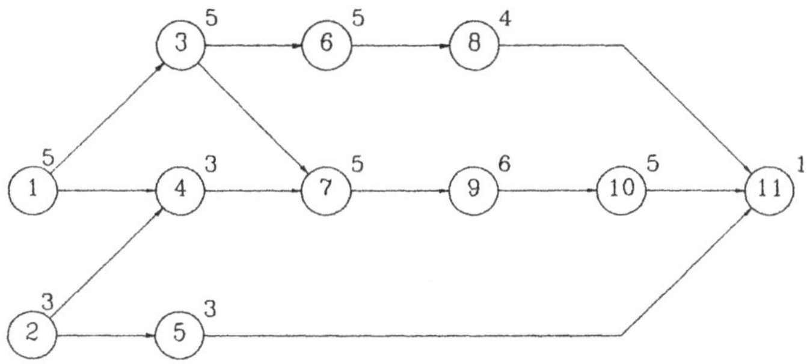

Figure 4.23. Example Problem

An initial lower bound LE is derived from any feasible solution. Assume that the solution $S_1 = \{1,2,3\}$, $S_2 = \{4,5,7\}$, $S_3 = \{6,9\}$, and $S_4 = \{8,10,11\}$ is known for the combination $(m,c) = (4,13)$ with $LE = 45/52 \cdot 100\% = 86.54\%$. The initial upper bound is $UE = 100\%$ due to the combinations (5,9) and (3,15) which fulfill the equation $t_{sum} = m \cdot c$.

The procedure starts with $E = UE = 100\%$ which corresponds to $BR = 0\%$ in Figure 4.24 (note that $BR = 100\% - E$). Assume that a horizontal line is added to the balance delay diagram for $BR = 0\%$. All (m,c)-combinations on this line are potential optimal combinations. Therefore, the feasibility of these combinations is examined. Since solving SALBP-F instances with few stations is often easier than solving such with many stations, the respective combinations are considered in order of increasing values of m (from the left to the right). If one of these combinations is feasible, the algorithm terminates, because the respective solution is optimal. Otherwise, the horizontal line is moved upwards (to larger BR and smaller UE values, respectively) until at least one further combination is reached. The process is repeated until a feasible (= optimal) combination is found or the line approaches the lower bound LE at the position 100% - LE on the y-axis. In the latter case, the initial solution is optimal.

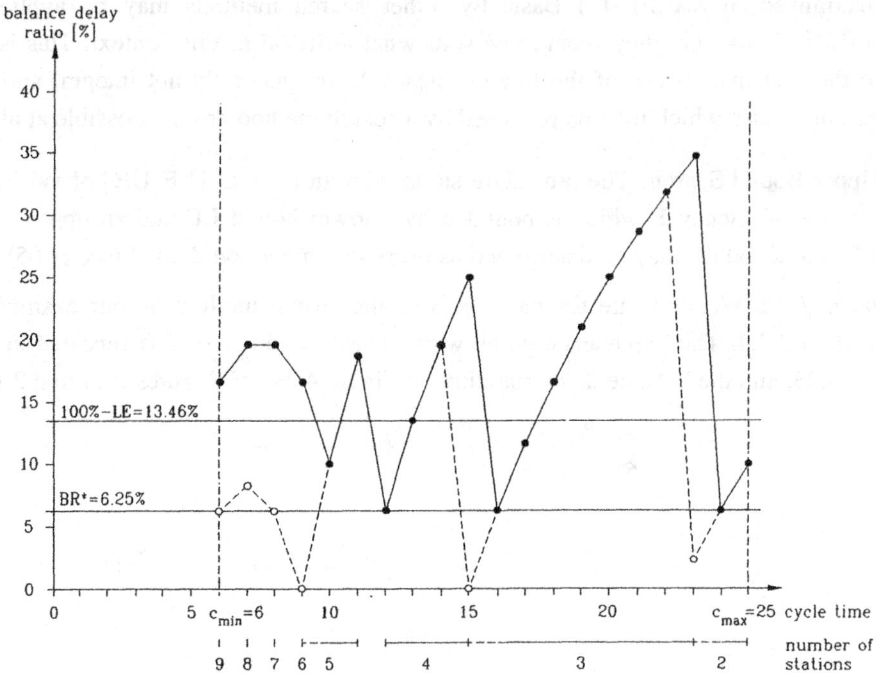

Figure 4.24. Balance Delay Function

Example. The combinations considered at first are (3,15) and (5,9), none of which is feasible. The line is moved to E = 97.83 % (BR = 2.17 %), and the combination (2,23) is examined without success. After moving the line to E = 93.75 % (BR = 6.25 %), five combinations are available which are considered from the right to the left. A feasible solution is found for the first combination (2,24).

Step 1. Compute theoretical minimal cycle times $c := \lceil t_{sum}/m \rceil$ for all $m \in [m_{min}, m_{max}]$. Sort those (m, c) - combinations whose line efficiency is larger than LE into a list L according to non-decreasing values of $m \cdot c$ (tie breaker: increasing m).

Step 2. Select the first combination (m, c) of list L and remove it from the list. Update the upper bound: $UE := 100\,\% \cdot t_{sum} / (m \cdot c)$.

Step 3. Examine the feasibility of the chosen combination (m,c). Stop, if (m,c) is feasible.

Step 4. Increase the cycle time of combination (m,c) to the next possible value c'. Insert the combination (m,c') into the sorted list L if $100\% \cdot t_{sum}/(m \cdot c') > LE$ and $c' \le c_{max}$. If the list L is not empty, then continue with step 2; otherwise, terminate.

Figure 4.25. Upper Bound Search for SALBP-E

Algorithm. Figure 4.25 gives an algorithmic description of the approach outlined above. It uses a sorted list L which contains the combinations to be evaluated. Only those combinations are stored which may improve the current lower bound LE. In each iteration, the first combination in L corresponds to the current upper bound UE. For each value of m, at most one combination is in the list at a time. The current combination (m,c) for a particular m is changed when (m,c) is shown to be infeasible. Then the next possible value c' for m stations is determined (usually $c' := c+1$, see below). The resultant combination (m,c') is sorted into the list L if it may improve LE.

Similar procedures (based on SALBP-1) are proposed by Zäpfel (1975, pp. 53), Klenke (1977, pp. 47) as well as Rosenblatt and Carlson (1985) for assembly line balancing problems with profit oriented objectives.

Table 4.12. Sorted List of (m,c)-Combinations

Posit.	m	c	E (%)	Action
1	3	15	100	UE = 100, infeasible, insert (3,16) with E = 93.75
2	5	9	100	UE = 100, infeasible, insert (5,10) with E = 90.00
3	2	23	97.83	UE = 97.83, infeasible, insert (2,24) with E = 93.75
4	2	24	93.75	UE = LE = 93.75, feasible, terminate
5	3	16	93.75	
6	4	12	93.75	
7	6	8	93.75	
8	8	6	93.75	
9	7	7	91.84	
10	5	10	90.00	
11	9	6	83.33	E < LE, not stored

Example (precised). Table 4.12 contains a sorted list of those (m,c)-combinations which are considered by the procedure. Note that list L only contains a part of the entries at a time. After step 1, the list is L = <(3,15), (5,9), (2,23), (4,12), (6,8), (8,6), (7,7)]. The combination (9,6) is not contained in L, because its efficiency is lower than LE. In the first two iterations, the combinations (3,15) and (5,9) are chosen and deleted from L. Since both are not feasible, the combinations (3,16) and (5,10) are built and added to the sorted list L. In the third iteration, the infeasible combination (2,23) is examined, and (2,24) is generated. This combination gets the first position in the current list L and results in a feasible solution. Hence, the algorithm is terminated with an optimal solution.

In the framework of the general search procedure described above, either SALBP-1 or SALBP-2 procedures may be used to examine the feasibility of the considered SALBP-F instances.

Application of SALBP-1 Procedures. If exact procedures for SALBP-1 are used, the cycle time c of a considered SALBP-F instance (m,c) is fixed, and a solution with at most m stations has to be found (see pp. 145 for the respective modifications of SALBP-1 procedures). If such a solution exists, the SALBP-F instance is feasible, and the search procedure terminates. Otherwise, the cycle time of the considered combination is increased to $c' = c+1$.

This approach shows the drawback that no feasible solution is found at all if the search procedure takes more computation time than available. Therefore, several modifications which attempt to obtain feasible solutions before the procedure is terminated are considered.

Complete SALBP-1 Solutions. Instead of considering SALBP-F, a SALBP-1 instance may completely be solved in step 3 of the procedure, i.e., the minimal number $m^*(c)$ of stations is determined for the given cycle time c. If $m^*(c)$ is less than or equal to m, the solution is optimal for SALBP-E. Otherwise, a feasible solution, which may improve the lower bound LE, is found if $m^*(c) \leq m_{max}$. In the first case, the procedure terminates; in the latter case, step 4 is omitted and the procedure continues with step 2.

Flexible Fathoming. However, completely solving a SALBP-1 instance often requires much more computation time than solving a corresponding SALBP-F instance. Therefore, it may be advantageous to restrict the above approach. That is, only some parts of the enumeration tree which cannot provide a solution feasible for the current SALBP-F instance (m,c) are additionally considered instead of generating the complete tree. If the procedure SALOME-1 or a similar algorithm is used, a possible strategy may be based on the following reflections. As long as the difference between UE and LE is rather large, it is desirable to search for a solution which improves LE. Furthermore, only a small amount of computation time is required to complete a current partial solution if only few unloaded stations remain, i.e., the difference between m and the current level k of the enumeration tree is small. These two aspects are combined in the condition

$$(UE-LE)/(m-k) \geq \beta, \tag{4.18}$$

where UE and LE denote the current values of the upper and lower bound, respectively, and k denotes the number of stations currently loaded, i.e., the level of the node. If the condition (4.18) is true, a node is not fathomed even though it cannot result in a solution feasible for the current SALBP-F instance. Numerical results indicate that the parameter β may be set to $0.2 \cdot UE'/m$, where UE' denotes the initial upper bound.

Application of SALBP-2 Procedures. If SALBP-2 procedures are only used to solve SALBP-F instances, their is no fundamental difference to a SALBP-1 based approach.

Enlarged Cycle Time Increments. While the modified cycle time c' is always set to c+1 when using SALBP-1 procedures, it may get a larger value for SALBP-2 based approaches. A larger increment of c can be obtained if the cycle time c+1 is not possible, i.e., no solution can exist with that cycle time. That prevents considering unnecessary SALBP-F instances and is achieved by completely applying SALOME-2 until the global lower bound LC is increased in the root node (cf. Section 4.2.4.1, pp. 159). The new value of LC represents the nearest possible cycle time c'. Furthermore, feasible solutions are obtained, because bound enlargements are allowed in all nodes but the root.

Complete SALBP-2 Solutions and Flexible Fathoming. Similar to SALBP-1 based approaches, SALBP-2 instances may be considered in each iteration of the search procedure instead of SALBP-F instances. Furthermore, the fathoming strategy with condition (4.18) can be applied to SALOME-2, too, in order to achieve feasible solutions without solving SALBP-2 instances completely.

4.4 Exact Solution Procedures for UALBP

The procedures for the different versions of SALBP described in the preceding sections can be transferred to the corresponding UALBP version. The adaptations necessary are explained in the following.

4.4.1 Exactly Solving UALBP-1 by ULINO

The procedure ULINO (U-line Optimizer; cf. Scholl and Klein (1997 b)) is an extension of SALOME-1 to the U-line assembly line balancing problem UALBP-1 (see Section 2.3.4). Due to the modified precedence constraints, the enumeration is only possible in forward direction, i.e., stations are considered in order of increasing numbers (from left to right), and a modified definition of assignable tasks (cf. Definition 4.2) is needed.

Definition 4.11. Within UALBP, an unassigned task j is *forward (backward) assignable* to a station k if all predecessors (successors) have already been assigned to one of the stations 1,...,k and the current idle time $I(S_k) = c - t(S_k)$ of station k is not smaller than the operation time t_j.

Since ULINO is a straightforward extension of SALOME-1, we concentrate on the differences and refer to the description of SALOME-1 in Section 4.1.4.

Branching. ULINO starts with station 1 at the root node at level 0 of the enumeration tree to be built. At any other node at level k of the tree the first k stations are already loaded defining a partial solution. Branching at a level k is performed by systematically constructing all *maximal* loads for station k+1 the order of which is defined by the enumeration scheme and the Local Lower Bound Method. Since both, forward and backward assignable tasks, are candidates for being included into a station load, the task renumbering is different from that of SALOME-1. In order to define a precedence-feasible renumbering of tasks each task gets a smaller number than all its successors and/or all its predecessors. The renumbering is obtained by successively shrinking the precedence network from both sides as follows. In iteration i = 1,...,n, the number i is given to a task which is a source or a sink of the residual network. Among several candidate tasks, the one with largest operation time is numbered and removed from the network together with its arcs (tie-breaker: most direct predecessors or successors, second-level tie-breaker: smallest original task number).

Example. We explain the renumbering and the branching process of ULINO by means of the precedence graph in Figure 4.26 with the cycle time assumed to be c = 10. The renumbering defines the ordering 1, 10, 4, 12, 2, 3, 6, 5, 7, 8, 9, 11.

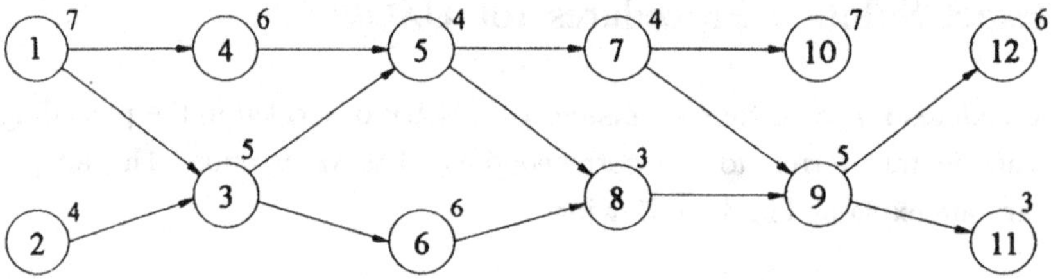

Figure 4.26. Example Precedence Graph

Figure 4.27 shows the enumeration tree built by the Local Lower Bound Method of ULINO following this renumbering of tasks. The nodes are numbered in their generation order; the shaded area encloses all nodes with local lower bound 6.

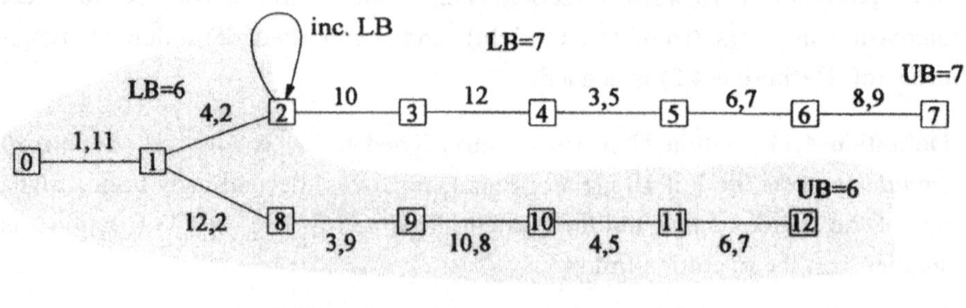

Figure 4.27. Enumeration Tree of ULINO

The initial bound is $LB = LB_1 = 6$ (cf. Section 2.3.4). The total idle time is $BD(6) = 6 \cdot 10 - 60 = 0$. At node 2, no first-class load with zero idle time exists and LB is increased to 7 leading to BD(7) = 10. Due to the increased available idle time, first-class loads for stations 3 to 7 are easily obtained and a feasible solution with UB = 7 stations is found at node 7. Nodes 3 to 7 are fathomed due to $LLB_k = 7$. After building the second subproblem of node 1, an optimal solution with 6 stations is found, because all loads on the way to node 12 have zero idle time.

Bounding. As mentioned in Section 2.3.4, the bound arguments LB_1, LB_2, LB_3, and LB_6 are valid for UALBP-1 and, thus, are applied within ULINO.

Logical Tests. ULINO applies the maximum load rule, the Jackson dominance rule, the tree dominance rule and the task time incrementing rule. Among those, the Jackson dominance rule must be adapted to UALBP-1 using the following definitions of potential dominance:

- A task j is *potentially forward dominated* by another task h if $t_j \leq t_h$ and all successors of j are also successors of h.
- A task j is *potentially backward dominated* by another task h if $t_j \leq t_h$ and all predecessors of j are also predecessors of h.

With these definitions a *forward version* of the rule can be formulated as:

A load S need not be considered (the corresponding node of the tree is not built) if there exist a task $j \in S$ and an unassigned task h such that: The predecessors of j and h are already assigned, j is potentially forward dominated by h, and $t_h - t_j \leq I(S)$.

In the *backward version* of the rule, the successors of j and h must already be assigned and j must be potentially backward dominated by h.

The potential dominances can easily be determined and stored before starting the enumeration. Whenever a load is generated, the forward and the backward rule are applied. If in either case a dominance is recognized this load is skipped.

Example. The following pairs (h,j) of potential forward dominances are present in our example instance: (1,2), (4,6), (7,8), (10,11), (12,11). Potential backward dominances are (12,11), (10,6), (6,4), (4,2), (1,2). For example, the load $S_1 = \{2,11\}$ needs never be considered, because the forward rule identifies that considering the load $\{1,11\}$ is sufficient. Note that in this case also the backward rule fits which prefers the load $\{2,12\}$.

4.4.2 Search Procedures for UALBP-2

By analogy with SALBP-2, optimal solutions to UALBP-2 (minimize c given m) can be determined by solving a series of UALBP-1 instances with several trial cycle times being given. Starting with an interval [LC,UC] of possible cycle times, different search methods may be applied to find the minimal cycle time for which the corresponding UALBP-1 solution requires no more than m stations. As search methods we can, e.g., use the Lower Bound Search, the Binary Search or the Fibonacci Binary Search (see Section 4.2.2). In each iteration of the search, a slightly modified version of ULINO is applied to examine the UALBP-1 instance currently considered: Nodes are fathomed whenever their local lower bound exceeds the given number m of stations.

Example. Given $m = 6$ for our example instance, the initial search interval may be [10,13]. The Lower Bound Search and the Fibonacci Binary Search only consider the (optimal) trial cycle time 10, the Binary Search starts with $c = 11$, identifies a feasible solution and proceeds to $c = 10$.

4.4.3 A Search Method for UALBP-E

UALBP-E is to maximize the line efficiency $E = 100\% \cdot t_{sum} / (m \cdot c)$, or equivalently, to minimize the line capacity $T = m \cdot c$ with m and c being variable in certain ranges. For solving this problem, the Upper Bound Search described for SALBP-E in Section 4.3.2 can be used applying the above-mentioned modification of ULINO for solving the feasibility problems in each iteration. Note that in case of minimizing T, a Lower Bound Search is present.

Example. Let us consider our example instance with the restrictions $m \in \{4,5,6,7\}$ and $c \in \{9,...,23\}$. Figure 4.28 shows the best value of T which can feasibly be obtained for each possible cycle time c (solid points). The non-filled points represent combinations (m,c) for which the necessary condition $T \geq m \cdot c$ is true but no feasible solution exists. The discrete points are connected by lines in order to illustrate the non-linear character of the line capacity depending on c and m.

The search starts with the ordered list <(4,15), (5,12), (6,10), (7,9)]. The first three combinations have $T = t_{sum} = 60$, the fourth one has $T = 63$. At first, the combination (4,15) is selected and the UALBP-1 instance with $c = 15$ solved. Since 4 stations are not sufficient, the combination (4,16) with $T = 64$ is built and inserted into the list (as last element).

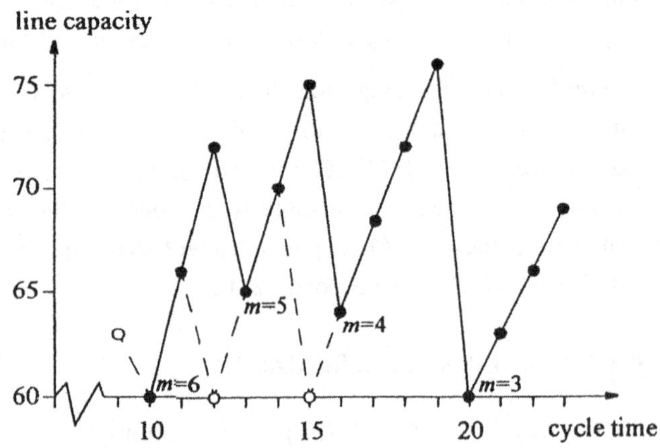

Figure 4.28. Line Capacity as a Function of c

Then (5,12) is chosen and proven to be infeasible. Therefore, (5,13) with $T = 65$ is appended to the list. The next combination (6,10) is feasible and optimal to UALBP-E with $T = 60$ ($E = 100\%$).

5 Heuristic Procedures for Single-Model Assembly Line Balancing Problems

In this chapter, we present heuristic procedures for simple assembly line balancing problems. General surveys and classifications of basic heuristic methods are given by Müller-Merbach (1981), Pearl (1984) as well as Zanakis et al. (1989). Recent heuristic principles are *simulated annealing* (cf. Aarts and Korst (1989)), *tabu search* (cf. Section 5.2.2), *genetic algorithms* (cf. Goldberg (1989), Michalewicz (1992) and Pesch (1994)), and *neural networks* (cf. Hopfield and Tank (1985) as well as Masson and Wang (1990)).

Most heuristic approaches for assembly line balancing described in the literature so far are designed for SALBP-1. Furthermore, most of them are based on simple priority rules and generate only one or a few feasible solutions. Little attention has been given to other problem types and improvement procedures. Even though some limited attempts have recently been made to apply new heuristic principles to some version of SALBP (cf. Gehring and Schütz (1994), Heinrici (1994), Suresh and Sahu (1994)), it seems to be necessary to investigate their usefulness for such problems in greater detail.

In order to close these gaps, we propose improved priority based methods for SALBP-1 and describe their application to solve SALBP-2 and SALBP-E. Furthermore, we develop improvement procedures which utilize the recent meta-strategy tabu search for SALBP-2 as well as SALBP-1.

5.1 Constructive Methods

As mentioned before, many heuristics have been proposed for SALBP-1 in the literature, mainly based on priority rules. Surveys and comparisons are given by Cauley (1968), Mastor (1970), Lutz (1974), Dar-El (1975), Kälberer (1976), Küpper (1982, pp. 166) as well as Talbot et al. (1986). In Section 5.1.1, we present traditional priority rule based procedures as well as extensions with respect to dynamic rules and bidirectional planning. Heuristics which construct solutions by some type of partial enumeration are described in Section 5.1.2. The application of heuristics for SALBP-1 within the framework of search methods for solving the problem versions SALBP-2 and SALBP-E is discussed in Section 5.1.3.

5.1.1 Priority Rule Based Procedures for SALBP-1

Most heuristics for finding initial solutions of SALBP-1 are based on static priority rules which are applied unidirectionally. Tasks are assigned to stations according to a task ranking.

5.1.1.1 Static Priority Rules

In the literature, a great variety of priority rules has been proposed. For surveys, see Talbot et al. (1986) as well as Hackman et al. (1989). In our presentation of such rules, we use the notations and definitions of Sections 2.1 and 4.1.1 as well as the notations given in Figure 5.1.

c	given cycle time
$\bar{m}$	upper bound on the number of stations
pw_j	positional weight of task j $(= t_j + \sum_{h \in F_j^*} t_h)$
pw_j^*	recursive cumulated positional weight of task j $(= t_j + \sum_{h \in F_j} pw_h^*)$
r_j	number of arcs in paths having j as its root $(= \lvert F_j \rvert + \sum_{h \in F_j^*} \lvert F_h \rvert)$
r_j^*	recursive cumulated number of arcs in paths with root j $(= \sum_{h \in F_j} (r_h^* + 1))$
E_j	earliest station for task j $(= \lceil (t_j + \sum_{h \in P_j^*} t_h)/c \rceil)$
L_j	latest station for task j to improve $\bar{m}$ $(= \bar{m} - \lceil pw_j / c \rceil)$
s_j	slack of task j $(= L_j - E_j + \epsilon)$

Figure 5.1. Notations

In Figure 5.1, different task *priorities* measuring task times and precedence relations are given. The *positional weight* combines the both characteristics, because it considers the operation times of a task and all successors in the precedence graph. As documented by the definition of the latest stations L_j, a close relationship exists between L_j and pw_j. However, the range of values is significantly smaller for the first measure such that the use of additional priorities is often necessary. An analogous definition may be given for the earliest stations E_j by means of a *reverse positional weight* which adds up the operation times of j and all its predecessors. The priority r_j considers the number of arcs in the subgraph built by task j and its successors. The *recursive cumulated positional weight* pw_j^* is a modification of the positional weight. It additionally incorporates the structure of the precedence relations between task j and its successors, because task times of transitive successors are contained repeatedly. Similarly, the priority r_j^* weights each arc corresponding to the number of successors which are connected to task j by this arc. The *slack* measures the degree of freedom for assigning a task to different stations. The parameter ϵ (small number) is added in order to avoid dividing by zero in some of the rules of Figure 5.2.

MaxTime	descending task time t_j
MaxPW ^{sm}	descending positional weight pw_j
MaxF ^{sm}	descending number of followers $\lvert F_j^* \rvert$
MaxIF	descending number of immediate followers $\lvert F_j \rvert$
MinE ^{m}	ascending earliest station E_j
MinL ^{m}	ascending latest station L_j
MinSlack	ascending slack s_j
MaxAvgPW	descending average positional weight $pw_j / \lvert F_j^* + 1 \rvert$
MinAvgL ^{sm}	ascending average latest station $L_j / \lvert F_j^* + 1 \rvert$
MaxTimeL	descending task time divided by latest station t_j / L_j
MaxFSlack	descending number of followers divided by slack $\lvert F_j^* \rvert / s_j$
MaxTimeSlack	descending task time divided by slack t_j / s_j
MaxA ^{sm}	descending total number of following arcs r_j
MaxCumA ^{sm}	descending cumulated number of following arcs r_j^*
MaxCumPW ^{sm}	descending cumulated positional weight pw_j^*
MinTaskNo ^{sm}	ascending task number j

Figure 5.2. Static Priority Rules

Figure 5.2 contains a comprehensive collection of known and new priority rules which build a task ranking according to ascending or descending values of some priority (cf., among others, Helgeson and Birnie (1961), Tonge (1961), Talbot et al. (1986), Hackman et al. (1989), Scholl and Voß (1994, 1996)). These rules are *static,* because they only need be computed once before a corresponding procedure is started. According to the following monotonicity property, the rules are subdivided into several groups.

Definition 5.1. A rule, which ranks tasks according to decreasing (increasing) values of a particular priority p_j, is called *monotonous* if $p_h \geq p_j$ ($p_h \leq p_j$) holds for all arcs (h,j) of the precedence graph. It is *strongly monotonous* if the condition $p_h > p_j$ ($p_h < p_j$) is true for all arcs (h,j).

In Figure 5.2, strongly monotonous rules are marked with ^{sm} and monotonous ones with ^{m}. The strong monotonicity of rules is important with respect to two planning strategies described in the following section.

5.1.1.2 Unidirectional Priority Based Procedures

Priority based procedures apply priority rules (main rule and tie breaking rules) to build a task ranking. In each of n iterations, a task with highest priority is chosen out of a set of (forward) available tasks and is assigned to a station.

We describe two basic planning strategies proposed in the literature. They differ with respect to the manner in which the tasks to be assigned are selected out of the set of available tasks.

Stationoriented Procedures. These procedures start with the first station (k = 1). The following stations are considered successively. In each iteration, a task with highest priority which is assignable to the current station k is selected and assigned. When no additional task is assignable to station k, it is closed, and the next station k + 1 is opened. That is, stations are loaded maximally before another one is considered.

Such procedures are, e.g., described by Talbot et al. (1986). For the rule MaxPW, this procedure is called *ranked positional weight technique* by Helgeson and Birnie (1961). The name *packing heuristic* is given to the same approach with the rule MinTaskNo by Nourie and Venta (1987).

Taskoriented Procedures. Among all available tasks, one with highest priority is chosen and assigned to the earliest station to which it is assignable.

Depending on whether the set of available tasks is updated immediately after assigning a task or after assigning all currently available tasks, taskoriented methods can be subdivided into *immediate-update-first* and *general-first-fit* methods (cf. Wee and Magazine (1982) as well as Hackman et al. (1989)). In the sequel, we assume that the immediate-update-first strategy is used due to its greater flexibility. Furthermore, the above two strategies are more efficient than the so-called *rank-and-assign* method which does not explicitly use a list of available tasks.

Example. The difference between stationoriented and taskoriented procedures is illustrated by means of two example problems provided in Figure 5.3.

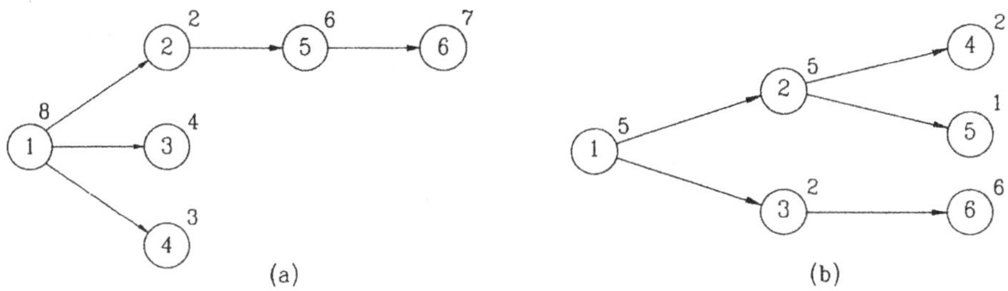

Figure 5.3. Example Precedence Graphs

The cycle time is set to c = 10 for the problem of Figure 5.3 (a). With MaxTime as priority rule, the task ordering <1,6,5,3,4,2> is obtained.

The stationoriented procedure performs the following steps: Open station 1 and assign the tasks 1 and 2; open station 2 and assign the tasks 5 and 3; open station 3 and assign the remaining tasks 6 and 4. The solution is $S_1 = \{1,2\}$, $S_2 = \{3,5\}$, $S_3 = \{4,6\}$ with m = 3 stations.

The assignments performed by the taskoriented procedure are: Task 1 to station 1 (1 to 1 for short), 3 to 2, 4 to 2, 2 to 1, 5 to 3, and 6 to 4. The resulting solution is $S_1 = \{1,2\}$, $S_2 = \{3,4\}$, $S_3 = \{5\}$, $S_4 = \{6\}$ with m = 4 stations.

For the problem of Figure 5.3 (b), we assume a cycle time of c = 7. Using MaxTime and the tie breaker MaxIF, the task ranking is <6,2,1,3,4,5>.

The stationoriented solution is $S_1 = \{1,3\}$, $S_2 = \{6\}$, $S_3 = \{2,4\}$, and $S_4 = \{5\}$ with m=4 stations, because task 6 is assigned too early for being combined with task 5.

Only 3 stations are needed for the taskoriented solution $S_1 = \{1,3\}$, $S_2 = \{2,4\}$, and $S_3 = \{6,5\}$.

Relationship between the Strategies. The above examples show that there is no general theoretical dominance for either procedure. Nevertheless, we can state an *equivalence property* for certain rules:

> In the case of *strongly monotonous rules*, both the taskoriented and the stationoriented procedure provide *identical* solutions. Furthermore, monotonous rules may be extended to be strongly monotonous by adding a suitable tie breaking rule (e.g., MinTaskNo).

The equivalence of stationoriented and taskoriented procedures with strongly monotonous rules follows from the subsequent reflections. As long as tasks can be assigned to the earliest station k which is currently not maximally loaded, both procedures perform the same assignments. A difference occurs if the task j with the highest priority does not fit into station k. While the taskoriented procedure assigns task j to a later station, the stationoriented procedure assigns another task to station k or closes it, if no further task can be added. The priority value of a task h which is assigned before task j by the latter procedure is not larger than that of j. Due to the strong monotonicity of the priority rule, all successors of task h have lower priorities than h as well as j. Therefore, the stationoriented procedure assigns the task j as the first one into the same station, to which it is assigned by the task-oriented procedure, as soon as this station is opened.

Both the stationoriented and the taskoriented procedures may be modified by *prefixing* tasks j with $E_j = L_j$ to station E_j before starting the iterations (cf. Section 4.1.4.2, p. 139). In this case, all predecessors of j have to be assigned to the stations $1,...,E_j$. If this is not possible, the procedure is not able to find a feasible solution with less than $\bar{m}$ stations. The initial upper bound $\bar{m}$ can be determined by a theoretical value (cf. Section 2.2.2.2).

5.1.1.3 Dynamic Rules and Bidirectional Planning

In what follows, we restrict our presentation to stationoriented procedures in order to avoid unnecessary explanations. Furthermore, our computational experiments strongly indicate that stationoriented procedures are superior (cf. Section 7.2.2.3).

Forward Procedures. The procedures described above develop solutions in ascending order of the stations. In that sense, they work unidirectionally from station 1 to subsequent stations in forward direction.

Backward Procedures. By applying a priority based procedure to the reverse precedence graph, a backward procedure is obtained, which considers the stations in descending order of their numbers. That corresponds with applying a *reverse rule*, which considers predecessors instead of successors, and vice versa, to the original problem (cf. the reverse positional weight rule mentioned above).

Bidirectional Procedures. Furthermore, the procedures can be extended to assign tasks *bidirectionally*, i.e., to apply a forward and a backward procedure simultaneously. This is achieved by building a forward and a backward ranking. In the backward case, the reverse rules have to be applied.

The bidirectional procedure starts with the stations $k_f = 1$ and $k_b = \bar{m} - 1$ as current stations in the forward and backward direction, respectively. In each iteration, one task with highest priority, which is either forward assignable to k_f or backward assignable to k_b, is chosen and assigned to the respective station. If no further task is assignable to k_f (k_b), the station is maximally loaded, and the next station $k_f + 1$

$(k_b - 1)$ is considered. The algorithm stops when all tasks are assigned. A feasible solution is obtained by combining the partial forward and backward solutions. A further stopping criterion is to terminate when $k_b < k_f$ holds, because this inequality indicates that no feasible solution which improves the upper bound $\bar{m}$ can be obtained with the present rule.

Direction Tie Breaker. If a chosen task is in the set of forward and backward available tasks simultaneously, an additional priority rule should be applied to decide whether it is assigned to its earliest or to its latest station. For example, one may prefer that station whose station time gets closer to c by assigning the task.

Example. Figure 5.4 shows the precedence graph of a problem with $n = 10$ tasks. The cycle time is assumed to be $c = 14$. A solution with $\bar{m} = 5$ stations may already be known. The problem is solved by applying a forward, a backward, and a bidirectional procedure, all of which use the primary rule MaxF with the tie breakers MaxTime and MinTaskNo. The forward task ranking is $<1,2,5,3,4,8,6,7,9,10>$, and the backward ranking is $<10,9,8,5,4,3,6,7,2,1>$. Note that in the backward case the second-level ties are broken in favour of the task with the largest original number, because the task numbering is assumed to be reversed, too.

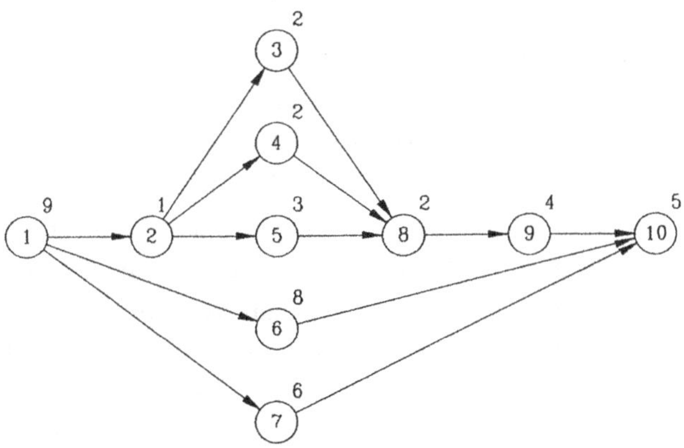

Figure 5.4. Example for Different Planning Directions

Forward Procedure. By chance, the tasks can be assigned in the order given by the task ranking. The resulting solution is $S_1 = \{1,2,5\}$, $S_2 = \{3,4,8,6\}$, $S_3 = \{7,9\}$, $S_4 = \{10\}$ with $m = 4$ stations.

Backward Procedure. Task 10 is the only backward available task. After assigning it to station 1, the tasks 6, 7, and 9 are backward available, and task 9 has the highest priority. It is assigned to station 1 together with the tasks 8 and 5. After opening station 2, the tasks 4, 3, and 6 are assigned in order of the ranking. Since task 7 does not additionally fit, task 2 is added to station 2. The two remaining tasks 7 and 1 cannot be combined and require the stations 3 and 4, respectively. The corresponding forward solution is $S_1=\{1\}$, $S_2=\{7\}$, $S_3=\{2,3,4,6\}$, $S_4=\{5,8,9,10\}$ with $m=4$ stations.

Bidirectional Procedure. The solution is obtained as follows. Task 1 and task 10 are the only forward and backward available tasks, respectively. The primary rule MaxF does not distinguish between the tasks due to $|F_1^*|=|P_{10}^*|$. (Note that predecessors of a task in the original graph are successors in the reverse graph, and vice versa.) Due to $t_1>t_{10}$, task 1 is chosen and assigned to station 1. Now, the tasks 2, 6, and 7 are forward available. Since task 10 has the highest priority among the tasks currently available, it is assigned to station $\bar{m}-1=4$. The tasks 6, 7, and 9 become backward available. Because of $|F_2^*|=|P_9^*|$ and $t_9>t_2$, task 9 is assigned to station 4, and task 8 becomes backward available. Now, task 2 is assigned to station 1, because it has the highest priority among all available tasks. After assigning task 8 to station 4, the remaining tasks 3,...,7 are forward and backward available simultaneously. Task 5 is selected and assigned to station 4, because this station gets the station time $t(S_4)=14$ (see the direction tie breaker above). The tasks 3 and 4 are assigned to station 1. Since the remaining unassigned tasks fit together into station 2, only 3 stations are required. By eliminating station 3 whose load is empty the solution $S_1=\{1,2,3,4\}$, $S_2=\{6,7\}$, $S_3=\{5,8,9,10\}$ with $m=3$ stations is obtained.

Dynamic Rules. All rules considered above are *static,* because priorities only depend on parameters not being influenced by the solution process (except the direction tie breaker). *Dynamic rules* utilize parameters which are modified through task assignments. For example, the values of E_j (L_j) can be modified by computing the earliest (latest) station to which a task j is *currently* assignable (i.e., already loaded stations are excluded). In the case of bidirectional planning, the number or sum of task times of *not yet* assigned successors or predecessors may be used instead of all successors or predecessors. Thus, dynamic counterparts can be defined for all rules of Figure 5.2 except MaxTime. Boctor (1995) defines a dynamic composite rule including the number of tasks getting available by assigning a task as a priority.

Even task-independent parameters like, e.g., the numbers of currently forward and backward available tasks, respectively, may be included in bidirectional rules in order to decide whether a forward or a backward step should be performed. Following the idea that the probability of making a disadvantageous assignment is low if the number of available tasks is small, the rule MaxTimeSlack can be extended to a dynamic bidirectional rule *MaxTSAvail:*

> Rank the tasks according to descending values of t_j/s_j divided by the number of forward and backward available tasks, respectively.

This rule represents a new class of priority rules for SALBP-1 which is particularly suitable for being applied within *dynamic bidirectional procedures.* Such procedures are successfully applied to the resource constrained project scheduling problem by Klein (1998).

Remark 5.1. Considering the extensive discussion of priority rules for other problem types, dynamic bidirectional rules have potential applications in various problem domains with network structures. For example, problems related with shortest path computations may be solved by bidirectional methods (cf. Lawler (1991)).

Furthermore, bidirectional procedures (including the bidirectional versions of the exact algorithms SALOME-1, SALOME-2, and TBB-2) are suitable for treating unpaced assembly line balancing problems with respect to the *bowl-phenomenon* (cf. Section 1.3.3). This is due to the tendency of bidirectional procedures to concentrate the (unavoidable) idle time in the middle of the line, instead of accumulating it at the end (or the start) like forward (backward) procedures do. Therefore, bidirectional procedures tend to produce a bowl-shaped station time distribution. In order to obtain a desired bowl distribution, these procedures may be modified with respect to different local cycle times of the stations (cf. Section 2.3.1.2).

5.1.1.4 Extended Priority Based Procedures

Priority rules may be used in a more general context than described above. In the sequel, we outline several possibilities to apply such rules more flexibly in order to obtain better solutions or to solve generalized problems.

Multi-Pass Heuristics. A simple extension of priority based approaches is obtained by applying such a procedure with different priority rules repeatedly (cf. Talbot et al. (1986)). Arcus (1963 a, 1966) proposes a stochastic version of a multi-pass heuristic, called COMSOAL, which generates a number of solutions such that a certain probability for finding an optimal solution is achieved. In each pass, the tasks to be assigned are randomly chosen among the respective available tasks. The selection probabilities are determined in proportion to priority values. A similar approach, NULISP, is described by Bonney et al. (1976) and Schofield (1979). COMSOAL and NULISP are furthermore extended to solve generalized problems with assignment restrictions, multiple manning, and parallel stations.

Flexible Rule Application. While most approaches inflexibly apply one or several rules, the procedure of Tonge (1965) randomly selects the used priority rule in each iteration of the assignment procedure. This idea is extended by Bennett and Byrd (1976) who present a procedure which stores experiences with certain priority rules for actual problem instances in a data base. These informations are utilized to choose priority rules for a new problem instance to be solved. Görke and Lentes (1976) describe a similar (multi-pass) approach. The probabilities for selecting a certain rule depend on the solution quality obtained for the same problem instance in former passes, i.e., with other rule combinations. For applications of that approach see Auwärter and Lentes (1977). Raouf et al. (1980) describe a priority based procedure with rule MaxTime which additionally takes into consideration whether or not tasks are members of a critical path in the precedence network.

Reduction Techniques. Baybars (1986 b) proposes a priority based procedure which involves heuristically reducing the problem size by some logical tests. The most simple reduction is the elimination of tasks each of which occupies one station by itself (cf. the task time incrementing rule, p. 117). Further reduction rules attempt to combine tasks and to decompose the precedence graph. In the last step, a heuristic solution is found by applying a backward stationoriented procedure with priority rule MaxIF and tie breaker MaxTime to the reduced precedence graph. Another procedure which involves reducing the problem size is described by Tonge (1960). A hierarchy of problems is constructed by successively combining groups of tasks which represent particular substructures (e.g., chains of nodes) of the precedence graph. A further heuristic method which is based on a repetitive problem reduction is proposed by Freeman and Swain (1986).

Generalized Problems. As mentioned above for the procedures COMSOAL and NULISP, priority rule based procedures can easily be modified for solving generalized problems. Hardeck (1974) describes an extension of the ranked positional weight technique for problems with assignment restrictions and parallel stations. Similar approaches (mostly based on MaxPW, too) are proposed by Heskiaoff (1968), Kovach (1969), Hautsch et al. (1972), Schad (1986), Bhattacharjee and Sahu (1988) as well as Boguschewski et al. (1990).

Combined Solutions. As stated in Remark 4.1 (see p. 132), SALBP-1 can be interpreted as a shortest path problem with exponential numbers of nodes and arcs. Each feasible solution can be represented by a path in such a graph. Therefore, Pinto et al. (1978) describe a two-stage solution approach. In the first step, a number of feasible solutions is determined by means of several priority rules (cf. the multi-pass heuristics). These solutions are used to construct a subgraph of the complete graph in the second step of the procedure. For this subgraph, a shortest path problem is solved. That is, the outlined approach tries to combine parts of several feasible solutions in order to obtain an improved complete solution (cf. Domschke et al. (1997, pp. 209)).

5.1.1.5 Worst Case Analysis

Since heuristic methods do not guarantee that optimal solutions are obtained, one is interested in the solution quality which is obtained on the average and in the worst case, respectively. In the case of SALBP, corresponding error bounds can be derived from properties of optimal solutions. For example, the (optimality) conditions (2.42) and (2.45) in Section 2.2.2.2 (see pp. 52) result in the following worst case bounds, where m and m* denote the number of stations in the heuristic and the optimal solution, respectively (cf. Hackman et al. (1989) as well as Domschke et al. (1997, pp. 203) for proofs):

$$m/m^* \leq \begin{cases} 2 - 2/m^* & \text{for } m = 2,4,6,... \\ 2 - 1/m^* & \text{for } m = 1,3,5,... \end{cases} \tag{5.1}$$

$$m < c \cdot m^* / (c - t_{max} + 1) + 1 \tag{5.2}$$

These bounds are valid for any reasonable heuristic which builds maximal station loads. The worst case bound (5.1) states that a heuristic solution never contains more than twice the number of stations as an optimal solution. For increasing parameters n (number of tasks) and c (cycle time), the worst case bound m/m* approaches the value 2.

While the worst case errors of any heuristic are bounded from above by the value 2, a lower bound is given by 3/2 for polynomial time heuristics unless P = NP (cf. Queyranne (1985) as well as Section 2.2.1.5). Average (expected) error bounds are discussed by Frieze and Baybars (1986).

5.1.2 Incomplete Enumeration Procedures for SALBP-1

A class of heuristic procedures consists of *incomplete enumeration techniques*. Generally, such methods are based on exact enumeration schemes, like those contained in branch and bound procedures, which are modified by heuristically restricting the search space.

Heuristic of Hoffmann (1963). The procedure works unidirectionally in a station-oriented manner. In each iteration k = 1,2,..., a load with minimal idle time is generated for station k. That is, a single branch of a stationoriented branch and bound procedure is constructed. In each node, the examined station loads are enumerated by the same technique which is used by EUREKA (cf. Section 4.1.2.2).

Example. For the problem instance of Figure 5.4 with c = 14, the heuristic of Hoffmann computes a solution with 4 stations as follows. In iteration 1, the load S_1 = {1,2,3,4} is considered first. Since it contains no idle time, it is assigned to station 1. For station 2, the loads {5,6,8}, {5,6}, {5,7,8}, {5,7}, {5,8,9}, {5,8}, and {5} are examined in the given order. The load S_2 = {5,6,8} with idle time 1 is chosen. In iteration 3, the tasks 7,9, and 10 remain. The station load with the least idle time is S_3 = {7,9}, and the only remaining task 10 is assigned to station 4.

Modifications. Even though the heuristic of Hoffmann only builds a single branch of an enumeration tree, it may require considerable computation times, because it has to examine all possible station loads of a current subproblem. Therefore, Gehrlein and Patterson (1975, 1978) propose a modification of the procedure which accepts a load for the currently considered station if a certain amount of idle time is not exceeded. The accepted portion of idle time depends on the balance delay time (total available idle time) for the theoretical minimum number of stations and can be controlled by a parameter.

Truncated Branch and Bound Procedures. Each branch and bound procedure can be applied as a heuristic by adding heuristic fathoming rules or imposing a restrictive time limit. For example, nodes of the enumeration tree may be fathomed if it is not likely that the current incumbent solution can be improved. This may be examined by comparing the average idle time of already loaded stations with the average idle time being available in an improved solution to be found. Furthermore, the number of descending nodes chosen for branching in a current subproblem can be restricted. Respective approaches are already mentioned in Section 4.1.2.3. Furthermore, see Nevins (1972), Dar-El (1973) as well as Haider and Moodie (1978).

5.1.3 Search Methods for SALBP-2 and SALBP-E

By analogy with exact methods, all above heuristic procedures for SALBP-1 can be applied to SALBP-2 and SALBP-E, respectively, within the framework of a suitable search method (cf. Sections 4.2.2 and 4.3.2). The SALBP-F instances, which have to be examined in each iteration of such search processes, are solved by heuristics instead of being solved by modified exact procedures for SALBP-1.

Search Methods for SALBP-2. Usually, the application of the Lower Bound Search is proposed (cf., among others, Mastor (1970), Dar-El (1973, 1975), Gehrlein and Patterson (1975), and Bonney et al. (1976)). The Binary Search is applied by Hardeck (1974).

While the different search methods are equivalent with respect to the solution quality if exact procedures are applied to the SALBP-F instances, differences may occur when applying heuristic procedures. This ensues from the fact that heuristics need not find solutions with monotonically decreasing numbers of stations for increasing cycle times. That is, a procedure may obtain a solution with m stations for a cycle time c and one with m+1 or more stations for cycle time c+1 (cf. Hackman et al. (1989) for an explicit example). In such a case, the Binary Search and the Upper Bound Search do not find a solution with cycle time c if the trial cycle time c+1 is examined prior to c. Therefore, only the Lower Bound Search guarantees that the smallest cycle time for which the respective heuristic obtains a solution with at most m stations is found.

Search Methods for SALBP-E. Applications of simple heuristic search procedures to SALBP-E are proposed by Kilbridge and Wester (1961 b), Wester and Kilbridge (1962), Nawara and Elezaby (1980), Bhattacharjee and Sahu (1988) as well as Shin and Min (1991 a, 1991 b). As is the case with exact procedures, different search methods are possible. However, most approaches apply some type of Upper Bound Search as described in Section 4.3.2. The procedure in Figure 4.25 (see p. 172) can be applied if the updating of the upper bound is eliminated from step 2, because the upper bound cannot be improved by a heuristic procedure.

5.2 Improvement Procedures

Improvement procedures are applied in order to improve a given initial solution which is determined by some constructive method (cf. Section 5.1). Usually, this is achieved by repeatedly re-arranging (small) parts of the solution. Traditional improvement procedures only accept alterations which result in an improvement of the current incumbent solution. Recent meta-strategies, like *tabu search* and *simulated annealing*, also allow interim deteriorations of the objective function value in order to avoid getting trapped in a local optimum.

In Section 5.2.1, we describe a pure improvement procedure for SALBP-2 which re-arranges solutions by shifting one or two tasks to another station. The improvement procedure is combined with tabu search in Section 5.2.3 after introducing the basic ideas of this meta-strategy in Section 5.2.2. A non-obvious application of this SALBP-2 approach to SALBP-1 is presented in Section 5.2.4.

5.2.1 A Steepest Descent Procedure for SALBP-2

As outlined above, improvement procedures try to improve a given feasible solution by iteratively *transforming* it into other feasible solutions. Such transformations are referred to as *moves*. Solutions which may be obtained from a given solution x by means of a single move are called neighbouring solutions or *neighbourhood* of x.

Traditionally, improvement heuristics try to find a sequence of moves which produces a trajectory of successively improved solutions. This may be achieved by two different strategies. A *steepest descent (best fit)* procedure chooses a move which leads to a maximum improvement of the current value of the objective function in each iteration. A *fastest descent (first fit)* procedure always realizes the move found first which improves the current solution. Whenever the evaluation of the complete neighbourhood is possible within a reasonable amount of computation time, the best fit strategy is preferred, because it is expected to produce the better solutions.

A basic steepest descent procedure for SALBP-2 is based on so-called shifts and swaps which can be explained using the following notation:

LP_j latest station to which a predecessor of task j is currently assigned

ES_j earliest station to which a successor of task j is currently assigned

The following definitions are required to describe the moves for transforming solutions to neighbouring solutions.

Definition 5.2. A station k is called *critical* if its station time determines the realized cycle time, i.e., $t(S_k) = \max\{t(S_i) \mid i = 1,...,m\}$. A current solution may contain several critical stations, each of which shows no idle time. Non-critical stations are called *slack stations*.

Definition 5.3. A *shift move* describes the movement of a task j from station k_1 to station k_2 with $k_1 \neq k_2$. This move is *feasible* if $k_2 \in [LP_j, ES_j]$. Shifts are denoted by (j, k_1, k_2). A *swap move* exchanges tasks j_1 and j_2, which are not related by precedence, between different stations k_1 and k_2. This move is feasible if the two corresponding shifts (j_1, k_1, k_2) and (j_2, k_2, k_1) are feasible. Swaps are abbreviated by (j_1, k_1, j_2, k_2).

In order to restrict our presentation to swap moves, we interpret shifts as swaps which exchange a real task j_1 and a fictitious task j_2 with zero operation time and no precedence relations.

The *neighbourhood* of a SALBP-2 solution consists of all transformed solutions which are obtained by a single feasible swap move. In order to improve the current solution, the steepest descent procedure selects a swap move concerning a critical station k_1 and a slack station k_2 which results in the largest reduction of the realized cycle time in each iteration. The new value $\bar{c}$ of the realized cycle time is determined by:

$$\bar{c} = \max\{t(S_{k_2}) + t_{j1} - t_{j2},\ t(S_{k_1}) + t_{j2} - t_{j1},\ \max\{t(S_k) \mid k \neq k_1 \text{ and } k \neq k_2\}\} \qquad (5.3)$$

If more stations than one are critical, the realized cycle time cannot be decreased by a single move. Therefore, also moves which do not change the realized cycle time have to be allowed. However, this requires additional mechanisms which avoid re-visiting of already considered solutions (cf. Section 5.2.2). The process stops when only moves which would increase the current realized cycle time exist.

A similar approach is proposed by Moodie and Young (1965) as well as Rachamadugu and Talbot (1991) for the related problem of improving the equality of work loads.

5.2.2 The Meta-Strategy Tabu Search

As already mentioned, tabu search is a meta-strategy for guiding heuristic improvement procedures to overcome local optimality. In the following, we give a short introduction to this recent methodology. The presentation is restricted to the basic components and some extensions which seem to be promising for an application of tabu search to SALBP. More detailed descriptions and further concepts can be found in Glover (1989, 1990), de Werra and Hertz (1989), Voß (1993 b) as well as Glover and Laguna (1997). Illustrative descriptions of the basic tabu search approach are given in Domschke et al. (1996 b,c). Successful applications of tabu search to problems somehow related to SALBP are reported by, e.g., Laguna and Glover (1993), Skorin-Kapov and Vakharia (1993), Hübscher and Glover (1994), Voß (1994), Scholl et al. (1997 a). A simple tabu search procedure for SALBP-2 is proposed by Heinrici (1994).

A further meta-strategy, simulated annealing, is not described here, because respective approaches are most often outperformed by procedures which are controlled by tabu search (cf., among others, Domschke et al. (1992), Voß (1993 b, pp. 179), Heinrici (1994)). A simulated annealing approach for a stochastic variant of SALBP-1 is proposed by Suresh and Sahu (1994).

5.2.2.1 Basic Ingredients of Tabu Search

Most often, steepest descent procedures get trapped in a local optimum which is not necessarily an optimal solution for the problem. In order to overcome this difficulty, a *steepest descent/mildest ascent procedure* is used. In each iteration, a move is performed which results in a best possible improvement or a least deterioration of the objective function value. That is, this approach selects the same moves as the steepest descent procedure does until the latter method stops in a local optimum. Then a deteriorating move is accepted, and the search proceeds.

Without additional mechanisms, this approach can cause a local optimal solution to be re-visited immediately after moving to a neighbouring solution, or in a later iteration of the search process, respectively. This difficulty already occurs if only non-deteriorating moves are allowed as is the case with the steepest descent procedure for SALBP-2 in Section 5.2.1.

In order to avoid this undesired repeated *cycling* between a number of solutions, tabu search excludes a subset of moves which may lead to previously visited solutions from being performed. This strategy is based on the following definition of attributes.

Definition 5.4. A move is characterized by one or more *attributes* which describe how a solution is changed by the move. Two attributes e and $\bar{e}$ are called *complementary* to each other if their successive realization leads to the same solution as before.

Example. A solution of a zero-one programming problem is given by the values of the binary variables. Moves transform a solution into a neighbouring one by changing the value of one variable from 0 to 1, or vice versa. Then the move e which sets the value of a variable to 1 is complementary to the move $\bar{e}$ which sets the value of the same variable to 0.

The attributes of all moves are stored in a so-called *running list*, representing the trajectory of solutions visited so far. The running list is used to build a second list, the *tabu list*. It contains attributes which are complementary to some attribute of the running list and might lead back to a previously visited solution. Consequently, all moves which consist of such attributes are forbidden *(tabu),* i.e., they must not be performed in the current iteration of the search process. Thus, the tabu list restricts the search to a subset of *admissible moves* in order to avoid re-visiting of solutions.

Tabu List Management. The method of constructing the tabu list out of the running list is the most important component of tabu search procedures and is referred to as *tabu list management.* It concerns updating the tabu list, i.e., deciding on how many and which attributes have to be set tabu within any iteration of the search. Two main approaches may be distinguished, static and dynamic methods.

Static Tabu List Management. The *tabu navigation method* (TNM) adds the attributes which are complementary to an attribute of a current move to the tabu list. The attributes stay tabu for a certain number of iterations until the probability of causing a re-visit of a solution is small. The efficiency of the algorithm depends on the choice of the *tabu status duration,* i.e., the *length* TL of the underlying tabu list. The TNM and similar static methods are quite simple and require only few computations but have several drawbacks. On the one hand, moves may be set tabu even though they do not lead to already visited solutions. On the other hand, a cycling may occur after removing the tabu status of attributes. That is, the length TL of the tabu list is a critical parameter, because a long tabu list may exclude more moves and a short tabu list less moves than necessary. Therefore, some efforts have been made to include dynamic aspects like, e.g., a random variation of TL within a specified interval, into TNM. Another approach may consist of activating or deactivating parts of the tabu list randomly or periodically (see p. 196).

Dynamic Tabu List Management. In order to (partly) eliminate the disadvantages outlined above, dynamic methods may be used. They decide on the tabu status of an attribute based on the current state of the solution process. Two respective methods are the *cancellation sequence method* and the *reverse elimination method* (cf. Glover (1990), Dammeyer and Voß (1993) as well as Voß (1993 b, pp. 57)). Since these methods are more elaborate, they require more expensive data structures and computations than static methods.

Basic Tabu Search Procedure. A general outline of a tabu search procedure for solving a minimization problem is given in Figure 5.5 (cf. Voß (1993 b, pp. 41) for a similar procedure). It starts with a feasible initial solution x* with objective function value z*. The same variables represent the final solution and the respective objective function value. The search can be terminated by different stopping criteria like a maximal number of iterations or a maximal computation time, respectively. Furthermore, the process is terminated when the objective function value of a solution found is equal to a lower bound.

```
Procedure Tabu Search (x*, z*);
x := x*;  z := z*;
repeat
    select the best admissible move which transforms x into x' with objective
    function value z' and add its attributes to the running list;
    compute attributes to be set tabu (tabu list management);
    perform the selected move:  x := x';  z := z';
    if z < z*  then z* := z;  x* := x  (store incumbent solution);
until stopping criterion is fulfilled;
```

Figure 5.5. Basic Tabu Search Method

5.2.2.2 Extensions of the Basic Approach

The basic tabu search procedure described in the previous section may be complemented with some additional components.

Aspiration Level Criteria. Even though a move is set tabu, it may be advantageous to select this move, because it does not necessarily lead to the re-visit of a solution. Obviously, moves which immediately result in an improvement of the best known solution should be performed independently of their tabu status *(global aspiration* *criterion)*. Furthermore, the tabu status of a move may be overridden if it leads to the best objective function value known for solutions which have particular values of one or more variables *(local aspiration criterion)*.

Varying Size of the Tabu List. The choice of an appropriate length TL of the tabu list is often very critical with respect to the quality of solutions obtained. Thus, the value of TL may be changed after a certain number of iterations without improving the best known solution. This can be done randomly or systematically, e.g., the size of the list may successively be reduced by a constant length.

Dynamic Parts of the Tabu List (Moving Gap). One may use a tabu list which is subdivided into a static and a dynamic sublist. While the static sublist is always active, only a part of the dynamic sublist is utilized at a time (cf., e.g., Hübscher and Glover (1994)).

Intensification Strategies. Intensification aims at concentrating the search to a specific region of the solution space by means of a so-called *frequency based memory* (cf. Glover (1990)). For this purpose, the number of iterations during which a problem variable shows a particular value is stored. Respective frequencies are determined by dividing these numbers by the total number of currently performed iterations. In the case of a high frequency, the value of a respective variable is fixed for some iterations.

Diversification Strategies. In the case of a restricted static tabu list, the search cannot definitely be prevented from cycling in a certain part of the solution space. Further, it may be advantageous to examine large parts of the solution space rather than intensively searching in a restricted part. In order to leave a region, variable values which have occurred in former solutions often are forbidden for a number of future iterations. This is achieved by means of the frequency based memory mentioned above.

5.2.3 A Tabu Search Procedure for SALBP-2

We propose a tabu search procedure for SALBP-2 which utilizes the move definition of the steepest descent procedure described in Section 5.2.1 (cf. Definition 5.3, p. 192). That is, in each iteration, a shift or a swap move concerning a critical station as well as a slack station which leads to the smallest realized cycle time is selected. Due to its simplicity, the TNM is chosen as tabu list management strategy. Hence, the tabu list has a given length TL. Attributes which are set tabu remain in the list for TL iterations.

Attributes. A shift (j,k_1,k_2) is characterized by the attributes "task j leaves station k_1" and "task j enters station k_2" which are symbolized by $(j,\bar{k}_1)$ and (j,k_2), respectively. In order to avoid reversing the move, the attribute (j,k_1), which is complementary to $(j,\bar{k}_1)$, is set tabu. Since swaps are composed of two shifts, they are characterized by the attributes $(j_1,\bar{k}_1)$, (j_1,k_2), $(j_2,\bar{k}_2)$, and (j_2,k_1). The attributes (j_1,k_1) and (j_2,k_2) are set tabu. Moves containing one or two of the attributes which are in the tabu list are forbidden. In each iteration, the algorithm selects an (admissible) move that is not tabu and leads to the best possible realized cycle time.

Stopping Criteria. The procedure may be stopped by one or more of the following criteria:

- A certain number of iterations (MaxIt) has been performed,
- a prespecified time limit is reached,
- a number of iterations has been performed without improving the best solution, or
- the solution is identified to be optimal, because its cycle time is equal to a known lower bound.

Conflict Management Strategies. A *conflict* occurs when no admissible move is available, i.e., the attributes stored in the tabu list prevent all possible moves from being performed. The following strategies may be used to resolve a conflict and to continue the search:

Strategy C1. The oldest entries of the tabu list are successively removed from the list until the list is empty or an admissible move exists.

Strategy C2. The neighbourhood of the current solution is enlarged by additionally considering moves which shift immediate predecessors (immediate successors) of tasks in critical stations to earlier (later) stations. That is, a most improving/least deteriorating move which concerns this set of tasks is performed. The rationale behind this (diversifying) approach is to enlarge the degree of freedom for those (critical) tasks being trapped within a critical station.

Strategy C3. Any move which either shifts a predecessor of a critical task to an earlier station or shifts a successor to a later station is considered. In this case, the criterion for evaluating a move is not primarily based on the objective function but on the closeness of a respective task to a critical station (measured in the difference of the respective station numbers). Ties are broken in favour of smaller realized cycle times.

In cases where conflicts cannot be resolved by C2 or C3, respectively, the strategy C1 is additionally applied.

Modifications of the Basic Approach. Some additional procedures may improve the basic approach described above:

Global Aspiration Criterion. All moves which directly improve the best known solution are performed even if they are tabu.

Size of the Tabu List. Three different strategies for determining the length TL of the tabu list are proposed.

- FTL: A fixed value of TL is chosen. A list length commonly used is 7.
- RTL: The list size is modified in a certain range by randomly increasing and decreasing TL by 1, respectively, after performing a certain number of iterations without improving the current incumbent solution.
- DTL: It systematically decrements TL after performing a number of non-improving iterations, i.e., a certain list length is used as long as improvements are possible before a smaller length is tried. The number of non-improving iterations allowed for a new value of TL is determined by equally subdividing the remaining iterations (MaxIt minus current iteration) among the remaining values of TL. The strategy starts with $TL = \lfloor (m+n)/3 \rfloor$ and successively reduces the list length by the constant 3 until TL would become smaller than $\max\{\lfloor n/6 \rfloor, 3\}$. Note that the sizes of the neighbourhoods to be evaluated are influenced by the number m of stations and the number n of tasks.

Moving Gap. The tabu list is subdivided into a static part of length $\lfloor TL/3 \rfloor$, which is always active, and a dynamic part containing the remaining (older) list entries. The dynamic part itself is equally subdivided into four subparts two of which are deactivated at a time. This is done by cyclically moving a window simultaneously covering two parts of the dynamic sublist every $\lfloor MaxIt/20 \rfloor$ iterations.

Intensification and Diversification. In order to intensify the search in certain regions or to direct the search into yet unvisited parts of the solution space, a frequency based memory is used. In this memory, the relative number of iterations, any task j belongs to any station k, is stored (denoted as z_{jk}). Five phases, each of which is active for $\lfloor MaxIt/5 \rfloor$ iterations, are distinguished: In the first phase, the information for the frequency based memory is collected. In the second phase, an intensification is achieved by fixing tasks which have mainly been in a certain station up to the current iteration. More precisely, all tasks j with $z_{jk} \geq 0.75$ are fixed to station k for $z_{jk} \cdot \lfloor MaxIt/5 \rfloor$ iterations. The third phase, without any intensification and diversification, is used to update the frequency based memory again. A diversification is achieved in the fourth phase by forbidding tasks j with $z_{jk} \geq 0.75$ to enter station k until z_{jk} falls below the value of 0.75 or an unresolvable conflict occurs. In the final phase, all task-station-combinations are free again.

5.2.4 A Tabu Search Approach for SALBP-1

In order to develop a tabu search procedure for SALBP-1, a suitable move definition is required. An obvious approach is to apply a modification of the neighbourhood used in our procedure for SALBP-2. Because of the cycle time being given, no critical stations exist and shift and swap moves may be allowed for all stations. This move definition, however, does not provide a clear distinction between different solutions, because only three different values are possible for the objective function in a transformed solution:

- A move improves a given solution with m stations to a neighbouring solution with m - 1 stations if the only task assigned to a station is shifted to another already used station.
- If one of several tasks currently being assigned to a station is shifted to a not yet used station, a solution with m + 1 stations is obtained.
- If neither a station is emptied nor an additional one is opened, the neighbouring solution has the same number of stations as the current one.

Within these three groups, no obvious further distinction is available between the moves. Furthermore, when m gets closer to optimality, the available idle time of the line reduces and the size of the neighbourhood (regarding feasible moves) reduces, too. Then in most iterations, the class of improving moves is empty and one has to choose among the moves of the third class or even the second one without having a respective ranking measure within the classes.

To summarize, there is no obvious measure to find 'good' moves, because the overall best objective function value eventually changes only after a large number of iterations when one station with zero station time can be achieved. Thus, it is difficult to apply tabu search to SALBP-1 in a direct manner.

Owing to these problems, we propose an *indirect* approach which is based on the dual relationship between SALBP-1 and SALBP-2 (cf. Section 2.2.1.3). By analogy with search procedures for SALBP-2 (cf. Section 4.2.2), instances of SALBP-1 can be solved by iteratively solving feasibility problems with the given cycle time c and different trial values for the number of stations. In each iteration, the tabu search procedure for SALBP-2 is applied in order to find a feasible solution.

In principle, all general search methods described in Section 4.2.2 may be used. Following our argumentation in Section 4.1.2.2, we prefer a *Lower Bound Method*, i.e., the search starts with a lower bound on the number of stations. If the tabu search procedure for SALBP-2 finds a feasible solution with its cycle time not exceeding this lower bound, the problem is solved optimally. If no solution is achieved within a certain number of iterations (or within a certain amount of computation time), the trial number of stations is increased by 1, and the process is repeated. As an additional stopping criterion, an upper bound may be calculated before applying the search procedure. Note that the optimality of a solution can only be proven during the first iteration.

The proposed method is a new general purpose approach which we call *dual strategy*, because it utilizes the (dual) relationship between different problems having the same feasibility problem. For example, the same relationship exists between the bin packing problem and the problem of scheduling identical parallel machines with the objective of minimizing the makespan (cf. Hochbaum and Shmoys (1987)). Therefore, the dual strategy can be applied to the bin packing problem which shows the same difficulties with respect to the definition of appropriate moves as SALBP-1. That is, instances of the parallel machine problem are solved by a tabu search procedure for several trial makespans (capacities of the bins) in order to find good solutions for the bin packing problem. An opposite approach with priority based procedures for the bin packing problem is proposed by Coffman et al. (1978). Note that the bin packing problem and the parallel machine problem are obtained from SALBP-1 and SALBP-2, respectively, by eliminating the precedence relations (cf. Section 2.2.1.4). Therefore, these problems can directly be solved by any procedure for the SALBP versions.

By the dual strategy, the complete search is transferred to the region of those feasible SALBP-2 solutions which are *infeasible* for SALBP-1. As soon as a feasible SALBP-1 solution is obtained, the search is stopped. This method enables tabu search to utilize the neighbourhood definition for SALBP-2 which usually provides a clear distinction between different neighbouring solutions. Note that the dual strategy is different from a *strategic oscillation* which only temporarily searches in the infeasible region, e.g., some precedence relations or cycle time constraints are violated for a certain number of iterations. In this case, the effect on the objective function value of neighbouring solutions is restricted, while the dual strategy allows any "distance" from feasibility. Therefore, it provides a higher degree of freedom which helps to diversify the search. This is especially advantageous in cases where the cycle time and the precedence constraints are very restrictive.

6 Solution Methods for Mixed-Model Balancing and Sequencing Problems

The present chapter is devoted to solution concepts for problems arising in mixed-model assembly line production systems. Section 6.1 considers procedures for solving mixed-model balancing problems. In Section 6.2, a summary of solution approaches for mixed-model sequencing is given and new exact and heuristic procedures are developed.

6.1 Procedures for Mixed-Model Balancing

A survey of different solution approaches for the mixed-model assembly line balancing problem (MALBP) is given. See, additionally, Suliman et al. (1985) or Domschke et al. (1997, chapter 4.4.1). We use the terms and notations of Section 3.2 (cf., especially, Figure 3.1, p. 80).

MALBP and its versions are defined on the basis of the average model, i.e., the mixed-model problem has the same underlying feasibility problem as SALBP (cf. Section 3.2.2.1). Therefore, fundamentally, any procedure for SALBP can be applied to obtain feasible MALBP solutions.

6.1.1 Procedures for Relaxed MALBP Versions

By relaxing the assumption that identical tasks have to be assigned to the same station for all models, the mixed-model problem is decomposed into M (number of models) independent single-model problems. A respective generalization of the shortest-path formulation of SALBP-1 (cf. Remark 4.1, p. 132) is presented by Roberts and Villa (1970) as well as Rao (1971). However, the assignment of identical tasks to different stations is usually not desired with respect to additional facility requirements, loss of specialization effects, complicated production control, and setup inefficiencies, respectively. Only in the case of multi-model production, where batches of models are processed, this relaxation of MALBP may be useful. An approach which *successively* solves SALBP-1 problems for the different models is described by Görke (1978, pp. 71). Whenever a task is assigned to a station, it is fixed for the remaining models. Another method for synchronizing the task assignments of the models is proposed by Macaskill (1969).

Further modifications of MALBP concern the cycle time constraints. A rather restrictive problem is obtained by imposing the cycle time restrictions for every model. That is, the operation times of all models in all stations must not exceed the cycle time. This, however, may lead to a poor efficiency of the line, because compensation effects between the models cannot be utilized. Due to these restrictive capacity constraints, no sequencing problem occurs. A branch and bound algorithm for deterministic and stochastic versions of this problem is proposed by Deutsch (1971); see, furthermore, Bedworth and Bailey (1982, pp. 281).

Another modification of MALBP with respect to the cycle time constraints is proposed by Link (1992). In order to obtain a task assignment which is independent of a particular model mix, the common cycle time constraints are replaced by different upper (and lower) bounds on the station time of every model. For a given number of stations, the objective 2 is considered. That is, the maximal deviation of any station time τ_{mk} from the average station time $\bar{t}_m$ of the respective model m is minimized (see (3.5), p. 88). Besides, the objective function (3.4) is used as a secondary objective in order to equalize those station loads which cannot be controlled by the primary objective. As a consequence of the model-dependent "cycle time" restrictions, the actual capacity requirements of the stations depend on the short-term model mix. These different requirements have to be met by a dynamic assignment of workers to the stations (cf. Decker (1993 a, chapter 2)). Link (1992) proposes a branch and bound procedure for solving the outlined problem.

6.1.2 Procedures for Unrelaxed Versions of MALBP

Usually, the versions MALBP-1 and MALBP-2 are solved by some method for the corresponding single-model problems SALBP-1 and SALBP-2, respectively. Such modifications are included in many single-model procedures (cf., among others, Arcus (1966), Kovach (1969), Magad (1971), Macaskill (1972), Schofield (1979), Driscoll and Abdel-Shafi (1985), Milas (1990)). Priority based and shortest path heuristics for a cost oriented version of MALBP are described by Chakravarty and Shtub (1985, 1986 b).

Methods for Smoothed Station Loads. The above approaches, however, do not attempt to equalize variations in the utilization of the stations depending on the models. That is, no additional objectives which distinguish between several solutions optimal for single-model objectives are considered. In the sequel, we outline a heuristic procedure which incorporates a respective secondary objective. Furthermore, we discuss how exact and heuristic procedures for versions of SALBP have to be modified in order to solve such problems. Note that primary objectives ("Minimize the number of stations", "Minimize the cycle time", "Minimize the line efficiency") correspond to the average model or, equivalently, to the joint (aggregate) model (cf. Definition 3.2, p. 83). On the contrary, the secondary goals refer to the operation times of the different models (cf. the objectives in Section 3.2.2.3).

The Heuristic of Thomopoulos. A heuristic greedy procedure for MALBP-F or MALBP-1, respectively, with the secondary objective function (3.3) is proposed by Thomopoulos (1970). Furthermore, it is applicable to a modified problem, where the cycle time constraints are complemented by a lower bound on the station times of the aggregate model. The secondary objective aims at levelling the station utilizations by minimizing the sum of absolute differences between the total station times τ'_{mk} and the average total station times $\bar{t}'_m$ for all models and stations (cf. p. 87). We outline the procedure for MALBP-F, i.e., the cycle time c and the number K of stations are given.

The heuristic procedure works unidirectionally in a stationoriented manner, i.e., the stations are loaded in order of increasing station numbers. It proceeds similar to the heuristic of Hoffmann (1963) which is described in Section 5.1.2. For each station (in each iteration) k = 1,...,K, a station load with a minimal value of

$$\Delta_k := \sum_{m=1}^{M} | \tau'_{mk} - \bar{t}'_m |$$

is selected. Even loads which are not maximal for the aggregate model have to be considered, because their Δ_k values may be smaller than those of corresponding maximal loads. In each iteration, the assigned tasks are removed from the precedence graph. The load for the currently considered station is obtained by systematically enumerating and evaluating a user-specified number of feasible subsets of the reduced precedence graph.

Modification of SALBP Procedures. Basically, each solution procedure for a version of SALBP can easily be applied to a respective version of MALBP with an arbitrary secondary objective for smoothed station loads (cf. Section 3.2.2.3).

For *exact procedures,* the following modifications are necessary. First, all optimal solutions have to be considered (a respective procedure is described by Dar-El and Rubinovitch (1979)). That is, branch and bound algorithms must not fathom a node if its local lower bound is equal to the current upper bound (concerning the primary objective). Secondly, the search cannot be restricted to maximal station loads, because non-maximal loads may be preferred by the secondary objective function. Thirdly, dominance and reduction rules which utilize the fact that only a single optimal solution is required for SALBP have to be omitted or relaxed. Whenever a solution which is equivalent to the incumbent solution with respect to the primary objective is found, the secondary objective decides on the solution being preferred.

In general, the described modifications of exact procedures will not result in effective algorithms, because all solutions being optimal for the primary objective have to be examined. Therefore, it seems to be promising to generalize the Local Lower Bound Methods of SALOME-1 for SALBP-1 and SALOME-2 for SALBP-2, respectively (cf. Sections 4.1.4.1 and 4.2.4.1). These methods systematically increase the local lower bound of each node by generating station loads in a respective order. This is achieved by successively enlarging intervals of the station times of the aggregate model. The approach can be complemented by subdividing the interval for a current (primary) local lower bound into subintervals according to increasing values of the secondary objective function. That is, the station loads available for branching in a particular node are ordered according to non-decreasing values of an appropriately defined combined objective function. Note that the objective 3 already combines a primary and a secondary goal (cf. Section 3.2.2.3 as well as Domschke et al. (1996 a)).

Similar modifications may be performed for *heuristic procedures*. In priority based methods, the secondary goal may be reflected by a respective tie breaking rule (cf. Section 5.1.1). In improvement procedures, the moves have to be evaluated by a combined objective function. As a positive consequence, the secondary objective enables a tabu search procedure to distinguish between solutions which are equivalent with respect to the primary goal. Therefore, the definition of a neighbourhood may be easier for MALBP-1 relative to SALBP-1 (cf. the discussion in Section 5.2.4).

6.2 Procedures for Mixed-Model Sequencing

In this section, we describe existing exact and heuristic solution procedures for several versions of the mixed-model sequencing problem (MSP). Surveys are provided by Suliman et al. (1985), Koether (1986), Yano and Bolat (1989), Ziegler (1990) as well as Domschke et al. (1997, chapter 4.4.2). For a special version of the problem which is to minimize the total work overload (MSP-W), we propose exact and heuristic solution approaches.

In order to facilitate the reading, we repeat the notations of Section 3.3, which are required to describe solution procedures, in Figure 6.1.

M	number of models, index: $m = 1,...,M$
d_m	demand for model m during the planning period
I	length of the sequence, total demand $(= \Sigma_m d_m)$, index: $i = 1,...,I$
c	fixed launch interval (cycle time, in TU)
K	number of stations, index: $k = 1,...,K$
l_k	length of station k (in TU; = physical length / conveyor speed)
τ_{mk}	processing time per unit of model m in station k $(\tau_{mk} \leq l_k)$
$\bar{t}_m$	average operation time per station and unit of model m $(= \Sigma_k \tau_{mk}/K)$
τ'_k	sum of operation times in station k per period $(= \Sigma_m d_m \cdot \tau_{mk})$
Δ_{mk}	displacement of operator in station k by a unit of model m $(= \tau_{mk} - c)$
x_{mi}	binary variable, $= \begin{cases} 1 & \text{if a unit of model m is assigned to seq. position i} \\ 0 & \text{otherwise} \end{cases}$
s_{ki}	start position of operator in station k for the i-th unit (in TU)
e_{ki}	end position of operator in station k after processing the i-th unit (in TU)
w_{ki}	utility work in station k to complete the i-th unit of the sequence (in TU)

Figure 6.1. Notations for Mixed-Model Sequencing (cf. Figure 3.7)

6.2.1 Exact Procedures

We survey existing exact procedures for several mixed-model sequencing problems and propose a new branch and bound approach for MSP-W.

6.2.1.1 Survey of Literature Approaches

Most versions of MSP (including MSP-W) are NP-hard optimization problems. They are usually considered as being computationally intractable for exact approaches due to their lack of structural properties which can be utilized for bound arguments. Therefore, only a few exact procedures are proposed in the literature up to now. They either solve very restrictive problem versions or are intended to serve as reference procedures for evaluating heuristic methods. As a consequence of the recent development of powerful computer systems, all known exact procedures but one were proposed very recently. In the sequel, we outline these approaches but do not attempt to compare them, because they solve quite different problem versions.

Procedure of Dar-El and Cucuy. A rather restrictive sequencing problem is solved by an iterative procedure (cf. Dar-El and Cucuy (1977)). By assuming that the operation times of each model are identical in all stations (i.e., $\bar{t}_m = \tau_{mk}$ for k = 1,...,K and m = 1,...,M), the problem is reduced to a single-station problem. As objective the minimization of the line length (i.e., the station length) subject to zero idle time and utility work is considered. The procedure starts with a lower bound on the line length and iteratively increases this value by 1 until a feasible solution is obtained. In each iteration, the following two steps have to be performed.

Step 1. All subsequences, which start and end in the initial state (cf. Definition 3.3, p. 100) and do not exceed the given station length, are systematically constructed by an enumeration procedure. That is, the worker starts at the left border of the station and returns to it after finishing work on the respective subsequence.

Step 2. The subsequences of step 1 can be appended in arbitrary order, because they start and end in the initial state, respectively. In the sequel, such subsequences are called *pattern* (cf. Definition 6.1, p. 214). Thus, a solution is sufficiently described by the quantities with which the different patterns are contained. A respective integer linear program is formulated and solved in order to find out whether or not the available patterns can be combined such that the given demands of all models are satisfied (cf. Section 6.2.3). When such a combination is found, the procedure terminates with an optimal solution. Otherwise, the station length is increased by 1 and the process is repeated.

Branch and Bound Procedures. Recently, a few branch and bound procedures for different versions of MSP were proposed.

Bard et al. (1994) describe the application of a standard branch and bound procedure for general mixed-integer programming to a version of MSP. The objective is to minimize a linear combination of the total line length and the amount by which production differs from strict proportion with demand. That is, the goals "minimize the line length" and "level the part usage rates" are considered simultaneously. The branch and bound procedure is based on a formulation of the problem as a zero-one mixed-integer linear program. It performs a depth-first-search (DFS, p. 116) and branches by setting binary variables x_{mi} to 1 and 0, respectively, indicating whether or not model m is assigned to sequence position i. Lower bounds are obtained from the LP-relaxation of the program.

Bolat (1994) describes a branch and bound procedure for an MSP version which minimizes the sum of set-up costs and the costs of utility work. Branching is organized by the minimal-lower-bound strategy (MLB, p. 116), i.e., active nodes are stored in a candidate list in order of non-decreasing lower bounds. In each step, a node with minimal bound is chosen for branching. For a node at level i of the tree, alternative subproblems are obtained by assigning one of the models to position i+1 of the sequence at a time. That is, the sequence positions are successively filled along the branches of the tree. Lower bounds on the set-up costs are computed for a sequence in which all units requiring the same set-up state of the line are consecutively produced as a batch. Lower bounds on the utility costs are derived from independently considering the stations. That is, the restriction of an identical sequence for all stations is relaxed. A similar procedure is proposed by Bolat et al. (1994).

Ng and Mak (1994) consider MSP with the objective function (3.16), which minimizes the sum of squared differences between cumulated demands and output volumes of all models at all sequence positions (see p. 104). This non-linear problem is solved by a branch and bound procedure with the following components. Lower bounds are obtained by a Lagrangean relaxation of the non-linear integer programming formulation of Miltenburg (1989). An initial upper bound is computed by a simple heuristic. The branching is organized as a DFS, and the nodes descending from a current subproblem are considered in order of non-decreasing local lower bounds. As is the case with the procedure of Bolat described above, the level i of the enumeration tree corresponds to the position i+1 of the sequence.

6.2.1.2 A Branch and Bound Procedure for MSP-W

In this section, we propose a branch and bound algorithm for solving MSP-W which is to minimize the total work overload (cf. Section 3.3.3). Since work overload is assumed to be compensated by utility work, we use these terms synonymously.

Branching. The enumeration is organized as a *depth-first-search* (DFS). The initial problem serves as the *root node* (at level 0) of the enumeration tree. It is branched into m subproblems *(descending nodes)* by assigning one of the m models to the first position of the sequence, at a time. Every node at level i of the tree is branched by assigning models, the demands of which are not already satisfied by the current partial sequence, to the position i+1 of the sequence (this enumeration scheme is identical with that in the procedures of Bolat as well as Ng and Mak outlined above). When all positions are occupied, a feasible solution is obtained. The best known solution *(incumbent solution)* provides an *upper bound* UB on the minimal total work overload.

In the case of DFS algorithms, the order in which the descending nodes are considered is important. To obtain good solutions as early as possible, the models used for branching are ordered according to increasing values of the additional work overload $w_{i+1}(m)$ that results if a unit of model m is assigned to position i+1:

$$w_{i+1}(m) := \sum_{k=1}^{K} \max\{0, s_{k,i+1} + \tau_{mk} - l_k\} \tag{6.1}$$

Very often, several models show the same value of $w_{i+1}(m)$. In order to obtain an unequivocal ordering, one or more tie breaking rules (abbreviated by TB) are used. Ties may be broken in favour of the model with

- the smallest index (TB1),
- the largest remaining demand (TB2),
- the lowest sum $\Sigma_k s_{k,i+2}$ of operator start positions for the (i+2)-th unit (TB3), or
- the largest difference between the cumulated demand $(i+1) \cdot d_m / I$ and the cumulated production volume (number of already assigned units of model m). At the first position (i = 1), a model with largest sum of operation times per unit ($= \bar{t}_m \cdot K$) is preferred (TB4).

Bounding. Lower Bounds (LB) on the work overload are computed by solving relaxations of MSP-W. Like most sequencing problems, MSP-W is related to the well-known *traveling salesman problem* (TSP) which is to find a closed tour through all nodes of a graph such that each node is visited once and the tour has minimal length (cf. Lawler et al. (1985) as well as Domschke (1997, chapter 3)). Since MSP-W is complicated by the fact that the current state S_{i+1} (the start positions of the operators for the (i+1)-unit of the sequence) does not only depend on the model assigned to sequence position i but may be influenced by all models in the partial sequence $<1,...,i>$. Hence, the sequencing problem can be interpreted as a *traveling salesman problem with time-dependent distances* (cf. Domschke et al. (1997, p. 267)).

TSP-Relaxation. MSP-W can be relaxed to obtain a TSP by generating a directed graph as follows. For each model m, d_m nodes are defined, i.e., the graph contains $I = \Sigma_m d_m$ nodes. Between all pairs of nodes a directed arc is introduced. In order to facilitate the presentation, an arc is identified by the models m and n which correspond to its start and terminal node, respectively, even though $2 \cdot d_m \cdot d_n$ arcs are required to connect all nodes for models m and n (if m = n, $d_m \cdot (d_m - 1)$ arcs are needed). Each arc (m,n) means that units of m and n occupy neighbouring positions i and i+1 of a sequence. The arc weights (distances) δ_{mn} are computed as the minimal work overload which cannot be avoided for a unit of model n if the line is in the initial state before processing a unit of model m:

$$\delta_{mn} := \sum_{k=1}^{K} \max\{0, \max\{0, \min\{\Delta_{mk}, l_k - c\}\} + \tau_{nk} - l_k\} \quad \text{for } m,n=1,...,M \qquad (6.2)$$

A fictitious node 0 with $\delta_{0m} := 0$ and $\delta_{m0} := \Sigma_k \max\{0, \tau_{mk} - c\}$ for m = 1,...,M serves as start and terminal node of the "salesman's" tour. The arcs (0,m) and (m,0) represent the assignment of a unit of model m to the first and last sequence position, respectively.

Example. We repeat the example problem of Section 3.3.1 with $m=3$ models and the demands $d_1=2$, $d_2=2$, and $d_3=1$. The launch interval is $c=9$. Each of $K=4$ stations has the length $l_k=10$ TU. The total available time per station is $I \cdot c=45$ TU. The operation times τ_{mk}, the average station times $\bar{t}_m$ per model unit, and the total station times τ'_k are given in Table 6.1.

Table 6.1. Operation Times

m \ k	1	2	3	4	$\bar{t}_m$
1	10	7	10	6	8.25
2	8	10	9	8	8.75
3	9	8	7	10	8.5
τ'_k	45	42	45	38	42.5

The distance matrix of our example (with one entry for each model combination; the complete matrix contains d_m rows and d_m columns for each model m) is given in Table 6.2. An optimal tour is (0,1,2,3,1,2,0) with length 1. Therefore, LB = 1 is a lower bound on the total work overload.

Table 6.2. Distance Matrix

m \ n	0	1	2	3
0	∞	0	0	0
1	2	2	0	0
2	1	0	1	0
3	1	0	0	1

Unfortunately, TSP is an NP-hard problem, too. Therefore, it is usually not possible to find an optimal solution for the TSP-relaxation of MSP-W.

TPP-Relaxation. Another relaxation of MSP-W (and of TSP as well) is the linear *transportation problem* (TPP). The models 1,...,M and the fictitious model 0 (with $d_0=1$) are interpreted as suppliers and demanders, respectively. The supply and demand quantities of the models m = 1,...,M are both equal to d_m. The distances δ_{mn} serve as unit transportation cost between a supplier m and a demander n. The minimal total transportation cost determines a lower bound LB on the total work overload for MSP-W. Unlike MSP-W and TSP, TPP is solvable in polynomial time. Respective solution procedures are, e.g., described by Domschke (1995, chapter 6).

Example. The transportation problem is defined by the transportation costs of Table 6.2 as well as the supply and demand quantities (1,2,2,1). An optimal solution with a total cost of 1 (LB = 1) consists of transporting one unit from 0 to 1, two units from 1 to 2, one unit from 2 to 1, one unit from 2 to 3, and one unit from 3 to 0. In this case, the solution of the transportation problem can be transformed into the tour (0,1,2,1,2,3,0) for the related TSP. In general, it is not possible to transform a TPP solution to a feasible TSP solution, because TPP provides no means for avoiding subtours.

Local Lower Bounds. In nodes at level i of the tree, local lower bounds LB_i are obtained by relaxing the current subproblem, which is defined by the remaining demands d'_m of all models and the current state $S_{i+1} = (s_{1,i+1},...,s_{K,i+1})$ of the line. The distances δ_{0m} are modified as follows (models with $d'_m = 0$ can be eliminated):

$$\delta_{0m} := \sum_{k=1}^{K} \max\{0, s_{k,i+1} + \tau_{mk} - l_k\} \qquad \text{for } m = 1,...,M \qquad (6.3)$$

A local lower bound LB_i is now computed by summing up the work overload already contained in the partial solution < 1,...,i > and the minimal objective function value of the respective (TSP- or TPP-) relaxation.

Fathoming. A node is fathomed if its local lower bound is not smaller than the current upper bound UB. The algorithm is terminated when UB gets equal to the (global) lower bound LB_0 of the root node or when the complete tree has been considered explicitly or implicitly.

Example. The tree shown in Figure 6.2 is built for our example problem. The global lower bound as well as the local lower bounds are computed by means of the TPP-relaxation and are given as node weights. The edges represent the assigned models. If a partial solution contains a work overload, the respective amount is specified behind the number of the model assigned at last. The rule TB4 is used as tie breaker for the selection of the model to be assigned; second-level ties are handled by TB1. The nodes are numbered according to the order of their generation. An optimal sequence <2, 1, 3, 2, 1> with UB = 2 is found in the first branch.

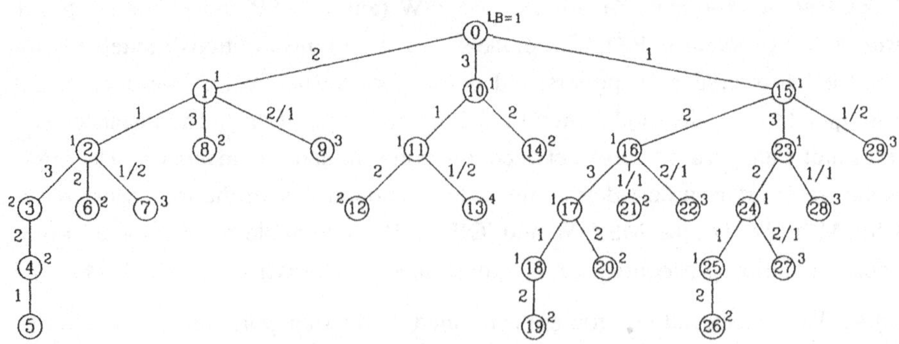

Figure 6.2. Branch and Bound Tree

The method of computing the lower bounds is already illustrated above. Now, we explain some of the branching steps performed by the procedure. In the root node 0, the models are assigned in the order of decreasing values of $\bar{t}_m$ in accordance with TB4. At first, model 2 is assigned to position 1. In the resulting node 1 at level $i=1$, the remaining demands are $d_1^{\prime}=2$, $d_2^{\prime}=1$, and $d_3^{\prime}=1$. The models $m=1,...,3$ are ordered according to decreasing differences between the proportional demand $((i+1)\cdot d_m/I)$ and the cumulated production $(d_m-d_m^{\prime})$. Since these differences are 0.8, -0.2, and 0.4, the models are considered for branching in the order 1, 3, 2. After assigning a unit of model 1 to position $i=2$ and three further assignments, a feasible solution with $UB=2$ is obtained in node 5. The search traces back to node 2, because the nodes 4 and 3 have a local lower bound of 2. This process is repeated until all nodes are completely branched or fathomed.

The example shows that the lower bounds are not very successful in reducing the enumeration tree, even though the optimal solution is found in the first branch. This is due to the compensation effects which are possible up to the last position of the sequence. That is, distances and transportation costs, respectively, are often very weak bounds on the actual work overload. Especially, in the case of long stations where large positive operator displacements can be equalized by succeeding small operation times, the branch and bound procedure tends to become a complete enumeration technique.

6.2.2 Heuristic Procedures in the Literature

Due to the computational complexity of MSP and its versions, most of the solution procedures described so far are heuristics. The majority of these procedures constructs an initial solution by means of priority rules, while only a few approaches attempt to improve an initial solution. In either case, the solution approaches reflect the relationship between MSP and TSP (cf. pp. 208). The procedures described below are originally designed for *different* versions of MSP. Nevertheless, we do not explicitly distinguish between the respective problem versions, because these procedures can basically be applied to any version.

6.2.2.1 Priority Based Methods

The positions of the sequence are successively considered, i.e., in each iteration $i=1,...,I$, a unit of some model m is selected by a priority rule and assigned to position i (cf. the priority based procedures for SALBP-1 in Section 5.1.1). In the literature, a large number of corresponding priority rules is proposed for different versions of MSP (cf. Section 3.3.2). Usually, these rules are *greedy*, i.e., they select a model which results in the smallest increase of the cumulated objective function value. In the following, we present some priority rules which select a model with remaining demand to be assigned to a sequence position i (cf. Domschke et al. (1997, pp. 269)):

- Select the model which causes the smallest increase of the cumulated idle time. Ties are broken in favour of the model with the smallest additional utility work (cf. Wester and Kilbridge (1964)).
- In order to equalize excess times, select a model with the largest and the smallest total operation time per unit, alternately (cf. Decker (1993 a, pp. 66)).
- The models are selected randomly according to certain probabilities. These selection probabilities may be defined as being equal for all models or proportional to the remaining demands of the models (cf. Wilhelm (1979, p. 107), Görke and Lentes (1981), Koether (1985, 1986, pp. 96) as well as Koether and Thaler (1986)). Another rule may randomly select the used priority rule in each iteration (cf. the flexible rule application for SALBP-1 in Section 5.1.1.4).
- Select the model which leads to the smallest increase of the labour costs including idle time costs and utility work costs. In the case of open stations, additional cost components are considered for working within overlap areas (cf. Thomopoulos (1967, 1968), Sahgal (1970)). Furthermore, costs for stopping the conveyor or for off-line completion of workpieces may be included in a cost oriented priority rule (cf. Görke and Lentes (1981)).
- Select the model which shows the largest difference between the cumulated requirement $i \cdot d_m / I$ and the cumulated production volume in order to level the part usage rates (cf., e.g., Miltenburg (1989) as well as the tie breaking rule TB4 in the branch and bound procedure of Section 6.2.1.2). Similar priorities are used in the procedures of Dar-El and Cother (1975), Ding and Cheng (1993) as well as Monden (1998). Furthermore, see Dar-El and Nadivi (1981), Sumichrast and Russell (1990) as well as Sumichrast et al. (1992).

Such greedy procedures tend to cumulate inefficiencies at the last positions of the sequence. Different strategies are proposed to overcome this drawback.

Look-ahead Procedures. They look ahead to later positions when assigning a model to the current position i. Therefore, the priority of a model m is modified by adding an estimated objective function value for the remaining partial sequence provided that a unit of model m is assigned to position i. Such estimates may be obtained by computing lower bounds or simulating sequences for the remaining positions (cf. Yano and Bolat (1989), Ziegler (1990, p. 206), Yano and Rachamadugu (1991) as well as Bolat (1994)).

Alternating Rule Application. Two or more rules which control different types of inefficiencies are alternately applied. For example, one may select a model with smallest increase of labour costs and one with largest total operation time per unit, alternately (cf. Macaskill (1973)).

A general *two-stage approach* is proposed by Felbecker (1980, chapter 8). In the first step, subsequences are generated by successively joining smaller subsequences such that the objective function value is minimally increased in each case. Initially, each unit builds a subsequence for its own. The joining of subsequences is stopped when they have reached a certain average length. In the second step, a TSP instance is defined by introducing a node for each subsequence and distances which estimate the increase of the objective function value when a subsequence is processed immediately after another one (cf. the TSP-transformation in Section 6.2.4.3, p. 227). This aggregated problem may be solved by any heuristic or exact procedure for TSP (cf., e.g., Lawler et al. (1985) and Domschke (1997, chapter 3)) resulting in a heuristic solution for MSP.

6.2.2.2 Improvement Methods

Due to the relationship between TSP and MSP, the same type of improvement procedures developed for the first can be used for the latter. Furthermore, MSP instances can heuristically be transformed to a corresponding TSP instance by defining appropriate distances and can directly be solved by a method for TSP as outlined above. For example, the well-known *2-opt procedure* may be applied to a transformed instance. In each iteration of this procedure, two arcs of the current TSP tour are exchanged for two not yet contained arcs such that the length of the tour is reduced (cf., e.g., Domschke and Drexl (1998, pp. 133)). Respective approaches for MSP are proposed by Schneeweiß and Söhner (1991, pp. 54) as well as Decker (1993 a, pp. 65).

Okamura and Yamashina (1979) present a steepest descent procedure for MSP which includes different types of moves. By a *shift* move, a single unit is transferred to another sequence position, while the positions of two subsequences with one or two elements are exchanged by a *swap* move. Another steepest descent procedure with shift moves is described by Ziegler (1990, pp. 227).

A tabu search approach with swap moves is proposed by Bard et al. (1994). In each iteration of this procedure, a most improving/least deteriorating move, which exchanges the positions of two units of different models, is performed. The neighbourhood is restricted by defining a maximal distance between corresponding positions. In order to avoid cycling between already visited solutions, a static version of tabu search with TNM is used (cf. Section 5.2.2.1). Furthermore, a global aspiration criterion allows performing a tabu move when the incumbent solution can be improved.

6.2.3 A Reformulation of MSP-W

MSP-W may be decomposed into the problem of finding a set of appropriate subsequences and that of combining such partial sequences to obtain a complete sequence (cf. the two-stage approach of Felbecker outlined above, p. 213). For a respective problem reformulation we need some definitions. Furthermore, see Definition 3.3 (p. 100) for the terms *state* and *initial state* of a line.

Definition 6.1. A partial sequence (subsequence) is called *pattern* if the line returns to the initial state after processing the units contained. That is, all operators start at the left border of their stations (or even later), perform their work on all units of the subsequence, and return to the initial positions in-time before subsequent units enter their stations. A pattern is *elementary* if a decomposition into smaller patterns with the same amount of work overload does not exist.

Remark 6.1. Each partial sequence of length $i-1$ which does not end in the initial state can be transformed into a pattern by introducing additional utility work of duration $s_{1i}+...+s_{Ki}$, where $S_i=(s_{1i},...,s_{Ki})$ denotes the state after processing the subsequence. Such patterns are called *artificial.*

With the above definitions, MSP-W may be reformulated as the problem of finding a pattern which minimizes the total work overload and contains all requisite units. If the optimal pattern is not elementary itself, it consists of smaller elementary patterns which can be combined to a complete sequence in an arbitrary order, because the patterns start and end in the initial state.

Formulation. If we assume that all (necessary) elementary patterns are known, the problem of combining them to meet the demands can be formulated as an integer linear program with the following *parameters* (cf. Dar-El and Cucuy (1977) for a similar formulation of a restricted problem with one station):

- P number of elementary patterns
- π_p elementary pattern, index: $p=1,...,P$
- u_p utility work (work overload) of pattern π_p
- a_{mp} number of times model m is contained in pattern π_p

With *integer variables* y_p for $p=1,...,P$, representing the quantities with which the patterns π_p are chosen, the problem can be formulated by (6.4) - (6.6).

$$\text{Minimize } F(y) = \sum_{p=1}^{P} u_p \cdot y_p \tag{6.4}$$

subject to

$$\sum_{p=1}^{P} a_{mp} \cdot y_p = d_m \qquad \text{for } m = 1,...,M \tag{6.5}$$

$$y_p \geq 0 \text{ and integer} \qquad \text{for } p = 1,...,P \tag{6.6}$$

Basically, optimal sequences for MSP-W can be constructed by appending the chosen patterns in an arbitrary order. Nevertheless, it may be advantageous to repeat identical patterns ($y_p \geq 2$) immediately in order to utilize learning effects and to have a regular material flow as long as possible.

Example. We consider an instance of MSP-W with $M=4$ models and $K=4$ stations. The demands are $d=(100,45,75,80)$, the cycle time is $c=10$, and the sequence length is $I=300$. Hence, the total available production time is $I \cdot c = 3{,}000$. The station lengths $l_1,...,l_4$ are given by the vector $(11,10,14,12)$, and the operation times τ_{mk} are defined by Table 6.3.

Table 6.3. Operation Times

m \ k	1	2	3	4	$\bar{t}_m$
1	9	8	7	10	8.5
2	11	9	13	11	11
3	10	10	8	11	9.75
4	8	9	12	8	9.25
τ'_k	2785	2675	2845	2960	2816.25

If we assume that the elementary patterns $\pi_1 = <3,2,1,4,1>$, $\pi_2 = <3,4,3,4,1>$, $\pi_3 = <2,3,4,1>$, $\pi_4 = <3,4,1,4,1>$, and $\pi_5 = <3,1,3,4,1>$ are known, an optimal solution of MSP-W without work overload can be determined by solving the problem (6.4) - (6.6). The $P=5$ patterns are described by the column vectors $a_{\cdot 1} = (2,1,1,1)$, $a_{\cdot 2} = (1,0,2,2)$, $a_{\cdot 3} = (1,1,1,1)$, $a_{\cdot 4} = (2,0,1,2)$, and $a_{\cdot 5} = (2,0,2,1)$. The optimal values of the integer variables are $y_1 = 30$, $y_2 = 11$, $y_3 = 15$, $y_4 = 6$, and $y_5 = 1$. Hence, the complete sequence may be composed of 30 copies of π_1, followed by 11 copies of π_2, 15 copies of π_3, 6 copies of π_4, and 1 copy of π_5.

A Related Problem. The reformulation of MSP-W is similar to the **one-dimensional cutting stock problem** (CSP-1, cf. Gilmore and Gomory (1961)). This problem considers the cutting of a *standard length* into smaller pieces of several *order lengths*, which are required in specified quantities. The objective is to minimize the number of standard length pieces to be cut. A model formulation of CSP-1 is given by (6.4)-(6.6) with $u_p = 1$ for $p = 1,...,P$. The restrictions (6.5) guarantee that the demands d_m of the different order lengths $m = 1,...,M$ are satisfied. The coefficients a_{mp} denote the quantities with which an order length m is contained in a *cutting pattern* p. The cutting patterns have to fulfill the (knapsack) condition

$$\sum_{m=1}^{M} a_{mp} \cdot L_m \leq L \qquad \text{for } p = 1,...,P, \tag{6.7}$$

where the parameters L_m and L designate the extents of the order lengths $m = 1,...,M$ and the extent of the standard length, respectively.

Remark 6.2. Usually, restrictions (6.5) are given as '≥'-conditions in order to avoid wasting material. Then only *efficient* cutting patterns, the waste of which is smaller than the shortest order length, have to be considered.

6.2.4 A Hybrid Procedure for MSP-W

The reformulation of MSP-W immediately leads to a two-stage solution approach (cf. Scholl et al. (1995)). At the first stage, all (or a subset of all) elementary patterns have to be built. At the second stage, the problem (6.4) - (6.6) is solved by a standard all-integer LP-solver. This approach has the main difficulty that the number of elementary patterns is usually very large and that it will neither be possible to generate nor to store all of them because of limited computation time and storage space. Therefore, we develop a heuristic procedure which restricts the number of patterns to be generated. Due to the relationship between MSP-W and CSP-1, which becomes obvious by the reformulation stated above, our approach contains a part which is similar to a heuristic for CSP-1. Therefore, the main features of this procedure are outlined before we describe the new approach.

6.2.4.1 A Heuristic for CSP-1

A heuristic procedure for CSP-1 is proposed by Stadtler (1990). It starts with applying the first fit decreasing heuristic for the related bin packing problem (FFD, cf. Coffman et al. (1984)) in order to determine an initial feasible solution. If the optimality of this solution cannot be proved, the LP-relaxation of CSP-1 is solved. This is done by means of the column generation procedure described below. If necessary, the optimal LP-solution is transformed to an integer solution by an iterative rounding procedure. In each of its iterations, one of the basic variables with largest fraction is rounded up to the nearest integer. A corresponding equation is introduced into the LP-model which is restricted to the variables (cutting patterns) already in the basis. This model is re-optimized by a revised simplex method. The rounding procedure stops when all variables are integral. Otherwise, a further iteration is performed. After terminating this procedure, some oversupply of demands may exist which is eliminated by reducing the quantities of cutting patterns having the largest portion of order lengths in an oversupply. If this reduction step produces an undersupply in some order lengths, the remaining problem is solved by the FFD heuristic.

Further hybrid heuristic procedures for CSP-1 containing column generation and rounding steps are, e.g., proposed by Pierce (1963), Roodman (1986) as well as Wäscher and Gau (1996).

The most interesting part of the outlined approach for CSP-1 is the column generation procedure which avoids generating all cutting patterns prior to optimization. Therefore, it is explained in greater detail in the subsequent section.

6.2.4.2 The Column Generation Procedure

The LP-relaxation of the integer linear program for CSP-1, obtained by dropping the integrality conditions of the variables y_p, is solved. Since it contains M restrictions, M variables (cutting patterns) are in the basis in each iteration of the (revised) simplex algorithm. Only the corresponding columns of the matrix of coefficients need be known and stored. This fact is utilized by the so-called *column generation procedure* of Gilmore and Gomory (1961, 1963) which is outlined in the sequel. A more detailed description is given by Terno et al. (1987, chapter 3.3). Furthermore, see Eiselt et al. (1987, chapter 6.2).

Current Basic Solution. We consider an iteration of the revised simplex method (cf., e.g., Domschke and Drexl (1998, pp. 30)). Let B and N denote the matrix of current *basic columns* and the matrix of current *non-basic columns*. Each column corresponds to a cutting pattern. The M cutting patterns represented by the basic columns constitute a current LP-solution of (6.4) - (6.6). The matrix N implicitly contains all remaining cutting patterns which are currently not known. The basic variables which correspond to the basic columns of B build a vector Y_B; the non-basic variables are contained in the vector Y_N. The demands of the order lengths are described by the column vector $d=(d_1,...,d_m)$. In the *initial basic solution,* M simple cutting patterns are used, each of which cuts $\lfloor L/L_m \rfloor$ pieces of an order length m. That is, the initial basic matrix B contains non-zero coefficients only at the main diagonal. In standard matrix form, the restrictions (6.5) can be written as:

$$BY_B + NY_N = d \qquad (6.5')$$

The values of the basic variables are determined by transforming (6.5') to

$$Y_B = B^{-1}d - B^{-1}NY_N. \qquad (6.8)$$

The inverse of the initial basic matrix is easily determined by building the reciprocal values on the main diagonal. The inverse matrices of modified basic solutions are obtained from the initial B^{-1} by successively performing standard simplex iterations (see below).

The costs u_p of all cutting patterns $p = 1,...,P$ are equal to 1, because one standard length is required for each cutting pattern. With the M-dimensional row vector $U_B = (1,1,...,1)$ and the (P-M)-dimensional row vector $U_N = (1,1,...,1)$, the objective function is given by:

$$F(Y_B, Y_N) = U_B Y_B + U_N Y_N \tag{6.9}$$

If Y_B is substituted for (6.8), the objective function is transformed to:

$$F(Y_B, Y_N) = (U_N - U_B B^{-1} N) Y_N + U_B B^{-1} d \tag{6.9'}$$

Since all non-basic variables are zero, the basic vector Y_B is determined by $B^{-1}d$. The objective function value is given by $U_B B^{-1} d$.

Reduced Costs. The row vector $U_N - U_B B^{-1} N$ contains the *reduced costs* of the unknown non-basic variables. The current solution is optimal if no non-basic variable exists which has negative reduced cost. Otherwise, the variable with the most negative value is chosen to enter the basis. The reduced cost of a single non-basic cutting pattern $q = (q_1,...,q_M)$ is obtained by $u_q - vq$, where $u_q = 1$ denotes the cost of pattern q and the row vector $v = U_B B^{-1}$ contains the *simplex multipliers* (entries of the objective function row in the current simplex tableau).

Column Generation. A non-basic variable (cutting pattern) with the most negative reduced cost is obtained by maximizing the scalar product vq. In order to get a feasible cutting pattern, the knapsack condition (6.7) and the integrality of the variables q_m have to be observed. Therefore, the requisite cutting pattern can be determined by solving an *integer knapsack problem.* Even though this problem is NP-hard, it can often be efficiently solved by dynamic programming or branch and bound algorithms (cf. Martello and Toth (1990 a, chapter 3)). If the maximal value of vq is larger than $u_q = 1$, the computed cutting pattern q builds the column for an entering variable. Otherwise, the current basic solution is optimal, because no non-basic variable with negative reduced cost exists.

Simplex Iteration. When the entering column q has been determined, a usual simplex iteration is performed. This includes computing the corresponding new pivot column $B^{-1}q$, choosing the pivot row (a current basic variable), and transforming the simplex tableau accordingly (cf., e.g., Hillier and Lieberman (1988, chapter 5) or Domschke and Drexl (1998, chapter 2.4)).

6.2.4.3 The Hybrid Heuristic for MSP-W

We adopt the basic ideas of the outlined procedure for CSP-1 and develop a heuristic algorithm for MSP-W. This approach is especially devised for the master sequencing problem arising in the hierarchical planning system (cf. Section 3.4), because it is able to handle large demand values.

At first, the LP-relaxation of (6.4) - (6.6) is solved by a column generation approach. If the resulting LP-solution is not integral, it is treated by a rounding procedure until no fractional values of the variables remain. If all contained patterns are non-artificial, i.e., no additional utility work is necessary to return to the initial state, a feasible sequence for MSP-W is obtained by appending all patterns in an arbitrary order (see Remark 6.1, p. 214). Otherwise, the procedure tries to find a *sequence of the patterns* which minimizes the amount of unnecessary work overload (note that we use the terms utility work and work overload synonymously). This is done by a combination of a priority based procedure which generates an initial pattern sequence and an improvement procedure with tabu search which is subsequently applied to improve the initial solution.

Phase 1. Solving the LP-Relaxation

We apply the column generation approach of Gilmore and Gomory with the following modifications.

Initial Basic Solution. The initial basic columns are obtained from the M simple elementary patterns which contain one unit of one model in each case. Therefore, the basic matrix B and its inverse B^{-1} are equal to the identity matrix.

Column Generation. The entering column is determined by finding a pattern with most negative reduced cost. This can be done by solving the mixed-integer problem (6.10) - (6.16), which includes the same variables as (3.18) - (3.25) and the integer vector $q = (q_1,...,q_M)$ denoting the column to be found (see pp. 105 for the formulation of MSP-W).

The row vector $v = (v_1,...,v_M)$ of simplex multipliers is computed by $v = U_B B^{-1}$, where the vector U_B specifies the work overloads connected with the current basic patterns. The objective function (6.10) minimizes the reduced cost $u_q - vq$, where u_q denotes the utility work of the pattern represented by the entering column q. The restrictions (6.11) to (6.13) guarantee that the pattern contains at most d_m units of each model m. The conditions (6.14) ensure that only the sequence positions from 1 to the length of the pattern are occupied. By the restrictions (6.15), it is enforced that the line returns to the initial state after processing the pattern.

$$\text{Minimize } RC(x,s,w,q) = u_q - vq = \sum_{k=1}^{K} \sum_{i=1}^{I} w_{ki} \cdot \sum_{m=1}^{M} x_{mi} - \sum_{m=1}^{M} v_m \cdot q_m \tag{6.10}$$

subject to (3.21)-(3.25) and

$$q_m \le d_m \qquad \text{for } m=1,...,M \tag{6.11}$$

$$\sum_{i=1}^{I} x_{mi} = q_m \qquad \text{for } m=1,...,M \tag{6.12}$$

$$\sum_{m=1}^{M} x_{mi} \le 1 \qquad \text{for } i=1,...,I \tag{6.13}$$

$$\sum_{m=1}^{M} x_{mi} \ge \sum_{m=1}^{M} x_{m,i+1} \qquad \text{for } i=1,...,I-1 \tag{6.14}$$

$$s_{ki} \le l_k \cdot \sum_{m=1}^{M} x_{mi} \qquad \text{for } k=1,...,K \text{ and } i=1,...,I \tag{6.15}$$

$$q_m \ge 0 \text{ and integer} \qquad \text{for } m=1,...,M \tag{6.16}$$

The problem (6.10)-(6.16) can be solved by a slight modification of our branch and bound algorithm of Section 6.2.1.2. This is illustrated by substituting q according to (6.12) and re-arranging the terms of the objective function:

$$RC'(x,s,w) = \sum_{i=1}^{I} \sum_{m=1}^{M} x_{mi} \cdot \left(\sum_{k=1}^{K} w_{ki} - v_m\right) \tag{6.10'}$$

Hence, the cost of assigning model m to position i is determined by the difference of the resulting work overload and the simplex multiplier v_m. Since it is not known with which quantities the models are contained in the requisite pattern, the lower bound arguments described in Section 6.2.1.2 cannot be applied. Furthermore, feasible solutions (i.e., patterns) are obtained in each node instead of being available only in the leaf nodes of the enumeration tree. Note that the formulation as well as the search is not restricted to elementary patterns, because it is difficult to identify the non-elementary ones.

Unfortunately, it may be necessary to construct the complete enumeration tree in order to determine the entering column. Consequently, it may occur that solving the LP-relaxation is much harder than solving the original MSP-W. This difficulty may be dealt with by the following strategies.

Pattern Selection Rule. Instead of determining a pattern with the most negative reduced cost *(best fit)*, we may choose the first one with a negative value of (6.10) which is found during the enumeration *(first fit)*. Then only in the last iteration, the complete enumeration tree has to be checked. Due to rather small improvements per simplex iteration, the first fit strategy usually results in a large number of simplex iterations, which may cause numerical problems in a computer program.

Thus, the search may be continued after finding the first candidate pattern until no improvement of (6.10) has been found during a user-specified time interval. This intermediate strategy is called *average fit.*

Cut-Off. For large problem instances, the column generation process can be very expensive in terms of computation time. The improvements of the objective function value are often very small in the last iterations, while the computation time per iteration tends to increase (cf. Gilmore and Gomory (1963)). Furthermore, small improvements often result in highly fractionated solutions with values of basic variables remarkably smaller than 1. That is, the LP-solution tends to deviate from integrality. Therefore, it may be advantageous to terminate the column generation process before the optimal solution of the LP-relaxation is determined with respect to both the computation time requirements and the quality of the integer solution generated in phase 2 of the heuristic (see pp. 223). Possible cut-offs to terminate the calculation are the following:

- A given time limit is exceeded.
- The average relative improvement of the objective function value per iteration falls below a given constant.

Restricted Pattern Length. The pattern length is the most critical factor for the computation time of the branch and bound algorithm used to find the patterns. Thus, the pattern length may be restricted to a user-defined maximal value. This restriction is an analogy to the knapsack condition (6.7) for the cutting patterns in CSP-1.

Besides the reduction of computation times, a restriction of the maximal pattern length may be advantageous with respect to the following reflections. The sensibility of the production system decreases with a reducing length of the patterns, because the line often returns to the initial state when short patterns are used. In this case, it is easy to include changed or additional patterns, which may be caused by short-term demand changes or production failures, respectively, into the current sequence. Furthermore, small patterns may be repeated several times which is advantageous with respect to learning effects and material flow aspects such as frequently discussed in the context of just-in-time production systems. Consequently, solutions should be composed of small patterns. However, the obtained solution of the LP - relaxation need not be optimal if the pattern length is restricted.

Remark 6.3. If a parametric solution is identified at the end of the column generation approach, i.e., the reduced cost of some pattern q is zero, the process may be continued for some further iterations. This proceeding may result in a solution with more integral values of the basic variables or may provide an idea which integer values are in question for some variable. However, one has to avoid a cycling of the solution process.

Example. We consider the example of Section 6.2.3 (see p. 215) again and restrict the pattern length to 5. In order to facilitate the reading, we repeat the problem data: M = 4, K = 4, d = (100,45,75,80), c = 10, I = 300, $l_1 = 11$, $l_2 = 10$, $l_3 = 14$, $l_4 = 12$, and the operation times τ_{mk} of Table 6.4.

Table 6.4. Operation Times (cf. Table 6.3)

m \ k	1	2	3	4	$\bar{t}_m$
1	9	8	7	10	8.5
2	11	9	13	11	11
3	10	10	8	11	9.75
4	8	9	12	8	9.25
τ'_k	2785	2675	2845	2960	2816.25

The column generation procedure starts with the simple patterns <1>, <2>, <3>, and <4> which are represented by the columns (1,0,0,0), (0,1,0,0), (0,0,1,0), and (0,0,0,1). All patterns but the first are artificial. Due to the cost vector U_B = (0,5,1,2), the current work overload is $WO = 100 \cdot 0 + 45 \cdot 5 + 75 \cdot 1 + 80 \cdot 2 = 460$. Table 6.5 shows a shortened version of the initial tableau of the revised simplex method. A detailed description of the method can, e.g., be found in Hillier and Lieberman (1988, chapter 5.2) or Domschke and Drexl (1998, chapter 2.4.3). The first four rows correspond to the current basic variables $Y_B = (y_{B_1}, y_{B_2}, y_{B_3}, y_{B_4})$, and the first four columns initially represent the respective patterns. Hence, the corresponding part of the tableau contains the basic matrix B which is equal to B^{-1}, because it is an identity matrix. The right hand side (RHS) specifies the values of the basic variables, i.e., the quantities with which the basic patterns are currently chosen. The last row represents the vector v of simplex multipliers and the current objective function value, respectively.

Table 6.5. Initial Simplex Tableau

	1	2	3	4	RHS
y_{B_1}	1	0	0	0	100
y_{B_2}	0	1	0	0	45
y_{B_3}	0	0	1	0	75
y_{B_4}	0	0	0	1	80
v	0	5	1	2	460

Since both B and B^{-1} are identity matrices, the vector of initial simplex multipliers is equal to the cost vector, i.e., $v = U_B$. Solving (6.10) - (6.16) with the maximal pattern length being restricted to 5 results in the pattern <2,1,2,4,1>. This pattern corresponds to the entering (pivot) column q=(2,2,0,1) with work overload $u_q = 2$ (in station 3) and reduced cost $RC = u_q - vq = -10$. Due to the usual selection rule for the pivot row (choose the row with minimal quotient of the RHS value and the coefficient of the transformed pivot column $B^{-1}q = (2,2,0,1)$), the current basic column (0,1,0,0) leaves the basis. Table 6.6 shows the modified tableau after the respective simplex iteration.

Table 6.6. Simplex Tableau After the First Iteration

	1	2	3	4	RHS
y_{B_1}	1	-1	0	0	55
y_{B_2}	0	0.5	0	0	22.5
y_{B_3}	0	0	1	0	75
y_{B_4}	0	-0.5	0	1	57.5
v	0	0	1	2	235

Note that the tableau always contains the inverse B^{-1} of the current basic matrix and the current values of the simplex multipliers, respectively. Usually, the last row is included in an extended basic matrix, too. Then the negative values of the simplex multipliers have to be entered. The vector $Y_B = (y_{B_1}, y_{B_2}, y_{B_3}, y_{B_4})$ represents the *current* basic variables, in each iteration.

After six further iterations, the final tableau of Table 6.7 is obtained. The optimal LP-solution without work overload contains $y_{B_1} = 32.5$ copies of the pattern <2,3,1,4,1>, $y_{B_2} = 5$ copies of <2,1,4,1>, $y_{B_3} = 7.5$ copies of <2,3,4,1>, and $y_{B_4} =$ 17.5 copies of <3,4,3,4,1>. Since only y_{B_2} is integral, no feasible solution is attained for MSP-W.

Table 6.7. Final Simplex Tableau

	1	2	3	4	RHS
y_{B_1}	1	-0.5	1	-1.5	32.5
y_{B_2}	0	0	-1	1	5
y_{B_3}	-1	1.5	0	0.5	7.5
y_{B_4}	0	-0.5	0	0.5	17.5
v	0	0	0	0	0

Phase 2. Transforming the LP-solution to an Integer Solution

If all variables of the LP-solution are integral, this step is omitted. Otherwise, one of the two following rounding procedures is applied to obtain an integer solution.

Procedure Round-One. In each iteration, the following steps are performed.

Step 1. All variables which already have integer values are fixed to these values.

Step 2. The remaining variables are considered in order of decreasing fractions. Ties are broken in favour of larger variable values. The first variable for which no oversupply (with respect to the values of already fixed variables) occurs for any model by rounding up its value to the nearest integer is fixed to the respective value. If no variable can be rounded up, all variables are rounded down to the nearest (smaller) integer.

Step 3. If all demands are satisfied by the fixed variables, the procedure terminates with an integer solution. Otherwise, a residual problem remains. If no variable was rounded up in step 2, this problem is solved by the branch and bound procedure of Section 6.2.1.2 resulting in one additional pattern, and the procedure stops. When a round up was performed, the LP-relaxation of the residual problem is solved by the column generation procedure of phase 1, and the process continues with step 1.

Procedure Round-All. In each iteration of this procedure, all variables are fixed to an integer value by the following steps.

Step 1. All variables which already have integer values are fixed to these values.

Step 2. The remaining variables get priorities which are computed by weighting their fractions with the lengths of the corresponding patterns. Ties are broken in favour of larger fractions. The variable with the smallest priority is rounded down resulting in undersupplies of some models (with respect to the current values of all variables). To eliminate these undersupplies, the remaining variables are considered in decreasing order of their priorities. Following this order, variables are rounded up to the nearest integer if the resultant increments of the model quantities do not exceed the current undersupplies (i.e., no oversupplies are allowed during the rounding process). Step 2 is repeated for the remaining variables with fractional values until all values are integral.

Step 3. The procedure terminates when no undersupplies remain. Otherwise, a residual problem has to be solved. If this problem is sufficiently small (i.e., the sum of remaining demands is not larger than the maximal pattern length), it is solved by the branch and bound algorithm resulting in one additional pattern, and the procedure terminates. Otherwise, the LP-relaxation of the residual problem is solved by the column generation procedure, and the procedure continues with step 1.

Example. We continue the above example with both rounding procedures. For the sake of simplicity, we number the patterns which are generated during the iterations consecutively. The optimal LP-solution (of the initial problem) contains the basic variables $y_1 = 32.5$, $y_2 = 5$, $y_3 = 7.5$, and $y_4 = 17.5$ (see Table 6.7) which denote the quantities of the patterns $\pi_1 = <2,3,1,4,1>$, $\pi_2 = <2,1,4,1>$, $\pi_3 = <2,3,4,1>$, and $\pi_4 = <3,4,3,4,1>$.

Round-One. The initial LP-solution is transformed to an integral one as follows.

Step 1. The variable y_2 is fixed to 5, i.e., pattern π_2 is selected with quantity 5.

Step 2. Due to the tie breaker, the quantity of pattern π_1 is rounded up to $y_1 = 33$.

Step 3. The residual problem is defined by the demands $d = (24,7,42,42)$. The optimal LP-solution for this problem contains the patterns $\pi_5 = \pi_3$, $\pi_6 = \pi_4$, $\pi_7 = <4,3,4,3>$, and $\pi_8 = <4,3,4,1>$ with quantities $y_5 = 7$, $y_6 = 17$, $y_7 = 0.5$, and $y_8 = 0$.

Step 1. The patterns π_5 and π_6 are fixed with $y_5 = 7$ and $y_6 = 17$. The pattern π_8 is eliminated.

Step 2. The only remaining non-integral variable y_7 cannot be rounded up. Hence, it is rounded down to zero, i.e., the pattern π_7 is eliminated, too.

Step 3. Finally, the residual problem with $d = (0,0,1,1)$ is solved by the branch and bound procedure. The resulting artificial pattern $\pi_9 = <4,3>$ requires a utility work of 1.

Result. The solution includes the five patterns π_1, π_2, π_5, π_6, and π_9 with quantities $y_1 = 33$, $y_2 = 5$, $y_5 = 7$, $y_6 = 17$, and $y_9 = 1$. All patterns but π_9 contain no work overload and are non-artificial. The objective function value is 1.

Round-All. The initial LP-solution is treated as follows.

Step 1. The pattern π_2 is fixed with quantity $y_2 = 5$.

Step 2. The remaining patterns π_1, π_3, and π_4 get the priorities 2.5, 2, and 2.5. Therefore, the quantity of pattern π_3 is rounded down to $y_3 = 7$. This leaves undersupplies of 0.5 for each model, and no other variable can be rounded up. Hence, step 2 is repeated: Pattern π_1 (or π_4) is considered and y_1 is rounded down to 32. Now, the current undersupplies of the four models are (1.5, 1, 1, 1). Pattern π_4 can be rounded up and fixed with quantity $y_4 = 18$, and step 2 is terminated.

Step 3. The residual problem is defined by the demand vector $d = (1,1,0,0)$ and is solved by the branch and bound procedure. The resulting artificial pattern $\pi_5 = <2,1>$ is fixed with quantity $y_5 = 1$.

Result. The solution includes the five patterns π_1, π_2, π_3, π_4, and π_5 with quantities $y_1 = 32$, $y_2 = 5$, $y_3 = 7$, $y_4 = 18$, and $y_5 = 1$. While the first four patterns do not contain any work overload and are non-artificial, the last one is artificial with a work overload of 1. Thus, the objective function value is 1.

Phase 3. Determining a Pattern Sequence

If the solution of phase 2 contains only non-artificial patterns, equivalent MSP-W sequences are generated by arbitrarily ordering the patterns. As mentioned above, it may be advantageous to repeat identical patterns immediately. Furthermore, one may schedule longer patterns first, because the planning certainty may decrease with increasing time horizon.

The Pattern Sequencing Problem (PSP). If artificial patterns are contained in the integer solution of phase 2, a sequencing problem arises. It consists of finding a *pattern sequence* with minimal reducible utility work (see below).

Definition 6.2. The utility work which results from transforming a subsequence into an artificial pattern is called *reducible* (cf. Remark 6.1, p. 214).

The problem is treated by a heuristic improvement procedure with tabu search strategy. For its presentation, we use the notations of Figure 6.3, additionally to those of Figure 6.1 (see p. 205).

P	number of patterns to be sequenced
π_p	pattern, index: $p = 1,...,P$
L_p	length of pattern π_p
y_p	quantity of pattern π_p
Ψ	length of the pattern sequence $(=\Sigma_p y_p)$, index: $\tau = 1,...,\Psi$
β_{kp}	earliest end position of operator in station k after processing pattern π_p
α_{kp}	latest start position for pattern π_p in station k
ω_p	reducible utility work of pattern π_p $(=\Sigma_k \beta_{kp})$
σ_p	reducible idle time of pattern π_p $(=\Sigma_k \alpha_{kp})$

Figure 6.3. Notations for Pattern Sequencing

Lower Bound for the Heuristic Solution. The non-reducible utility work contained within the patterns is excluded from PSP, because it cannot be avoided by the pattern sequence. Thus, the total non-reducible utility work provides a lower bound on the objective function value of the best solution which can be attained with the given patterns. However, it does not represent a lower bound for MSP-W itself.

Earliest End Positions. For each pattern π_p, we calculate the positions of the operators in all stations by starting in the initial state and applying the recursive formulae

(3.9) and (3.10) for $k=1,...,K$ and $i=2,...,L_p+1$ (see p. 99). The terminal state S_{L_p+1} defines earliest end positions of the operators after processing pattern π_p, i.e., $\beta_{kp}=s_{k,L_p+1}$ for $k=1,...,K$. Obviously, these end positions are zero for non-artificial patterns, because they terminate in the initial state. The *reducible utility work* ω_p of pattern π_p is equal to the sum of earliest end positions.

Latest Start Positions. Now, latest start positions α_{kp} are computed so that the earliest end positions can be realized. The sum of latest start positions is symbolized by σ_p and can be interpreted as *reducible idle time*. The latest start positions of a pattern π_p are easily computed in reverse order of the contained units by formulae analogous to (3.10).

TSP-Transformation. PSP may approximately be transformed into a traveling salesman problem by defining distances δ_{pq} between each pair of patterns (π_p,π_q) as follows.

$$\delta_{pq} = \sum_{k=1}^{K} \max\{0, \beta_{kp} - \alpha_{kq}\} \qquad \text{for } p, q = 1,...,P \tag{6.17}$$

The definition of a TSP instance is complemented by introducing a fictitious pattern π_0 with $y_0=1$, $\delta_{0p}=0$ and $\delta_{p0}=\Sigma_k\beta_{kp}$ for all $p=1,...,P$. For each pattern π_p, y_p node copies have to be defined, each of which is connected with all other nodes (including the other copies for the same pattern) by directed arcs.

The distances δ_{pq} provide upper bounds on the (reducible) utility work which is caused by assigning the patterns π_p and π_q to neighbouring sequence positions τ and $\tau+1$. If $\delta_{pq}>0$, the earliest end positions of pattern π_q cannot be realized in all stations, because the earliest end positions of pattern π_p cause violating the latest start positions of π_q in some stations. That is, even if pattern π_p is started in the initial state, the realized end positions of pattern π_q exceed its earliest end positions in at least one station. These position differences represent potential utility work for pattern π_q or patterns in later sequence positions.

Note that the TSP-transformation is no relaxation of the pattern sequencing problem, because the distances represent upper bounds on the utility work of two neighbouring patterns. Therefore, the length of each TSP tour provides an upper bound on the objective function value of a related PSP instance.

Initial Pattern Sequence. An initial solution is obtained by the following greedy procedure which utilizes the TSP-transformation.

The first sequence position is given to a pattern with smallest value of the reducible idle time σ_p, because idle time cannot be utilized for the compensation of utility work at this position. Each further sequence position $\tau=2,...,\Psi$ is filled by assigning a pattern with positive remaining demand which has the smallest distance to the pattern at sequence position $\tau - 1$ (cf. the *nearest-neighbour* heuristic for TSP in Domschke (1997, pp. 110)). Ties are broken in favour of the smallest value of the reducible idle time σ_p.

Improvement Procedures (cf. Section 5.2). A pattern sequence can be transformed into a *neighbouring solution* by exchanging the sequence positions τ_1 and τ_2 of two (copies of) different patterns π_p and π_q. The *moves* are abbreviated by $(\pi_p,\tau_1,\pi_q,\tau_2)$. A similar move definition is used by Okamura and Yamashina (1979) as well as Bard et al. (1994). Without loss of generality, we assume that $\tau_1 < \tau_2$. The objective function value of a transformed solution requires the re-calculation of the operator positions in the partial sequence from τ_1 to Ψ for the underlying original sequence of units. The re-calculation can be omitted up to sequence position τ_2 if the current and the new start positions of the pattern at sequence position $\tau_1 + 1$ are identical or do not violate its latest start positions both. If this exception is right for the pattern at sequence position $\tau_2 + 1$, too, no re-calculation is necessary at all.

Descent Procedures. In each iteration, a move which results in a reduction of the utility work is performed. With respect to the selection of moves we may distinguish two strategies. A *best fit* procedure always chooses a most improving move, whereas a *first fit* procedure performs the first improving move which is found.

Tabu Search. In each iteration, a most improving/least deteriorating move is selected. In order to avoid cycling between already visited solutions, a static tabu search strategy (tabu navigation method) is used (cf. Section 5.2.2.1). We define attributes (π,τ) which denote that a pattern π is set to a sequence position τ. A tabu list of fixed length TL (tabu status duration) temporarily stores the two attributes (π_p,τ_1) and (π_q,τ_2) for each move $(\pi_p,\tau_1,\pi_q,\tau_2)$ performed. These attributes avoid that a copy of pattern π_p (π_q) is set to the sequence position τ_1 (τ_2) during the next TL iterations.

Move Evaluation Strategies. As mentioned above, each move requires the re-calculation of at least a part of the underlying unit sequence. Therefore, evaluating a single move requires up to $K \cdot (I - 1)$ applications of the recursive formulae (3.10). The number of moves to be evaluated in each iteration of the tabu search procedure is of order $O(\Psi^2)$. Hence, $O(K \cdot I \cdot \Psi^2)$ computations have to be performed in each iteration. Even for problems with moderate sizes this approach is usually very expensive in terms of computation time. Hence, one may use first fit strategies or even a simplified move evaluation.

We distinguish two further evaluation strategies besides the *complete move evaluation* discussed above:

- *TSP-Evaluation:* The sequencing problem is transformed to a TSP as described above. In each iteration, a non-tabu move which shows the largest reduction/ least increase in the length of the underlying TSP tour is selected. By this method, the evaluation of a move requires no more than 4 additions and 4 subtractions, because two nodes are exchanged in a TSP tour by eliminating and inserting at most 4 arcs, respectively. Only for the selected move an exact re-calculation of the sequence is required.
- *Pre-Selection:* The above strategy compares moves by estimating the reduction/ increase of the utility work. Therefore, it may select moves which provide a large estimated reduction but really increase the objective function value. Thus, all non-tabu moves which do not increase the length of the TSP tour are additionally exactly calculated. Out of this subset of *promising* moves, the best one is chosen even though a non-promising move may really be better.

Variation of TL. Initially, the tabu list gets a certain *basic length.* If no improvement of the current solution is obtained after a given number of iterations, the *actual length* of the tabu list is increased by 1 in order to *diversify* the search. This is repeated until an improvement is found or a *conflict* occurs, i.e., no admissible move is available. When an improvement is achieved, TL is reset to the basic length. In case of a conflict, the basic length is repeatedly reduced by 1 until at least one move becomes admissible. Then the current basic length is used as actual length of the tabu list, and the process continues.

Stopping Criteria. The tabu search procedure may be stopped by one or more of the following criteria:

- The basic length of the tabu list is already zero, and no move is admissible for the actual length.
- An optimal solution for PSP without reducible utility work is found.
- A given maximal number of iterations has been performed.
- A given limit of the computation time has been exceeded.

Adjusting the Phases of the Hybrid Procedure

The hybrid heuristic procedure described above is mainly influenced by the given *maximal pattern length.* While the computation time required for the column generation process (phases 1 and 2) exponentially grows with an increasing value of this parameter, the complexity of the pattern sequencing problem (phase 3) decreases, and vice versa. Since the computation time is usually restricted, this complexity trade-off influences the solution quality. Furthermore, non-reducible utility work may be caused within the patterns if the pattern length is too large. Additionally, the rounding decisions of phase 2 tend to be more critical with respect to the objective function value than for small patterns. On the other hand, compensation effects within patterns cannot be utilized if the maximal pattern length is too small. For these reasons, the choice of an appropriate maximal pattern length is important for the performance of the hybrid procedure. In our computational experiments, which include MSP-W instances with up to $I=400$ and $M=20$, the best values for this parameter are 4 and 5 (cf. Section 7.3.3). However, the repeated application of the procedure with several values of the maximal pattern length is most promising.

If the maximal pattern length is set to I, the procedure mainly reduces to the branch and bound procedure, utilized for finding the patterns in phase 1. This is especially true when the optimal sequence is an elementary pattern itself. If the parameter is set to 1, the phases 1 and 2 are useless and a tabu search approach to the original MSP-W is present.

7 Computational Experiments

In this chapter, we describe computational experiments performed to evaluate the effectiveness of solution procedures for simple assembly line balancing and mixed-model sequencing.

The main attention is given to existing and new approaches for SALBP-1 and SALBP-2 (cf. Chapters 4 and 5), because most solution procedures for generalized balancing problems, including mixed-model problems, are based on these problem types (see Sections 2.3 and 6.1). In order to examine the effectiveness of such procedures for solving a generalized single-model problem, we include search methods for SALBP-E (cf. Section 4.3). For all problem types, different exact and heuristic procedures are comparatively evaluated on the basis of benchmark data sets which are obtained by collecting problems from the literature (including both real-world and constructed problems). Furthermore, the different UALBP approaches (Section 4.4) are investigated.

With respect to mixed-model sequencing, we perform comparisons of different exact and heuristic approaches for MSP-W (see Section 6.2). The tests are based on randomly generated problem instances, because no benchmark data sets are available in the literature. Furthermore, a random problem generation facilitates considering different types of problem instances in order to recognize specific strengths and limitations of procedures.

The present chapter is organized as follows. In Section 7.1, we describe the computational environment used to perform the tests. Section 7.2 reports on data sets, test settings, and respective results for SALBP and UALBP procedures, while Section 7.3 is devoted to comparing mixed-model sequencing procedures.

7.1 Hardware and Software Environment

All computations described in this chapter are performed on a single IBM-compatible personal computer in order to achieve comparable results. The following specifications of the used computer system influence its computational performance (note that the used product names are registered trademarks):

- Central processing unit: Intel 80486 DX2-66
- Operating system: Microsoft DOS 5.0
- Available memory: 582 kilo bytes of conventional memory in the real mode, 5.2 mega bytes of extended memory in the protected mode, access time: 70 nano seconds

In order to perform fair comparisons of different algorithms, influences of the programming language or the compiler, the way of coding or the data structures have to be eliminated. Therefore, all algorithms contained in the subsequent tests are coded by the author using Borland's Pascal 7.0 (cf. Borland (1992)). Furthermore, the basic data structures required are identical for all compared algorithms.

Assembly Line Balancing Procedures. In order to achieve a fast access, precedence relations (i,j) and task times t_j as well as temporary and solution data are stored in standard array types provided by Pascal. Whenever possible, the real mode is used, because the programs run about two times slower in the protected mode. Due to the two-dimensioned array of size n^2 required for storing the precedence relations, the protected mode must be used for the largest problem with $n = 297$ tasks. Since this is likewise true for all algorithms, no attempt is made to convert corresponding computation times into real mode equivalents.

Mixed-Model Sequencing Procedures. The size of a problem instance is determined by the number M of models, the number K of stations, and the length I of the sequence. The input data consist of an $(M \times K)$-matrix of processing times τ_{mk}, a demand vector $d = (d_1,...,d_M)$, a vector $l = (l_1,...,l_K)$ of station lengths, and the cycle time c (cf. Figure 3.7, p. 97). Solutions are represented by a vector of length I. The storage space requirements are mainly determined by the $(K \times I)$-matrix of operator start positions s_{ki} used for calculating the work overload of a current (partial) sequence. In order to use the real mode only, the problem size is restricted to $I = 400$, $K = 20$, and $M = 20$ in our tests.

7.2 Assembly Line Balancing Procedures

We describe computational experiments performed in order to evaluate exact and heuristic procedures for the different versions of SALBP. All tests are based on a collection of precedence graphs described in Section 7.2.1. In the subsequent sections 7.2.2, 7.2.3, and 7.2.4, procedures for the problem types SALBP-1, SALBP-2, and SALBP-E are considered, respectively. For each problem type one or more data sets are constructed from the collection of precedence graphs. These data sets are used to evaluate a number of exact as well as heuristic procedures with respect to solution quality and computation time requirements. All tests include procedures described in the literature as well as those proposed in the present work. Complete descriptions of all data sets are given in Scholl (1993) and can be downloaded from the web at: *http://www.bwl.tu-darmstadt.de/bwl3/forsch/projekte/alb/index.htm*

7.2.1 Collection of Precedence Graphs

We present precedence graphs for simple assembly line balancing problems which have been assembled from the literature. Some of these precedence graphs have been modified or complemented, respectively. We omitt informations on cycle times and numbers of stations which are added in subsequent sections to obtain problem instances for the different versions of SALBP.

Table 7.1. Characterization of Precedence Graphs

Name	Reference	n	t_{min}	t_{max}	t_{sum}	OS	TV	div	conv
Arcus 1	Arcus (1963 b)	83	233	3691	75707	59.1	15.8	0.73	0.73
Arcus 2	Arcus (1963 b)	111	10	5689	150399	40.4	568.9	0.63	0.63
Bartholdi 1	Bartholdi (1993)	148	3	383	5634	25.8	127.7	0.74	0.72
Bartholdi 2	Bartholdi (1993), modified	148	1	83	4234	25.8	83	0.74	0.72
Bowman	Bowman (1960)	8	3	17	75	75	5.7	0.88	0.8
Buxey	Buxey (1974)	29	1	25	324	50.7	25	0.74	0.78
Gunther	Gunther et al. (1983)	35	1	40	483	59.5	40	0.78	0.76
Hahn	Hahn (1972)	53	40	1775	14026	83.8	44.4	0.63	0.63
Heskiaoff	Heskiaoff (1968)	28	1	108	1024	22.5	108	0.68	0.69
Jackson	Jackson (1956)	11	1	7	46	58.2	7	0.77	0.77
Jaeschke	Jaeschke (1964)	9	1	6	37	83.3	6	0.73	0.73
Kilbrid	Kilbridge and Wester (1962)	45	3	55	552	44.6	18.3	0.67	0.69
Lutz 1	Lutz (1974)	32	100	1400	14140	83.5	14	0.76	0.82
Lutz 2	Lutz (1974), modified	89	1	10	485	77.6	10	0.75	0.75
Lutz 3	Lutz (1974)	89	1	74	1644	77.6	74	0.75	0.75
Mansoor	Mansoor (1964)	11	2	45	185	60	22.5	0.79	0.91
Mertens	Mertens (1967)	7	1	6	29	52.4	6	1	0.78
Mitchell	Mitchell (1957)	21	1	13	105	71	13	0.74	0.7
Mukherjee	Mukherjee and Basu (1964)	94	8	171	4208	44.8	21.4	0.51	0.51
Rosenberg	Rosenberg and Ziegler (1992)	25	1	13	125	71.7	13	0.74	0.69
Sawyer	Sawyer (1970)	30	1	25	324	44.8	25	0.83	0.83
Scholl	Scholl (1993)	297	5	1386	69655	58.2	277.2	0.7	0.69
Tonge	Tonge (1961)	70	1	156	3510	59.4	156	0.78	0.73
Warnecke	Warnecke (1971), modified	58	7	53	1548	59.1	7.6	0.72	0.81
Wee-Mag	Wee and Magazine (1981 b)	75	2	27	1499	22.7	13.5	0.85	0.62

In Table 7.1, we describe the collected precedence graphs by specifying the original reference, the number n of tasks, the minimal and maximal task time as well as the sum of task times, in each case. Furthermore, we provide single-valued complexity measures for the precedence graphs including the order strength OS (in %), the time variability ratio TV as well as the degrees of divergence (div) and convergence (conv), respectively (see p. 39 and p. 42). For computing div and conv, the precedence graphs have to be transformed to single-source and single-sink graphs, respectively.

7.2.2 Comparative Evaluation of SALBP-1 Procedures

First, we present the data sets used. Subsequently, comparative tests for exact and heuristic procedures are described and evaluated.

7.2.2.1 Data Sets for SALBP-1

We describe benchmark data sets, which are used in the literature to evaluate and compare algorithms, and provide optimal objective function values. Furthermore, we develop a new data set involving problem instances which are more challenging than those in former data sets.

Table 7.2. Data Set of Talbot et al.

Name	c	m*	c	m*	c	m*	c	m*	c	m*
Bowman	20	5^+								
Mansoor	48	4	62	3	94	2				
Mertens	6	6^+	7	5	8	5^+	10	3	15	2
	18	2								
Jaeschke	6	8^+	7	7^+	8	6^+	10	4	18	3
Jackson	7	8^+	9	6	10	5	13	4	14	4
	21	3								
Mitchell	14	8	15	8^+	21	5	26	5	35	3
	39	3								
Heskiaoff	138	8	205	5	216	5	256	4	324	4
	342	3								
Sawyer	25	14^+	27	13^+	30	12^+	36	10^+	41	8
	54	7^+	75	5						
Kilbrid	57	10	79	7	92	6	110	6	138	4
	184	3								
Tonge	176	21^+	364	10	410	9	468	8	527	7
Arcus 1	5048	16^+	5853	14^+	6842	12	7571	11^+	8412	10^+
	8898	9	10816	8^+						
Arcus 2	5755	27	8847	18^+	10027	16^+	10743	15^+	11378	14
	17067	9								

Data Set of Talbot et al. This data set ('Talbot data set' for short) contains 12 precedence graphs with 8 to 111 tasks. By adding various cycle times c, Talbot et al. (1986) generated 64 instances. In order to give a compact presentation, the instances for each precedence graph are grouped into five columns in Table 7.2. For each instance, we give the cycle time c and the minimal number of stations m*. If m* is not equal to the simple lower bound $LB_1 = \lceil t_{sum}/c \rceil$, the symbol '+' is added (see pp. 43).

Data Set of Hoffmann. The Talbot data set does not provide a hard test, because most instances are easily solved by common algorithms. Therefore, Hoffmann (1990, 1992) tried to find more strenuous instances by considering cycle times which would have a total idle time close to zero if the lower bound LB_1 could be realized. These instances are expected to have an optimal solution with more than LB_1 stations or only few optimal solutions with that number of stations. Hoffmann used the following cycle times to define instances:

$$c(m) := \lceil t_{sum}/m \rceil \quad \text{for } m = \lfloor t_{sum}/(2 \cdot t_{max}) \rfloor, \ldots, \lfloor t_{sum}/t_{max} \rfloor \tag{7.1}$$

Table 7.3 contains the data set with 50 instances ('Hoffmann data set' for short) which results from applying (7.1) to the larger precedence graphs of the Talbot data set. If m* is currently not available for an instance, the best known values of a lower and an upper bound on m* are provided (separated by a hyphen).

Table 7.3. Data Set of Hoffmann

Name	c	m*	c	m*	c	m*	c	m*	c	m*
Sawyer	27	13$^+$	30	12$^+$	33	11$^+$	36	10$^+$	41	8
	47	7	54	7$^+$						
Kilbrid	56	10	62	9	69	8	79	7	92	6
	111	5								
Tonge	160	23$^+$	168	22$^+$	176	21$^+$	185	20$^+$	195	19$^+$
	207	18$^+$	220	17$^+$	234	16$^+$	251	14	270	14$^+$
	293	13$^+$	320	11						
Arcus 1	3786	21$^+$	3985	20$^+$	4206	19$^+$	4454	18$^+$	4732	17$^+$
	5048	16$^+$	5408	15$^+$	5824	14$^+$	6309	13$^+$	6883	12$^+$
	7571	11$^+$								
Arcus 2	5785	27$^+$	6016	26$^+$	6267	25$^+$	6540	24$^+$	6837	23$^+$
	7162	22$^+$	7520	20-21	7916	20$^+$	8356	19$^+$	8847	18$^+$
	9400	17$^+$	10027	16$^+$	10743	15$^+$	11570	13		

Additionally, Hoffmann included the Talbot data set in his tests of EUREKA (cf. Section 4.1.2.2) and examined the precedence graphs *Tonge* for all cycle times between 156 and 1746 (23 to 3 stations) and *Sawyer* for all cycle times between 25 and 161 (14 to 3 stations). Only a few instances of both precedence graphs (especially those which are contained in the above table) appeared to be hard in terms of computation time.

Data Set of Scholl. Table 7.4 contains a data set with 168 instances which includes precedence graphs not being considered in other data sets so far. The cycle times are determined by (7.1). Contrary to the former data sets, this new data set contains problem instances for which the difference $m^* - LB_1$ is larger than 1. In those cases, the respective difference is specified following the '+' sign.

Table 7.4. Data Set of Scholl

Name	c	m*	c	m*	c	m*	c	m*	c	m*
Rosenberg	14	10^{+}	16	8	18	8^{+}	21	6	25	6^{+}
	32	4								
Buxey	27	13^{+}	30	12^{+}	33	11^{+}	36	10^{+}	41	8
	47	7	54	7^{+}						
Lutz 1	1414	11^{+}	1572	10^{+}	1768	9^{+}	2020	8^{+}	2357	7^{+}
	2828	6^{+}								
Gunther	41	14^{+2}	44	12^{+}	49	11^{+}	54	9	61	9^{+}
	69	8^{+}	81	7^{+}						
Hahn	2004	8^{+}	2338	7^{+}	2806	6^{+}	3507	5^{+}	4676	4^{+}
Warnecke	54	31^{+2}	56	29^{+}	58	29^{+2}	60	27^{+}	62	27^{+2}
	65	25^{+}	68	24^{+}	71	23^{+}	74	22^{+}	78	21^{+}
	82	20^{+}	86	19^{+}	92	17	97	17^{+}	104	15
	111	14								
Wee-Mag	28	63^{+9}	29	63^{+11}	30	62^{+12}	31	62^{+13}	32	61^{+14}
	33	61^{+15}	34	61^{+16}	35	60^{+17}	36	60^{+18}	37	60^{+19}
	38	60^{+20}	39	60^{+21}	40	60^{+22}	41	59^{+22}	42	55^{+19}
	43	50^{+15}	45	38^{+4}	46	34^{+}	47	32-33	49	32^{+}
	50	32^{+2}	52	31^{+2}	54	31^{+3}	56	30^{+3}		
Lutz 2	11	49^{+4}	12	44^{+3}	13	40^{+2}	14	37^{+2}	15	34^{+}
	16	31	17	29	18	28^{+}	19	26	20	25
	21	24								
Lutz 3	75	23^{+}	79	22^{+}	83	21^{+}	87	20^{+}	92	19^{+}
	97	18^{+}	103	17^{+}	110	15	118	14	127	14^{+}
	137	13^{+}	150	12^{+}						
Mukher-jee	176	25^{+}	183	24^{+}	192	23^{+}	201	22^{+}	211	21^{+}
	222	20^{+}	234	19^{+}	248	18^{+}	263	17^{+}	281	16^{+}
	301	15^{+}	324	14^{+}	351	13^{+}				
Barthol-di 1	403	14	434	13	470	12	513	11	564	10
	626	9	705	8	805	7				
Barthol-di 2	84	51	85	50	87	49	89	48	91	47
	93	46	95	45	97	44	99	43	101	42
	104	41	106	40	109	39	112	38	115	37
	118	36	121	35	125	34	129	33	133	32
	137	31	142	30	146	29	152	28	157	27
	163	26	170	25						
Scholl	1394	50	1422	50^{+}	1452	48	1483	47-48	1515	46-47
	1548	46^{+}	1584	44	1620	44^{+}	1659	42-43	1699	41-42
	1742	40	1787	39	1834	38	1883	37	1935	36
	1991	35	2049	34	2111	33	2177	32	2247	31
	2322	30	2402	29	2488	28	2580	27	2680	26
	2787	25								

Combined Data Set. The three data sets with 64, 50, and 168 instances described above build a combined data set with 269 instances, because 13 instances are contained in both the data sets of Talbot et al. and Hoffmann. Currently, optimal solutions are known for 263 instances.

7.2.2.2 Exact Procedures

In the sequel, we report on computational experiments with different versions of the branch and bound procedures FABLE, EUREKA, and SALOME-1 for the data sets defined above. These procedures are described in Section 4.1.2.2 and 4.1.4, respectively. Since SALBP-1 is NP-hard, it is not possible to solve all problem instances to optimality in a reasonable amount of time. Therefore, we use a time limit of 500 cpu seconds, i.e., the computations are terminated when the computation time exceeds this limit before a proven optimum is found. In such *time-out* situations, the quality of the solution achieved depends on the number of stations of the incumbent solution *(upper bound)*. Note that input and output operations are not included in the computation times.

Unidirectional Procedures. With a first experiment, we compare the enumeration schemes as well as the bounding and dominance rules. Therefore, the influence of the planning direction is eliminated, and FABLE (Fab for short) is compared to the forward procedure of EUREKA (EurF) and the basic forward version of SALOME-1 described in Section 4.1.4.1 (SL 1). All algorithms have to observe the time limit of 500 seconds. In the case of a time-out, the computations of FABLE and SALOME-1 are simply stopped, whereas EUREKA additionally applies the heuristic of Hoffmann (1963) in forward direction. This leads to slightly increased computation times (up to 10 seconds but less than 1 second on the average in our tests).

The procedures are applied to the data sets of Section 7.2.2.1. Table 7.5 shows a summary of the results for the combined data set as well as the three included ones. It provides following information:

- # opt: number of instances for which an optimal solution is found and proven
- av.rel: average relative deviation from optimality in %
- max.rel: maximum relative deviation from optimality in %
- av.cpu: average computation time in seconds

Remark 7.1. Relative deviations are computed by $(UB - m^*)/m^*$, where UB is the number of stations in the solution found and m^* is the optimal number of stations. If m^* is not available for an instance, the best known lower bound is used. The average computation times include time-out cases with 500 seconds.

Table 7.5. Comparison of Forward Procedures

Data Set	Combined			Talbot			Hoffmann			Scholl		
Algorithm	Fab	EurF	SL1	Fab	EurF	SL1	Fab	EurF	SL1	Fab	EurF	SL1
# opt.	179	165	191	64	63	64	48	46	48	80	69	92
av.rel	0.95	1.10	0.75	0.00	0.06	0.00	0.25	0.41	0.25	1.44	1.62	1.12
max.rel	7.69	14.29	7.89	0.00	3.70	0.00	7.69	7.69	7.69	7.14	14.29	7.89
av.cpu	175.9	226.4	153.9	0.2	9.0	0.8	48.8	88.1	37.6	267.1	333.4	235.2

The results of Table 7.5 show that Fab clearly outperforms EurF with respect to all provided measures and all data sets. SL 1 which is an improved combination of the first two algorithms clearly shows the best results, especially for our new challenging data set. For the Talbot data set, SL 1 is slower than Fab because of a higher initial effort which is irrelevant for the more difficult problems in the Scholl data set. Note that SL 1 computes modified versions of the lower bound LB_4 and the potential dominances. Since the enumeration method of SL 1 is very similar to that of EurF, its better results are mainly due to the used bounding and dominance rules. This is confirmed by an analysis of the efficiency of these rules which is presented below (see pp. 242). The improvements of SL 1 relative to Fab are mainly influenced by the different enumeration techniques and the modifications mentioned above.

The Extended Version. The extended version of SALOME-1 (SL 2 for short) described in Section 4.1.4.2 is compared to the basic version SL 1. In addition to SL 1, it includes prefixing, the simple permutation rule, and the bounding arguments LB_5, LB_6, and LB_7 to compute the global and the local lower bounds. Prefixing is applied before enumerating loads for a station. The permutation rule is used after the maximum load rule but prior to the Jackson dominance rule, because the first rules are less complex than the latter one. The bounds are applied immediately before branching a node.

Table 7.6. Results of the Basic and the Extended Version of SALOME-1

Data Set	Combined		Talbot et al.		Hoffmann		Scholl	
Algorithm	SL 1	SL 2	SL 1	SL 2	SL 1	SL 2	SL 1	SL 2
# opt	191	199	64	64	48	48	92	100
av.rel	0.75	0.67	0.00	0.00	0.25	0.25	1.12	1.00
max.rel	7.89	7.69	0.00	0.00	7.69	7.69	7.89	6.67
av.cpu	153.9	138.1	0.8	0.5	37.6	39.9	235.2	209.3

The results in Table 7.6 show that further improvements can be obtained by the extended algorithm. In order to facilitate reading, we repeat the results of SL 1 from Table 7.5. SL 2 additionally solves 8 hard problems of the Scholl data set and reduces the execution times by about 10% on average. For the Hoffmann data set, which contains only a few hard instances, the additional components lead to a slight increase of the computation times.

The Planning Direction. In the following, we consider the influence of the planning direction on the different procedures. Therefore, we compare the bidirectional version of SALOME-1 (Bi SL for short, cf. Section 4.1.4.3), the original unidirectional version of FABLE (Fab), and a version of EUREKA (EurFB) which successively performs a forward and a backward planning. In order to observe the time limit, EurFB is modified according to the proposal of Hoffmann (1992): The available computation time is equally split up between the forward and the backward procedure, i.e., both procedures have to observe a time limit of 250 seconds. In order to obtain feasible solutions when both the forward and the backward procedure run out of time, the heuristic is finally applied in forward and backward direction without a time limit. As in the case of the pure forward version EurF, this results in slightly increased computation times (up to 18 seconds in our tests).

Table 7.7. Influence of the Planning Direction

Data Set	Combined			Talbot et al.			Hoffmann			Scholl		
Algorithm	Fab	EurFB	BiSL	Fab	EurFB	BiSL	Fab	EurFB	BiSL	Fab	EurFB	BiSL
# opt.	179	194	224	64	64	64	48	47	48	80	96	125
av.rel	0.95	0.72	0.46	0.00	0.00	0.00	0.25	0.34	0.25	1.44	1.05	0.67
max.rel	7.69	12.00	7.69	0.00	0.00	0.00	7.69	7.69	7.69	7.14	12.00	4.55
av.cpu	175.9	199.1	98.6	0.2	5.3	0.2	48.8	78.4	40.4	267.1	293.9	145.9

The results are summarized in Table 7.7 (those of Fab are repeated from Table 7.5). Whereas the pure forward application of EUREKA performs worse than Fab, the present application in two directions provides clearly better results. EurFB solves more problems of the combined data set to optimality and obtains the smaller average deviations than Fab. However, the maximum deviations are larger, because only the heuristic gives feasible solutions in time-out cases. These results are due to the Scholl data set. For the other data sets (which were the basis of former comparisons, cf. Hoffmann (1992) and Johnson (1993)), both codes perform similarly. Despite of the smaller number of optimal solutions, the average computation times of Fab are much shorter than those of EurFB. Some problems which are solved by Fab in less than 1 second take some hundreds of seconds with EurFB. In these cases, the bounding and dominance rules of Fab are very successful in reducing the enumeration tree. This outcome confirms the results observed by applying the original Fortran implementations of Johnson and Hoffmann (cf. Scholl and Klein (1997 a)).

Bi SL clearly outperforms the other procedures for the Scholl data set for all measures. For the two former data sets, no significant differences occur, because they mainly include easy problems. In comparison to the extended forward procedure SL 2, the bidirectional method shows significant improvements (cf. Table 7.6).

The Problem Structure. In order to examine the influence of the precedence relations and task times on the performance of the procedures, we subdivide the combined data set into subsets which represent particular structures. This is achieved by means of two corresponding complexity measures, the order strength OS and the time variability ratio TV.

Nevertheless, the subsequent argumentation is rather limited, because only single problem characteristics are reflected by such measures (cf. Section 2.2.1.5). For example, the number of tasks is an insufficient complexity measure. While all instances of the precedence graph *Bartholdi 1* are solved within seconds by all procedures, most instances of *Bartholdi 2,* which has the same number of tasks and the same precedence relations, are rather hard (cf. Table 7.1, p. 233). Furthermore, their time variability ratios are very similar. That means, none of the simple measures contained in Table 7.1 is able to explain these different complexities. More versatile measures, like those based on stages, have the difficulty of not being operational for a comparative evaluation.

Notwithstanding these difficulties, some interdependencies between problem characteristics and the behaviour of algorithms can be detected, especially by considering a mix of instances with similar values of a particular complexity measure.

Order Strength. The precedence graphs contained in the combined data set have order strength values between 22.5 and 83.8 (cf. Table 7.1). We define three classes of instances with OS $\leq$ 45, 45 < OS $\leq$ 65, and OS > 65, respectively (OS in %). Note that all medium and large scale instances ($\geq$ 30 tasks) in the data sets of Talbot et al. and Hoffmann fall into the restricted interval [40,55] (cf. Johnson (1993)). The results of the procedures Fab, EurFB, and Bi SL are shown in Table 7.8. Since the classes contain different numbers of instances, the percentage of proven optimal solutions (# opt (%)) is specified instead of their number. Furthermore, the average relative deviations (av.rel) are provided.

Table 7.8. Influence of the Order Strength

Instance class		Fab	EurFB	Bi SL
OS $\leq$ 45	# opt (%)	67.54	64.91	80.70
(114 instances)	av.rel	0.94	0.53	0.51
45 < OS $\leq$ 65	# opt (%)	59.80	74.51	80.58
(103 instances)	av.rel	1.10	0.98	0.60
OS > 65	# opt (%)	80.39	86.27	94.23
(52 instances)	av.rel	0.69	0.64	0.10

Bi SL outperforms the other procedures with respect to all three classes. Mainly for the problems with high order strength, it provides very promising results, because it utilizes the precedence constraints in a profitable manner, especially by prefixing and the bidirectional branching method.

All procedures provide the worst solutions for the class with medium order strength. This shows the trade-off between different influences of the precedence relations on the complexity of problem instances (cf. the discussion in Section 2.2.1.5). While they complicate finding station loads with low idle times, they restrict the number of feasible solutions simultaneously. Due to the latter aspect, all procedures obtain a large percentage of proven optimal solutions for instances with a high order strength.

EurFB shows a discrepancy between the percentage of proven optimal solutions and the average deviations. While it obtains better deviation values for smaller values of OS, it proves more solutions to be optimal for larger values. This is due to the application of the heuristic (cf. Section 5.1.2) when the forward and backward enumeration procedures fail. This greedy heuristic performs best if SALBP-1 approaches the bin packing problem (OS = 0), i.e., the precedence constraints do not prevent from finding station loads with small idle times.

Fab is outperformed by the two other procedures for all classes, because it often enumerates unnecessary parts of the enumeration tree (cf. Section 4.1.2.2, pp. 121). Like Bi SL, Fab obtains its best solutions for instances with high order strength. This finding is opposite to that of Johnson (1988) who has concluded that Fab performs well for low order strength.

Time Variability Ratio. The combined data set contains instances with their TV values ranging from 5.7 to 568.9 (cf. Table 7.1). The results for three instance classes with small, medium and large TV values are given in Table 7.9.

Table 7.9. Influence of the Time Variability Ratio

Instance class		Fab	EurFB	Bi SL
TV ≤ 15	# opt (%)	59.77	57.47	88.51
(87 instances)	av.rel	0.92	0.90	0.26
15 < TV ≤ 80	# opt (%)	92.68	97.56	100.00
(82 instances)	av.rel	0.27	0.05	0.00
TV > 80	# opt (%)	51.00	64.00	65.00
(100 instances)	av.rel	1.53	1.10	1.02

All procedures provide the best results for medium TV values, whereas the instances with small and large values, respectively, are much harder. If TV is small, the procedures have difficulties to find station loads with a low idle time, because all task times show similar values and/or small times are not available (cf. the discussion of TV in Section 2.2.1.5, p. 39). For such instances, dominance rules like the Jackson dominance rule are important. Furthermore, it seems to be necessary to spend more time for searching "good" loads rather than branching inferior ones. Since EurFB does not use any dominance rule and Fab does not attempt to consider promising loads first, Bi SL clearly dominates both other approaches.

In case of large TV values, all procedures show the worst results. This is mainly due to some very large tasks contained in respective instances which can hardly be combined with other tasks. Furthermore, this class contains the largest instances of the data set.

Analysis of the Efficiency of SALOME-1. We try to examine the influence of the various components of SALOME-1 on its good performance based on the 269 instances of the combined data set.

Global Lower Bound. At first, we examine the profits and the efforts of the bounds LB_2, LB_3, LB_4', LB_5, LB_6, and LB_7 when applied in the root node to improve the initial bound (cf. Section 2.2.2.1). Obviously, such an improvement is only possible if the minimum number of stations m* exceeds the trivial bound LB_1. This is true for 139 instances of the combined data set, while for 117 instances $m^* = LB_1$ holds (for 13 further instances it is not clear if m* is equal to or larger than LB_1). Hence, the examination is restricted to the first class with 139 instances, for 76 of which the initial bound can be increased to the optimal value m* by one or more of the bound arguments. Table 7.10 compares the efficiency of the bound arguments LB_i ($i=2,3,4',5,6,7$) for the 139 instances with respect to:

- av.inc: average relative increase of the simple bound LB_1 by LB_i in % (relative increase: $(LB_i - LB_1)/LB_1 \cdot 100\,\%$)
- # max: number of instances for which $LB_i = \max\{LB_k \mid k = 2,...,7\}$
- # best: number of instances for which LB_i is better than all other bounds (in brackets: LB_i is better than all other bounds if LB_4' is excluded)
- tot.cpu: total computation time for the whole data set (all 269 instances)

Table 7.10. Profits and Computation Times of Bound Arguments

Bound	LB_2	LB_3	LB_4'	LB_5	LB_6	LB_7
av.inc	5.09	1.20	8.60	1.20	5.53	5.16
# max	19	8	128	76	82	80
# best	0 (0)	0 (0)	32 (-)	0 (15)	5 (6)	3 (3)
tot.cpu	0.05	0.05	37.42	0.44	0.05	0.16

Since LB_4' provides the best bound values for many instances, its average computation time of 0.14 seconds per instance (at most 0.93 seconds) is justified. LB_2 and LB_3 result in good bound values in some special cases only. Besides, LB_2 is contained in the computation of LB_4' and LB_6, LB_3 in that of LB_4' (cf. Remark 2.2, p. 47). Hence, it is not necessary to compute them when LB_4' or LB_6 are present. Though LB_5 does not result in large improvements of the bound on the average, it shows the best values for 15 instances when LB_4' is absent. Since LB_4' is only applied in the root node, LB_5 which can easily be determined is useful in all other nodes. LB_6 and LB_7 provide the best bound values only in a few cases but take negligible computation times.

Due to the above computation times in the root node, it is justified that all bounds except LB_4' are applied to determine *local lower bounds* in each other node of the enumeration tree.

Fathoming and Dominance Rules. In order to examine the efficiency of various methods for reducing the size of the enumeration tree, the numbers of nodes, which are either branched or eliminated by one of the bounding and dominance rules, are counted. Table 7.11 shows a corresponding comparison of the algorithms SL 1, SL 2 as well as Bi SL. The first column contains the average number of generated nodes per instance. The further columns show the average portions of nodes fathomed by the maximum load rule (Maxload for short), the simple permutation rule (Perm), the Jackson dominance rule (Jackson), or by the application of local bound arguments (Bound) as well as the part of nodes which have to be branched.

Table 7.11. Profits of Fathoming and Dominance Rules

	Av. no. of nodes	Relative number of nodes (%)				
		Maxload	Perm	Jackson	Bound	Branched
SL 1	159,496	20.53	-	44.27	3.87	31.32
SL 2	82,354	31.43	8.07	22.33	11.58	26.59
Bi SL	23,133	5.64	4.42	27.05	21.59	41.31

Remark 7.2. The rules are applied in the given order. In case of SL 1, the used local bounds are LB_2 and LB_3. The other algorithms contain LB_3, LB_5, LB_6 and LB_7 (see above).

We observe that the more versatile algorithms consider significantly less nodes than the basic method SL 1. That means, SL 2 and Bi SL spend more time for searching suitable loads instead of branching disadvantageous ones. This is mainly due to prefixing and the flexible planning direction. The significant reduction of generated nodes by Bi SL shows that especially the bidirectional decision rule, which chooses the planning direction dynamically, is very successful in reducing the number of unnecessary branching operations.

Furthermore, Table 7.11 shows that the Jackson dominance rule as well as the maximum load rule are the most important ones. In case of SL 1, both rules together are able to fathom about 65 % of all nodes. For SL 2 and Bi SL, the influence of the Jackson dominance rule declines due to prefixing, because no prefixed task can be dominated by any not prefixed one. Note that this fact immediately follows from the definition of the potential dominance (cf. Definition 4.5', p. 138). Since a large quantity of (inferior) nodes built by SL 1 is avoided by SL 2, the relative number of nodes fathomed by the maximum load rule and the local lower bounds increases. In the latter case this is, however, mainly due to the additional bound arguments LB_5 to LB_7. The simple permutation rule makes only a small contribution in both algorithms, but its application is very simple. In Bi SL, the influence of the maximum load rule is significantly reduced, because the bidirectional branching rule tends to choose a station having only a few or large assignable tasks. The same is true for the permutation rule. Consequently, a larger part of the nodes passes through the remaining rules resulting in their larger relative success. Just so, a larger part of the nodes remains for branching.

Local Lower Bounds. A further measure for evaluating the different algorithms is the relative number of branched nodes in which the local lower bound has to be increased by the enumeration procedure (when all loads feasible for the old bound have been examined without success). The respective values for SL 1, SL 2, and Bi SL are 0.411 %, 0.278 %, and 0.007 %. This shows that the bidirectional procedure is very successful in avoiding unnecessary enlargements of the idle time, i.e., the size of the enumeration tree is often almost as small as possible.

Forward and Backward Steps. Another interesting characteristic is the relative number of forward and backward steps of Bi SL. For the combined data set, almost 75 % of the steps are forward ones indicating that the forward direction is better suited to this set of problems. In case of the Talbot and Hoffmann data sets, 90.1 % and 99.7 % are forward steps, respectively, showing that the contained problems have special properties which are extremely advantageous for forward algorithms. This is impressively underlined by applying FABLE to the reverse problems, because it only solves 58 and 22 of the instances within the time limit, whereas it finds 64 and 48 optimal solutions in forward direction, respectively (cf. Table 7.5, p. 238). For the Scholl data set, Bi SL performs about 72 % forward steps. Though this implies a slight dominance of forward planning, FABLE solves 110 instead of 80 instances in backward direction. These results indicate that a bidirectional version of FABLE should be significantly more successful than the present unidirectional procedure.

The Tree Dominance Rule and Dynamic Renumbering. As recent results show, applying the *tree dominance rule* (see p. 129) and a *dynamic renumbering* of tasks considerably improves the performance of SALOME-1 (cf. Scholl and Klein (1999)). The renumbering is called dynamic, because it is applied in each node of the enumeration tree instead of being performed only once in the root node. By doing so, it is possible to consider properties of the respective partial solution when numbering the tasks (cf. the dynamic priority rules in Section 5.1.1.3). A detailed description of the dynamic renumbering scheme is given in Scholl and Klein (1999). The improved bidirectional version of SALOME-1 which additionally includes the two above-mentioned components is abbreviated by Bi STR.

Since the tree dominance rule is the key component of the procedure OPTPACK of Nourie and Venta (1991), we give in Table 7.12 additional results not only for Bi STR but also for OPTPACK (using the original Pascal implementation of Nourie and Venta).

Table 7.12. Results of Improved SALOME-1 and OPTPACK

Data Set	Combined		Talbot et al.		Hoffmann		Scholl	
Algorithm	BiSTR	OPTPACK	BiSTR	OPTPACK	BiSTR	OPTPACK	BiSTR	OPTPACK
# opt	245	172	64	64	48	48	146	73
av.rel	0.22	0.60	0.00	0.00	0.25	0.25	0.28	0.88
max.rel	7.69	7.69	0.00	0.00	7.69	7.69	3.23	5.88
av.cpu	56.1	188.8	0.20	0.11	23.3	23.8	82.9	295.2

Compared to the results of Bi SL in Table 7.7 it becomes obvious that Bi STR provides further significant improvements concerning each measure. For the challenging Scholl data set, 21 additional instances are solved to optimality within the given time limit of 500 seconds and the average deviation from optimality is reduced from 0.67% to 0.28%. These improvements are due to both, the tree dominance rule and the dynamic renumbering (see Scholl and Klein (1999)).

OPTPACK gets better average and maximal deviations than EUREKA and FABLE but cannot prove as many optimal solutions as the latter are able to (cf. Table 7.7). The satisfying deviation values are obtained owing to the heuristics OPTPACK initially applies to each instance. Though the tree dominance rule significantly reduces the size of the enumeration tree (as the results of Bi STR document), the lack of further dominance rules and versatile bounds is responsible for not being able to find the optimum or to prove it within the computation time available.

Summary of the Results. The results of this section can be summarized as follows. Both the new enumeration technique (bidirectional Local Lower Bound Method) and the application of various bounding and dominance rules are powerful and make considerable contributions to the good performance of SALOME-1. Especially, the bidirectional strategy, the tree dominance rule and the dynamic renumbering are very successful.

The better results of EUREKA compared to FABLE are only due to the fact that it uses both planning directions and a heuristic. The comparison of the forward versions of these algorithms (cf. Table 7.5, p. 238) indicates that the renunciation of dominance and bounding criteria by EUREKA is not justified at all (especially that of the maximum load rule) though the enumeration technique of the Lower Bound Method is clever. FABLE contains versatile fathoming techniques which make it possible to solve a large number of problem instances very efficiently. Though OPTPACK sometimes performs better than FABLE and EUREKA, it is often not able to find and prove the optimum within a reasonable amount of computation time due to its lack of manifold dominance and bounding criteria.

The findings outlined above may contribute to the recent discussion between Hoffmann (1992, 1993), Johnson (1993) as well as Nourie and Venta (1996) in a clarifying manner.

The data sets of Talbot et al. and Hoffmann can no longer serve as benchmark data sets for SALBP-1, because most of the problem instances contained are easy due to their particular precedence structures. The new Scholl data set is much more challenging and contains very different problem structures (cf. Scholl (1993)).

7.2.2.3 Priority Based Heuristics

In the sequel, we report on computational experiences with priority based procedures for SALBP-1. Different ways of assigning tasks as well as several priority rules and planning directions are considered (cf. Section 5.1.1).

In all procedures, initial values of the upper bound on the number of stations are obtained by computing theoretical upper bounds as described in Section 2.2.2.2. Furthermore, LB_1 and LB_2 are used for computing initial lower bounds (see pp. 43).

Taskoriented versus Stationoriented Procedures. The first test is concerned with the comparison of the taskoriented and stationoriented planning strategies (cf. Section 5.1.1.2). For each method, we explore all rules of Figure 5.2 (see p. 181) and take MaxIF as tie breaking rule in all cases except MaxF and MaxIF. For the latter two rules, MaxTime serves as tie breaker. Second-level ties are broken by MinTaskNo.

The results for the combined data set with 269 instances are summarized in Table 7.13 and Table 7.14. The following information is provided:

- # f.opt: number of instances for which an optimal solution is found but not necessarily proven

- av.rel: average relative deviation from optimality in %
- max.rel: maximum relative deviation from optimality in %
- av.abs: average absolute deviation from optimality (additional stations)
- max.abs: maximum absolute deviation from optimality (additional stations)

Computation times are not provided, because they are less than 0.3 cpu seconds in each case and do not considerably differ between stationoriented and taskoriented methods. Nevertheless, different priority rules have different time and space complexities (cf. Hackman et al. (1989)).

Table 7.13 shows that the *successor related* rules MaxPW, MaxF, MaxCumPW, and MaxIF provide the best results in taskoriented procedures, while those of MinE and MinTaskNo are rather poor (see Figure 5.2 for a description of the rules). The latter rule serves as a reference rule, because it only depends on the initial numbering of the tasks and does not try to utilize the structure of the problem. The best rules obtain average relative deviations which are about a half of the average deviation of the reference rule. Our results essentially confirm the findings of Hackman et al. (1989) who have concluded a similar ranking for a subset of the rules considered here.

Table 7.13. Taskoriented Procedures

Rule	Max Time	Max PW	Max F	Max IF	Min E	Min L	Min Slack	Max Avg PW	Min Avg L	Max Time L	Max F Slack	Max Time Slack	Max A	Max Cum A	Max Cum PW	Min Task No
# f.opt	103	132	129	114	75	109	96	105	114	104	116	108	107	110	129	72
av.rel	4.92	3.97	3.83	4.41	7.87	4.94	5.36	5.05	4.84	4.60	4.98	4.57	5.19	5.20	4.00	7.32
max.rel	33.3	50	33.3	33.3	50	50	50	33.3	50	33.3	50	33.3	50	50	50	50
av.abs	0.93	0.71	0.70	0.87	1.65	0.88	1.01	0.99	0.87	0.90	0.91	0.90	0.92	0.91	0.71	1.38
max.abs	4	6	6	4	7	6	7	4	6	4	7	5	6	6	6	6

Table 7.14. Stationoriented Procedures

Rule	Max Time	Max PW	Max F	Max IF	Min E	Min L	Min Slack	Max Avg PW	Min Avg L	Max Time L	Max F Slack	Max Time Slack	Max A	Max Cum A	Max Cum PW	Min Task No
# f.opt	127	132	129	124	75	110	99	101	114	138	116	134	107	110	129	72
av.rel	4.16	3.97	3.83	4.09	7.97	4.92	5.24	5.33	4.84	3.59	4.95	3.74	5.19	5.20	4.00	7.32
max.rel	33.3	50	33.3	33.3	50	50	50	33.3	50	33.3	50	33.3	50	50	50	50
av.abs	0.82	0.71	0.70	0.79	1.67	0.88	1.01	1.07	0.87	0.68	0.90	0.71	0.92	0.91	0.71	1.38
max.abs	4	6	6	5	7	6	7	7	6	3	7	3	6	6	6	6

The best rules within the stationoriented procedure are MaxTimeL and the new rule MaxTimeSlack, followed by MaxPW, MaxF, MaxCumPW, MaxTime, and MaxIF (cf. Table 7.14). This ranking of rules is similar to that found by Talbot et al. (1986) who considered a subset of the rules used here and a stationoriented procedure. Their results implied that *task time related* priority rules (MaxTimeL, MaxTime) are superior to successor related ones (see above).

By performing a rule-by-rule comparison of the stationoriented and taskoriented procedures, we observe that the results are identical for the rules MaxPW, MaxF, MinAvgL, MaxA, MaxCumA, MaxCumPW, and MinTaskNo. This is due to the strong monotonicity of those rules (cf. Definition 5.1 and Figure 5.2, pp. 181), because this property leads to identical solutions for both planning strategies. The results of the monotonous rules MinE and MinL, respectively, are slightly different, because the first-level tie-breaker MaxTime is not monotonous.

As mentioned above, the two planning strategies lead to different rankings of rules. The best rules for the taskoriented procedure are successor related ones. Except MaxIF, those (strongly monotonous) rules obtain the same results within the stationoriented procedure. But in the latter case, the task time related rules perform remarkably better. Applying MaxTimeL or MaxTimeSlack within the stationoriented procedure gives the overall best results (see, especially, the values for the average relative deviation (av.rel) and the maximal absolute deviation (max.abs) in Table 7.14). With respect to the remaining rules, the results of both strategies are quite similar. Only in case of MaxAvgPW, the taskoriented procedure is slightly better.

As a summary of the results, we conclude that the stationoriented procedure should be preferred to the taskoriented one. For further comparisons underlining the dominance of stationoriented procedures we refer to Scholl and Voß (1994).

Unidirectional and Bidirectional Procedures. In the sequel, we report on computational experiments with priority based methods utilizing different planning directions. Where computation times are given, they refer to the times excluding input and output operations as well as the computation of the transitive closure (cf. Definition 2.2, p. 23). According to our above results, we choose the stationoriented planning strategy and the four static rules MaxTimeL, MaxTimeSlack, MaxF, and MaxPW. These rules are applied in a forward and a backward procedure, respectively. Furthermore, these and the rule MaxTSAvail are used as dynamic rules in a bidirectional procedure (cf. Section 5.1.1.3). Ties are broken in the same way as described above.

Table 7.15 shows a summary of the results obtained by applying these 13 rules to the combined data set with 269 instances (the first four columns are repeated from

Table 7.14). Besides the information contained in the former tables, it provides the following measures:

- # best: number of instances for which the respective rule provides the best solution found by any of the rules applied in the test
- av.cpu: average computation time per instance (in seconds)
- max.cpu: maximal computation time per instance (in seconds)

With respect to the average relative deviation from optimality (av.rel) and the number of found optimal solutions (# f.opt), the best rules are Bi-MaxTSAvail, F-MaxTimeL, Bi-MaxTimeSlack, Bi-MaxPW and Bi-MaxF (F = forward, B = backward, Bi = bidirectional). In all cases but MaxTimeL, the bidirectional rules outperform the respective unidirectional ones. In case of MaxTimeL, the forward procedure performs best, but B-MaxTimeL is the worst rule in the entire test indicating that MaxTimeL is sensitive with respect to data changes. Since the average deviation of the bidirectional version of this rule is better than the mean value of the deviations of both unidirectional rules, it may generally be the better choice.

Table 7.15. Different Planning Directions

	Forward				Backward				Bidirectional				
Rule	MaxF	Max PW	Max TimeL	Max Time Slack	MaxF	Max PW	Max TimeL	Max Time Slack	MaxF	Max PW	Max TimeL	Max Time Slack	Max TS Avail
# f.opt	129	132	138	134	131	124	120	121	134	134	132	136	138
# best	201	199	204	198	207	196	170	173	213	204	190	194	197
av.rel	3.83	3.97	3.59	3.74	3.71	3.77	4.27	4.12	3.65	3.63	3.80	3.53	3.21
max.rel	33.3	50	33.3	33.3	33.3	33.3	33.3	33.3	50	33.3	33.3	25	25
av.abs	0.70	0.71	0.68	0.71	0.67	0.71	0.88	0.87	0.66	0.71	0.79	0.75	0.70
max.abs	6	6	3	3	4	4	5	4	5	6	4	4	3
av.cpu	0.03	0.03	0.03	0.03	0.02	0.04	0.04	0.03	0.16	0.15	0.13	0.14	0.15
max.cpu	0.28	0.28	0.22	0.28	0.27	0.27	0.33	0.28	0.88	0.88	0.88	0.83	0.83

The average deviations in the numbers of stations (av.abs) are almost equal for most of the rules, only the backward versions of MaxTimeL and MaxTimeSlack show significantly larger values. The maximal absolute deviations (max.abs) are noteworthy, because three rules (F-MaxF, F-MaxPW, and Bi-MaxPW) require six additional stations in the worst case, while the best rules take no more than three unnecessary stations. With respect to the number of best solutions, the best rules are based on MaxF. Considering this measure, it becomes apparent that very often several rules obtain solutions with identical objective function values.

The results are summarized as follows. Forward rules may outperform backward rules, and vice versa, showing their data dependency, because backward procedures act as forward procedures on reverse precedence graphs. This disadvantage is removed by bidirectional rules which are able to choose the planning direction by dynamically evaluating the structure of the precedence graph.

The computation times of bidirectional procedures are much larger than those of unidirectional ones, because the dynamic priorities must permanently be updated after assigning tasks. An average factor of five holds. However, this additional effort seems to be negligible, because all times are smaller than 1 cpu second. In the case of unidirectional procedures, the application of dynamic rules is not profitable (cf. Scholl and Voß (1996)).

Summary of the Results. The tests show that priority based methods with suitable priority rules are able to compute solutions which are considerably better than solutions derived from an arbitrary task numbering in a very small amount of computation time. Furthermore, we can conclude that stationoriented approaches are more promising than taskoriented ones. Especially together with task time related rules like MaxTimeL and the new rule MaxTimeSlack, the stationoriented procedure performs well. Further improvements can be achieved by bidirectional planning including the new dynamic rule MaxTSAvail.

7.2.2.4 The Tabu Search Approach for SALBP-1

In the sequel, we report on numerical results with our indirect tabu search approach to SALBP-1 *(dual strategy)*. This approach is compared and combined with different exact procedures in order to find out which type of algorithm provides the best results for several limitations of the available computation time.

As described in Section 5.2.4, the dual strategy applies the tabu search procedure for SALBP-2 within the framework of a Lower Bound Method. It starts with the lower bound $LB = \max\{LB_1, LB_2\}$. For each trial station number, a priority based procedure for SALBP-2 with Bi-MaxF as priority rule and FBS as search method is used to compute an initial solution for the tabu search procedure (cf. Section 5.1.3). The tabu search procedure itself includes the global aspiration criterion, the conflict management strategy C3, the list size modification scheme DTL, and the moving gap strategy (cf. Section 5.2.3). The chosen specification of the contained procedures reflects the results of Sections 7.2.3.3. and 7.2.3.4, where this combination is shown to be effective.

Two different methods are used to obtain initial feasible solutions (for SALBP-1) before the search process of the dual strategy starts.

- PrioTabu: An initial solution is determined by a stationoriented priority based method for SALBP-1 with the rule Bi-MaxF and tie breakers MaxTime and MinTaskNo (this rule is used, because it is already applied within the search process).

- EurTabu: As recommended by Hoffmann (1992), the algorithm EUREKA is applied with a time limit of 3 seconds for the forward and backward enumeration procedure, respectively, before the contained heuristic of Hoffmann (1963) is applied in either direction, if necessary.

These two versions of our tabu search approach for SALBP-1 are compared with the exact algorithms FABLE, EUREKA, and SALOME-1. These reference procedures are selected, because no efficient heuristic improvement procedure is known for SALBP-1. Furthermore, each exact algorithm is reduced to a heuristic when it is terminated by a time limit before a proven optimum is found. Therefore, we perform a comparison of the three (truncated) branch and bound procedures with the new tabu search approach for different time limits in order to examine which procedures should be used if a certain amount of computation time is available. The following limits are used: LIMIT = 50, 100, 250, and 500 seconds.

When the time limit is exceeded, the station number of the best known solution provides an upper bound. The branch and bound procedures are applied as described in Section 7.2.2.2 (cf. Fab, EurFB, and Bi SL). In the tabu search procedures, a total of 100 · LIMIT iterations is performed for each trial station number m, before m is increased by 1.

Table 7.16. Comparison of Branch and Bound Procedures with Tabu Search Procedures

Method	FABLE	EUREKA	SALOME-1	PrioTabu	EurTabu
LIMIT = 50					
# f.opt	193	212	217	191	212
av.rel	1.10	0.81	0.61	1.14	0.67
max.rel	8.82	12.00	7.69	14.29	7.69
av.cpu	19.7	23.5	13.0	12.4	10.8
LIMIT = 100					
# f.opt	195	215	219	197	213
av.rel	1.07	0.76	0.58	0.94	0.65
max.rel	8.82	12.00	7.69	14.29	7.69
av.cpu	38.2	44.4	23.7	22.5	16.7
LIMIT = 250					
# f.opt	199	215	221	201	215
av.rel	0.98	0.76	0.54	0.88	0.62
max.rel	7.69	12.00	7.69	14.29	7.69
av.cpu	91.5	103.6	54.2	52.5	34.2
LIMIT = 500					
# f.opt	199	218	227	200	214
av.rel	0.95	0.72	0.46	0.86	0.63
max.rel	7.69	12.00	7.69	7.69	7.69
av.cpu	175.9	199.1	98.6	101.8	62.6

Table 7.16 summarizes the results of the different procedures for the time limits used. It provides some of the measures contained in previous tables. Note that the number of optimal solutions which are *found* but are not necessarily proven is provided even for the branch and bound procedures (in contrast to Table 7.7, p. 239).

The results show that SALOME-1 is still the best procedure with increasing advantage relative to the other approaches for larger time limits. In contrast to FABLE and EUREKA, it is able to prove the optimality of almost all optimal solutions found (compare Table 7.7 and Table 7.16 for LIMIT = 500). In case of EUREKA, this outcome shows that the contained heuristic is rather powerful. This is underlined by the fact that EUREKA provides relatively good solutions for small time limits where the enumeration procedure is less relevant. The initial solution of EurTabu, which is obtained by a short-time application of EUREKA, has an average relative deviation of 0.95 % which is reached by FABLE only for LIMIT = 500.

Both tabu search approaches obtain considerable improvements of the initial solutions for SALBP-1 by the iterative application of the tabu search procedure for SALBP-2. Even for the time limit of 50 seconds, the average relative deviation is reduced from 3.65 % (cf. the column Bi-MaxF in Table 7.15) to 1.14 % with PrioTabu. EurTabu reduces its initial average relative deviation (0.95 %) to 0.67 % (for LIMIT = 50) or 0.62 % (for LIMIT = 250), respectively. Considering the average number of iterations (trial station numbers) of the dual strategy within both procedures PrioTabu and EurTabu, the decision to take a Lower Bound Method is reasonable, because in most cases only one iteration has to be performed (less than 1.5 on average).

While PrioTabu only outperforms FABLE, EurTabu additionally provides better average deviations than EUREKA for all time limits taking roughly a third of the computation time. However, EUREKA finds more optimal solutions. The improvements of PrioTabu for enlarged time limits are considerable, while those of EurTabu are rather small. Nevertheless, PrioTabu does not achieve solutions as good as those of EurTabu for LIMIT = 50 even if it gets a tenfold time limit. Surprisingly, the results of EurTabu for LIMIT = 500 are even worse than those for LIMIT = 250. This is due to the different numbers of iterations performed within the tabu search procedure together with the list size modification strategy DTL.

The results of EurTab indicate that it may be profitable to combine a truncated branch and bound procedure with a tabu search approach. Therefore, we consider a combination of SALOME-1 and tabu search. For LIMIT = 500, two procedures are applied. In the first, SALOME-1 gets a time limit of 100 seconds, while the subsequent tabu search approach gets 400 seconds. The second procedure restricts the computation time of both parts to 250 seconds. The initial solutions obtained by

SALOME-1 have average relative deviations of 0.58 % and 0.54 %, respectively (cf. Table 7.16 for LIMIT = 100, 250). The final solutions show deviations of 0.50 % and 0.48 %, respectively. Even though tabu search gains a reduction in both cases, the results of the pure application of SALOME-1 for LIMIT=500 are still better (0.46 %). The same is true for the number of found optimal solutions (221 and 223 for the combined approaches versus 227 for SALOME-1). Only the computation times are reduced by the combined procedures (61.5 sec and 76.5 sec versus 98.6 sec). Nevertheless, both combined approaches obtain better solutions than SALOME-1 for 5 instances, while the latter gets better solutions for 7 instances.

Summary of the Results. The new branch and bound procedure SALOME-1 performs best even if only short computation times are available. The tabu search approach based on the dual strategy is competitive with truncated branch and bound procedures, especially in combination with such a procedure. As reported by Scholl and Voß (1996, appendix), tabu search is well suited to problem instances with high order strength and small time variability ratio, like SALOME-1. In the first case, tabu search provides an inherent diversification possibility, based on the conflict management strategy C3. In the latter case, the tendency of having tasks with similar operation times increases the possibilities of exchanging tasks.

7.2.3 Comparative Evaluation of SALBP-2 Procedures

We define new data sets for SALBP-2 and perform tests in order to examine different exact and heuristic procedures with respect to their computational efficiency.

7.2.3.1 Data Sets for SALBP-2

In the literature, no benchmark data set is available for SALBP-2. We present two new data sets consisting of precedence graphs which are contained in data sets for SALBP-1. For each graph, a range $[\underline{m},\overline{m}]$ of station numbers is defined resulting in $\overline{m}-\underline{m}+1$ different instances. Generally, the ranges are restricted by $\underline{m} \geq 3$ and $\overline{m} \leq \min\{\lceil t_{sum}/t_{max}\rceil + 2, \lfloor n/2 \rfloor\}$. Additionally, some easy instances are eliminated, and the number of instances is limited to 28 per precedence graph in order to avoid one problem structure dominating the whole data set. Therefore, in some cases only a subrange of $[\underline{m},\overline{m}]$ is used (cf. Scholl (1993)).

Data Set 1. The first data set for SALBP-2 includes all problems with $n \geq 30$ tasks of the data sets of Talbot et al. and Hoffmann for SALBP-1 (cf. Section 7.2.2.1). Additionally, it contains the precedence graphs *Buxey, Gunther, Lutz 1,* and *Lutz 2* (cf. Table 7.1, p. 233) resulting in a total of 128 problem instances.

For each precedence graph, the different instances are grouped into four columns. In each case, the minimum cycle time c* is specified. If c* is not yet known, an interval of possible values is provided. Instances for which c* is larger than the theoretical minimum cycle time $LC_1 = \max\{\lceil t_{sum}/m \rceil, t_{max}\}$ are marked with '+' (see p. 55). If the difference $c^* - LC_1$ is larger than 1, the respective value is added after the '+' sign.

Table 7.17. Data Set 1 for SALBP-2

Name	m	c*	m	c*	m	c*	m	c*
Buxey	7	47	8	41	9	37^{+}	10	34^{+}
	11	32^{+2}	12	28^{+}	13	27^{+2}	14	25
Gunther	6	84^{+3}	7	72^{+3}	8	63^{+2}	9	54
	10	50^{+}	11	48^{+4}	12	44^{+3}	13	42^{+2}
	14	40	15	40				
Kilbrid	3	184	4	138	5	111	6	92
	7	79	8	69	9	62	10	56
	11	55						
Lutz 1	8	1860^{+92}	9	1638^{+66}	10	1526^{+112}	11	1400
	12	1400						
Lutz 2	9	54	10	49	11	45	12	41
	13	38	14	35	15	33	16	31
	17	29	18	28^{+}	19	26	20	25
	21	24	22	23	23	22	24	21
	25	20	26	19	27	19^{+}	28	18
Sawyer	7	47	8	41	9	37^{+}	10	34^{+}
	11	31^{+}	12	28^{+}	13	26^{+}	14	25
Tonge	3	1170	4	878	5	702	6	585
	7	502	8	439	9	391^{+}	10	352^{+}
	11	320	12	294^{+}	13	271^{+}	14	251
	15	235^{+}	16	221^{+}	17	208^{+}	18	196^{+}
	19	186^{+}	20	177^{+}	21	170^{+2}	22	162^{+2}
	23	156	24	156	25	156		
Arcus 1	3	25236	4	18927	5	15142	6	12620^{+2}
	7	10826^{+10}	8	9554^{+90}	9	8499^{+87}	10	7580^{+9}
	11	7084^{+201}	12	6412^{+103}	13	5864^{+40}	14	5441^{+33}
	15	5104^{+56}	16	4850^{+118}	17	4516^{+62}	18	4317^{+111}
	19	4068^{+83}	20	3882^{+96}	21	3691	22	3691
Arcus 2	3	50133	4	37600	5	30080	6	25067
	7	21486	8	18800	9	16711	10	15040-15041
	11	13673	12	12534	13	11570	14	10747^{+4}
	15	$10035\text{-}10036^{+8}$	16	$9411\text{-}9414^{+11}$	17	$8854\text{-}8861^{+7}$	18	$8376\text{-}8377^{+20}$
	19	$7920\text{-}7928^{+4}$	20	7520-7526	21	$7181\text{-}7203^{+19}$	22	$6839\text{-}6859^{+2}$
	23	$6551\text{-}6579^{+11}$	24	$6273\text{-}6289^{+6}$	25	$6085\text{-}6106^{+69}$	26	$5844\text{-}5856^{+59}$
	27	5689						

Data Set 2. The second data set for SALBP-2 consists of 174 instances with precedence graphs having $n \geq 53$ tasks (cf. Table 7.1, p. 233).

Table 7.18. Data Set 2 for SALBP-2

Name	m	c*	m	c*	m	c*	m	c*
Hahn	3	4787^{+111}	4	3677^{+170}	5	2823^{+17}	6	2400^{+62}
	7	2336^{+332}	8	1907^{+132}	9	1827^{+52}	10	1775
War-necke	3	516	4	387	5	310	6	258
	7	222	8	194	9	172	10	155
	11	142^{+}	12	130^{+}	13	120	14	111
	15	104	16	98^{+}	17	92	18	87^{+}
	19	84^{+2}	20	79^{+}	21	76^{+2}	22	$72\text{-}73^{+}$
	23	69^{+}	24	66^{+}	25	64^{+2}	26	64^{+4}
	27	60^{+2}	28	59^{+3}	29	56^{+2}		
Wee-Mag	3	500	4	375	5	300	6	250
	7	215	8	188	9	167	10	150
	11	137	12	125	13	116	14	108
	15	100	16	94	17	89	18	84-87
	19	$84\text{-}85^{+5}$	20	77^{+2}	21	72	22	69
	23	66-67	24	66^{+3}	25	66^{+6}	26	65^{+7}
	27	$63\text{-}65^{+7}$	28	$63\text{-}64^{+9}$	29	63^{+11}	30	56^{+6}
Lutz 3	3	548	4	411	5	329	6	275^{+}
	7	236^{+}	8	207^{+}	9	184^{+}	10	165
	11	151^{+}	12	138^{+}	13	128^{+}	14	118
	15	110	16	105^{+2}	17	98^{+}	18	93^{+}
	19	89^{+2}	20	85^{+2}	21	80^{+}	22	76^{+}
	23	74						
Muk-herjee	3	1403	4	1052	5	844^{+2}	6	704^{+2}
	7	621^{+19}	8	532^{+6}	9	471^{+3}	10	424^{+3}
	11	391^{+8}	12	358^{+7}	13	325^{+}	14	311^{+10}
	15	288^{+7}	16	268^{+5}	17	251^{+3}	18	239^{+5}
	19	226^{+4}	20	220^{+9}	21	208^{+7}	22	200^{+8}
	23	189^{+6}	24	179^{+3}	25	172^{+1}	26	171
Bart-holdi 1	3	1878	4	1409	5	1127	6	939
	7	805	8	705	9	626	10	564
	11	513	12	470	13	434	14	403
	15	383						
Bart-holdi 2	27	157	28	152	29	146	30	142
	31	137	32	133	33	129	34	125
	35	121	36	118	37	115	38	112
	39	109	40	106	41	104	42	101
	43	99	44	97	45	95	46	93
	47	91	48	89	49	87	50	85
	51	84						
Scholl	25	2787	26	2680	27	2580	28	2488
	29	2402	30	2322	31	2247	32	2177
	33	2111	34	2049	35	1991	36	1935
	37	1883	38	1834	39	1787	40	1742
	41	1699-1706	42	1659-1667	43	$1621\text{-}1627^{+}$	44	1584
	45	$1549\text{-}1560^{+}$	46	1515-1522	47	1483-1494	48	1452
	49	$1423\text{-}1434^{+}$	50	1394	51	1386	52	1386

Combined Data Set. The distinction of both data sets is somewhat arbitrary. However, they have been used separately in former computational experiments. In the sequel, we use both data sets as a combined set with 302 instances. Currently, the optimal solutions are known for 276 instances.

7.2.3.2 Exact Procedures

We report on results of computational experiments comparing SALBP-1 based exact search methods with the direct branch and bound approaches TBB-2 and SALOME-2 (cf. Section 4.2).

As in our tests for SALBP-1 procedures, a time limit of 500 seconds is taken to restrict the available computation time for all algorithms. In time-out cases, only a heuristic solution is obtained which is characterized by the current upper bound UC and the current (global) lower bound LC. Note that lower bounds become more relevant for SALBP-2 relative to SALBP-1, because the range of possible objective function values is much larger (cf. the footnote on p. 274). Therefore, we measure the solution quality by the relative deviations

$$\text{rel.UC} := (\text{UC} - c^*)/c^* \cdot 100\% \quad \text{and} \quad \text{rel.LC} := (c^* - \text{LC})/c^* \cdot 100\%$$

of the upper and the lower bound from the optimal cycle time c*, respectively. If the optimal cycle time is unknown, the best value of a lower bound is taken as c*.

All procedures are applied to the combined data set and are compared with respect to the following measures (based on 302 instances):

- # opt: number of instances for which an optimal solution is found and proven
- # fail: number of instances without improvement of the initial feasible solution
- av.rel.UC: average relative deviation of UC from optimality in %
- max.rel.UC: maximum relative deviation of UC from optimality in %
- av.rel.LC: average relative deviation of LC from optimality in %
- av.cpu: average computation time in seconds (500 seconds in time-out cases)

Comparison of SALBP-1 Based Search Methods. We compare several search methods for SALBP-2 which iteratively solve SALBP-F instances (cf. Section 4.2.2). The following methods are examined: Lower Bound Search (LBS), Binary Search (BS), Fibonacci Binary Search (FBS), and Binary Search with Prespecified Entry Point (EBS).

All procedures proceed from an initial search interval [LC,UC] which is bounded by the simple lower bound LC_1 and an upper bound UC. The latter bound is determined as described in Remark 2.5 (see pp. 60). However, this initial upper bound is often far away from optimality. Therefore, the Upper Bound Search is excluded from the test. Within each of the four tested search methods, a version of FABLE which is modified to solve SALBP-F instances is applied.

The results in Table 7.19 show that LBS and FBS solve the most instances to a proven optimum within the available computation time. This is mainly due to the fact that 156 out of the 302 instances of the combined data set have minimal cycle times being equal to the simple lower bound LC_1 (for 15 further instances it is currently not clear if LC_1 is optimal). For these instances, LBS and FBS have to examine only a single trial cycle time, while the other methods require more iterations depending on the initial size of the search interval. On the other hand, LBS fails to find feasible solutions at all when it runs out of time (152 instances). In these cases, only the initial feasible solution with a worst case error of almost 100 % is available. FBS is faced with the same basic difficulty when a time-out appears in the Fibonacci Search phase. This problem has a minor influence on EBS, because it may arise only when solving the SALBP-F instance for the first trial cycle time takes too much time or this cycle time is not feasible. Only BS finds feasible solutions for all instances. Consequently, it shows the best average and maximum deviations of UC from optimality. The worse values of the other methods result from the instances for which they are not able to improve the initial solutions. With respect to the average computation times, LBS and FBS are sligthly better than the two other search methods, because they find more optimal solutions and need fewer iterations for the respective instances. Furthermore, they result in better lower bound deviations, because they systematically increase the lower bounds.

Table 7.19. Comparison of Search Methods

	LBS	BS	FBS	EBS
# opt	150	141	150	140
# fail	152	0	126	32
av.rel.UC	24.17	1.25	15.95	7.10
max.rel.UC	99.96	11.84	99.96	99.86
av.rel.LC	0.22	0.40	0.28	0.40
av.cpu	264.06	282.50	264.24	281.50

The results can be summarized as follows. Only BS obtains acceptable solutions for all instances. While performing well for many instances, the search methods LBS and FBS fail to improve the initial feasible solutions for harder problems. The choice of the particular first trial cycle time by EBS seems not to be advantageous, because this cycle time is not appropriate in some cases.

Different Planning Directions of SALOME-2. We compare the versions UniFor, UniBack, BiFor, BiBack, and BiFlex of SALOME-2. The first two algorithms load the stations in forward and backward direction, respectively. The procedures BiFor and BiBack work in a static bidirectional manner, and BiFlex in a flexible bidirectional manner (cf. Section 4.2.4.3). The initial bounds LC and UC are determined as described above (cf. p. 256). A summary of the results is presented in Table 7.20.

The results of UniFor are clearly worse than those of its mirror image UniBack even though they only differ with respect to the order in which stations are considered. The difference is illustrated by the fact that both procedures obtain the same objective function value for 162 instances only, while UniBack finds better solutions for 76 and UniFor for 64 instances. This underlines the importance of the planning direction and its data dependency stated for SALBP-1 in Section 7.2.2.2.

Table 7.20. Comparison of Planning Directions

	UniFor	UniBack	BiFor	BiBack	BiFlex
# opt.	164	184	171	182	217
# fail	1	0	1	2	0
av.rel.UC	1.09	0.90	1.24	1.51	0.56
max.rel.UC	10.05	11.73	53.22	53.22	10.03
av.rel.LC	0.33	0.21	0.24	0.19	0.12
av.cpu	242.74	208.71	228.99	210.92	157.03

The static bidirectional algorithms BiFor and BiBack are not able to outperform the unidirectional procedures. On the contrary, they perform worse for some instances (cf. the large maximal deviations). BiFlex clearly obtains the best results with respect to all criteria because of its dynamic flexible branching strategy. It is able to obtain solutions which are at least as good as the best solutions of the four static algorithms for 260 of the 302 instances. That means, it usually identifies an advantageous planning direction.

Search Methods versus Direct Procedures. We compare search based procedures with the two direct approaches SALOME-2 and TBB-2. All procedures have to observe the time limit of 500 seconds and start with the same initial upper bound UC as described above. Following our previous results, SALOME-2 is represented by its flexible bidirectional version BiFlex. Similarly, the used version of TBB-2 contains the flexible bidirectional strategy for selecting stations (see p. 150), because it clearly provides better results than other selection rules, especially better than those which perform a task selection. We do not provide corresponding numerical results but refer to Scholl (1994) who compared procedures with station selection and task selection for data set 1 (see pp. 253).

Due to our above results, we use the Binary Search as search method starting from the initial interval [LC,UC]. Three different procedures are obtained by applying FABLE, EUREKA, and SALOME-1 within the search process, respectively. These branch and bound procedures for SALBP-1 are modified for solving SALBP-F instances and are used in their best versions (cf. Table 7.7, p. 239). Since EUREKA contains several parts, the computation time remaining in a current iteration of the search process is equally split up between the forward procedure and the backward one. When the time limit is reached, only the heuristic procedure is applied for all remaining trial cycle times. This leads to an average additional computation time of 7 seconds.

The five procedures are applied to the combined data set with 302 instances. The results are summarized in Table 7.21 (cf. Tables 7.19 and 7.20 for the results of FABLE (column BS) and SALOME-2 (column BiFlex), respectively).

Table 7.21. Search Methods and Direct Approaches

	Binary Search			Direct	
	FABLE	EUREKA	SALOME-1	TBB-2	SALOME-2
# opt	141	178	215	201	217
av.rel.UC	1.25	0.62	0.53	1.30	0.56
max.rel.UC	11.84	11.86	15.25	17.86	10.03
av.rel.LC	0.40	0.55	0.11	0.21	0.12
av. cpu	282.50	293.34	164.43	188.36	157.03

The algorithms SALOME-1 (with Binary Search) and SALOME-2 show quite similar results. While the latter obtains more proven optimal solutions and a clearly smaller maximal deviation, the first shows a slightly smaller average deviation. Both procedures clearly outperform the other approaches with respect to all measures except the maximal deviation of SALOME-1. The similarity of the results of SALOME-1 and SALOME-2 is based on their close relationship reflecting the relatedness of SALBP-1 and SALBP-2. Both procedures apply Local Lower Bound Methods within which SALBP-F instances arise for temporarily fixed local lower bounds on the number of stations and the cycle time, respectively. Due to this relationship, they contain common components like the procedure for enumerating station loads, most of the dominance and reduction rules and the flexible bidirectional branching strategy. The main differences are: SALOME-2 avoids parts of the enumeration tree being repeatedly enumerated and only considers possible cycle times. SALOME-1 is only used for solving SALBP-F instances, i.e., each node is fathomed whenever its local lower bound has to be enlarged.

TBB-2 does only well with respect to the number of optimal solutions and the lower bound deviations. The large average and maximal deviations of UC from optimality are due to the enumeration scheme which concentrates on improving the current UC instead of trying to hold a current lower bound as the SALOME procedures do. Therefore, its efficiency strongly depends on the value of the initial upper bound (in Section 7.2.3.4, we show that it obtains much better solutions in combination with a tabu search procedure, p. 266). Furthermore, it assigns only a single task in each branching step such that the application of expensive logical tests is required in order to avoid constructing poorly utilized station loads. Since FABLE has the same basic drawbacks, it achieves rather poor results, too.

The results of the search procedures confirm the findings of Section 7.2.2.2 with respect to the procedures FABLE, EUREKA, and SALOME-1 (cf., especially, Table 7.7 and the corresponding interpretation, p. 239).

7.2.3.3 Priority Based Heuristics

In this section, we examine the performance of search methods for SALBP-2 which apply priority rule based heuristics for SALBP-1. The initial search interval is determined as in Section 7.2.3.2 for exact search methods. Results are summarized with respect to following measures:

- # f.opt: number of instances for which an optimal solution is found but not necessarily proven
- # best: number of instances for which the respective rule provides the best solution found by any of the rules applied in the test
- av.rel: average relative deviation from optimality in % (= av.rel.UC)
- max.rel: maximal relative deviation from optimality in % (= max.rel.UC)
- av.cpu: average computation time per instance (in seconds)
- max.cpu: maximal computation time per instance (in seconds)

Different Priority Rules and Planning Directions. The first test concerns the influence of different priority rules on the quality of SALBP-2 solutions obtained. Therefore, we compare the same set of 13 priority rules tested for SALBP-1 in Section 7.2.2.3 (cf. Table 7.15, p. 249) within the framework of Binary Search. The results are summarized in Table 7.22.

Table 7.22. Priority Rules and Planning Directions

	Forward				Backward				Bidirectional				
Rule	MaxF	Max PW	Max TimeL	Max Time Slack	MaxF	Max PW	Max TimeL	Max Time Slack	MaxF	Max PW	Max TimeL	Max Time Slack	Max TS Avail
# f.opt	22	22	31	28	25	25	29	29	28	25	34	31	33
# best	90	71	86	81	92	73	67	69	117	92	90	84	93
av.rel	2.57	2.57	2.74	2.79	2.47	2.52	3.28	3.25	2.29	2.58	2.78	2.76	2.63
max.rel	12.50	12.50	11.67	11.67	11.67	10.71	12.12	12.12	11.67	12.99	11.84	13.33	10.67
av.cpu	0.26	0.25	0.23	0.23	0.25	0.25	0.26	0.26	1.43	1.42	1.23	1.26	1.36
max.cpu	1.92	1.87	1.87	1.70	1.87	1.76	1.93	1.98	8.79	8.85	7.80	8.40	8.79

With respect to the average relative deviation values, all procedures perform well. This is illustrated by considering a reference procedure which randomly chooses the tasks to be assigned. It results in an average (maximal) deviation of 5.6 % (18 %). Even though the different rules show similar results, we make some interesting observations. MaxF and MaxPW outperform the other rules with respect to the average deviations for all directions. Both rules prefer tasks with many successors which results in an enlargement of the degree of freedom for choosing tasks in later iterations. This may be advantageous within a search process for SALBP-2, because only feasible solutions have to be found for SALBP-F instances instead of optimal ones in case of SALBP-1. Although the rules MaxTimeL and MaxTimeSlack show larger average deviations, they find more optimal solutions underlining their efficiency in solving SALBP-1 (cf. Table 7.15, p. 249). The dynamic rule Bi-MaxTSAvail appears to be a good compromise with respect to both measures, because it outperforms MaxF, MaxPW, and MaxTimeSlack for the number of optimal solutions and MaxTimeL and MaxTimeSlack for the average solution quality. Similarly to SALBP-1, the bidirectional rules need computation times which are up to six times larger than those for unidirectional rules.

Search Methods. The next test examines different search methods. It includes the Lower Bound Search (LBS), Binary Search (BS), Fibonacci Binary Search (FBS) and Binary Search with Prespecified Entry Point (EBS). Based on our above results, the bidirectional dynamic rule Bi-MaxTSAvail is used as the priority rule for all search methods.

Table 7.23 summarizes the results for the combined data set. In addition to the preceding table, it informs on the average and maximum number of iterations (av. and max. # iter) of the search procedure, i.e., the number of trial cycle times.

Table 7.23. Comparison of Search Methods

Search method	LBS	BS	FBS	EBS
# f.opt	33	33	33	33
# best	301	252	259	248
av.rel	2.29	2.63	2.51	2.62
max.rel	9.26	10.67	10.67	10.67
av. # iter	29.47	7.93	7.05	6.78
max. # iter	286	15	19	14
av.cpu	4.92	1.36	1.37	1.21
max.cpu	43.50	8.78	10.55	9.01

Of course, LBS shows the smallest average deviations but requires the largest computation times due to the large average number of iterations per instance. The other methods successively subdivide their search interval. Among these, FBS is slightly dominating with respect to solution quality, because it combines a lower bound oriented part and a Binary Search. The results of EBS and BS are almost identical with a smaller average number of iterations for EBS, because it chooses a first trial cycle time which is closer to the optimum especially for large search intervals. Since FBS provides good results in short times, it should be the best choice to compute initial feasible solutions for improvement procedures.

The above differences in the solution qualities show that the objective function values obtained by a priority rule based procedure for SALBP-1 need not monotonically decrease for increasing cycle times (cf. Section 5.1.3 as well as Hackman et al. (1989)).

7.2.3.4 Tabu Search Procedures for SALBP-2

We report on computational experiments with tabu search procedures for SALBP-2 (cf. Section 5.2.3). Several possible components of such procedures are examined. Additionally, see Scholl and Voß (1996).

List Size Modification Strategies. The first test is concerned with the length TL of the tabu list. In Section 5.2.3 (p. 198), three strategies for determining the tabu list size are described. We compare FTL with the list sizes 7 (FTL7) and 15 (FTL15), RTL, and DTL. RTL modifies TL by using uniformly distributed random numbers as follows: TL is restricted to the interval [8,22]. Initially, TL = 15 is used. After $\lfloor MaxIt/10 \rfloor$ iterations without improving the currently best solution, TL is decremented or incremented by 1 or keeps the same value with probability 1/3 for each choice. If TL = 8 (TL = 22), the probability is 1/2 for keeping the value or incrementing (decrementing) TL by 1. DTL is applied as described in Section 5.2.3. For the combined data set, the initial value of TL varies between 12 for the smallest problem with 29 tasks and 7 stations and 116 for the largest one with 297 tasks and 52 stations.

Initial feasible solutions are computed by the search method FBS with the priority rule Bi-MaxF, because both appear to be effective in the tests of Section 7.2.3.3. The tabu search procedures stop when the currently best solution is equal to the lower bound LC_1, when MaxIt = 50,000 iterations have been performed, or when a time limit of 500 cpu seconds has been reached, whatever occurs first. The global aspiration criterion is applied in all cases. If in an iteration no admissible move can be found, i.e., all moves are set tabu, the conflict management strategy C3 is used. When it fails to resolve the conflict, the strategy C1 is applied. Furthermore, the tabu list is subdivided into a static and a dynamic part by the moving gap strategy which is shown to be effective by Scholl and Voß (1996).

A summary of the results is contained in Table 7.24 which provides the same information as former tables and additional information on:

- av. # GA: average number of iterations in which a move is chosen due to the global aspiration criterion
- av. # confl: average number of iterations in which all admissible moves are set tabu (conflict) and the conflict management strategy C3 fails

Table 7.24. List Size Modification Strategies

TL Strategy	FTL 7	FTL 15	RTL	DTL
# f.opt	144	162	168	165
# best	202	256	266	247
av.rel	0.66	0.43	0.40	0.41
max.rel	7.94	5.26	5.26	5.26
av. #GA	0.75	1.45	1.45	2.24
av. #confl	0.13	142.77	147.94	106.45
av.cpu	108.99	87.72	84.90	85.85

The results show that all versions of the tabu search procedure are able to obtain considerable improvements on the initial solutions, which have an average relative deviation from optimality of 2.28 %. However, the results indicate that TL is a very sensitive parameter for solving SALBP-2. For the combined data set, FTL 7 is remarkably worse than FTL 15 with respect to the average solution quality. The best results are obtained by RTL and DTL which vary the list size. The application of the global aspiration criterion is successful for all procedures. Especially for DTL, which initially considers large tabu list sizes, more than two tabu moves improving the incumbent solution are performed per instance on the average. Due to its small list size, FTL 7 shows the least number of iterations in which a conflict occurs that cannot be resolved by the conflict management strategy C3. Besides showing the worst solution quality, FTL 7 takes the longest average computation times. This is due to the larger number of admissible moves in each iteration, because only a few moves are set tabu.

Influence of the Initial Feasible Solution. We examine the influence of the quality and the structure of different initial feasible solutions on the performance of improvement procedures. Following the results of Section 7.2.3.3, the priority rule MaxF is applied within the framework of FBS as forward, backward, and bidirectional rule with the tie breakers MaxTime and MinTaskNo. A bidirectional random rule is used as reference procedure, i.e., forward and backward priorities are derived from uniformly distributed random numbers.

Table 7.25 shows the results of the initial procedures as well as those of a steepest descent procedure (cf. Section 5.2.1) which tries to improve the respective initial solutions. The random rule yields initial solutions with an average deviation which is about twice as large as those of the other rules. For the MaxF rules, the steepest descent procedure obtains only a reduction of about 0.1 %, in each case, and the poor random solutions can only be improved by 0.8 %. For most instances, no improvement is achieved at all. These results show that a pure steepest descent procedure is not a suitable improvement procedure for SALBP-2 whatever initial solution is given.

Table 7.25. Influence of the Initial Solution

	Initial rule	Forward	Backward	Bidirectional	Random
Initial Solution	av.rel	2.57	2.47	2.28	5.59
	max.rel	12.50	11.67	11.67	17.86
Steepest Descent	av.rel	2.46	2.36	2.18	4.79
	max.rel	12.50	11.67	11.67	17.86
Tabu Search	# f.opt	132	136	140	136
	# best	217	220	242	209
	av.rel	0.69	0.64	0.57	0.67
	max.rel	7.79	5.26	5.26	5.26
	av.#GA	1.79	2.14	1.86	4.86
	av.#confl	1.39	2.61	1.08	1.07
	av.cpu	8.71	8.47	10.47	9.38

The tabu search procedure with DTL yields results being much better than those of the steepest descent procedure. Already after 1,000 iterations, the average deviations are reduced to about 1 % for all initial rules. Table 7.25 contains results for MaxIt = 5,000 which are discussed as follows.

The influence of the initial solutions is rather limited, because only small differences of solution quality remain. For MaxIt = 50,000, the average deviations are reduced to values very close to 0.4 % whatever initial rule is used.

The bidirectional rule slightly outperforms the other rules. Apparently, this is based on the structure of solutions which are constructed bidirectionally. Forward (backward) procedures tend to cumulate idle times in later (earlier) stations, while a bidirectional one tends to generate a bowl-shaped work load distribution, i.e., idle times are concentrated in the middle of the line (cf. Remark 5.1, p. 187). Due to the precedence constraints, it is often easier to move tasks from early or late stations to stations in the middle of the line than to do so in the reverse direction. Therefore, bowl-shaped initial solutions may lead to enlarged neighbourhoods (in the first iterations of the search process).

The difference between the results of the forward and backward initial rule are presumably due to the fact that our implementation of the tabu search procedure (arbitrarily) chooses among several most improving moves that one which is found first. That is, among the stations in question the earliest one is selected. That may subsequently result in a restricted neighbourhood, because the early stations tend to be heavily loaded. In the backward case, this search concentrates on stations with more idle times. Even though the initial solutions of the random rule are poor, the corresponding solutions of the tabu search procedure are sligthly better than those resulting from the forward rule. Possibly, this is due to the tendency of the random rule to produce solutions with the stations being loaded more equally.

Conflict Management Strategy. In the above tests, the conflict management strategy C3 is utilized. It systematically moves predecessors and successors of (critical) tasks which are trapped within critical stations to earlier and later stations, respectively. In Table 7.26, we present a comparison with the simpler strategies C1, which successively shortens the tabu list, and C2, which tries to move direct predecessors and successors of critical tasks (cf. p. 197).

Table 7.26. Different Conflict Management Strategies

Strategy	C1	C2	C3
# f.opt	160	164	165
av.rel	0.50	0.44	0.41
av. #confl	7789.89	4239.60	106.45
av.cpu	96.67	106.58	85.85

The results are obtained by applying the tabu search procedure with DTL and MaxIt = 50,000 (cf. Table 7.24). C3 outperforms the simpler strategies with respect to all measures. Especially, the number of unresolvable conflicts is much smaller than that of the other strategies. Note that this value counts the number of iterations in which the list size must be reduced. Even the computation times favour C3 due to the fact that fewer conflicts have to be resolved throughout the search.

Intensification and Diversification. A simple strategy is tested by Scholl and Voß (1996). However, this approach does not lead to improvements of the solution quality or the computation time relative to corresponding procedures without such strategies. Especially, it seems to be difficult to perform intensification and diversification in a rather schematic manner as usually proposed. In this respect, the conflict management strategy C3 provides a more flexible diversification mechanism.

Combining Tabu Search and Exact Procedures. We apply the tabu search procedure to obtain improved initial solutions for the branch and bound procedures TBB-2 and SALOME-2. It contains the global aspiration criterion and the strategies C3, moving gap, and DTL. For each list size considered within DTL, the search process is terminated when 1,000 iterations without improvement of the incumbent solution have been performed.

Table 7.27. Combination of Tabu Search with Exact Procedures

Strategy	pure TBB-2	pure SALOME-2	with Tabu Search TBB-2	with Tabu Search SALOME-2
# opt	201	217	214	218
# f.opt	207	220	224	223
av.rel	1.30	0.56	0.35	0.31
max.rel	17.86	10.03	7.27	7.27
av.cpu	188.36	157.03	165.46	159.03

Table 7.27 repeats the results of both exact procedures without tabu search given in Table 7.21 (see p. 259). These results are compared with those being obtained by starting the same methods with the initial upper bounds provided by the tabu search procedure. Both the numbers of proven (#opt) as well as found optimal solutions (#f.opt) are specified. The initial tabu search solutions are characterized by #f.opt = 146, av.rel = 0.73, max.rel = 7.27, and av.cpu = 14.04. The computation times are contained in the times of the combined procedures, the results of which are given in the columns on the right.

Both exact procedures obtain solutions significantly improved by the initial tabu search application. The results are even better than those of the best pure tabu search approaches (cf. Table 7.24, p. 263). Especially for TBB-2, the results are clearly improved underlining the importance of a sharp initial upper bound for this approach. In the case of SALOME-2, the upper bounds are used for prefixing.

Summary of the Results. The tests show that tabu search procedures are well suited to SALBP-2 due to the natural neighbourhood definition and the objective which usually provides a clear distinction between solutions. The difficulties arising from the precedence constraints are dealt with by a conflict management strategy which contains a diversification mechanism. Furthermore, a global aspiration criterion and a moving gap strategy help to improve solutions. Preliminary results indicate that further improvements can be obtained by dynamic tabu list management strategies like the cancellation sequence method (cf. p. 195).

The combination of exact and tabu search procedures is very promising, because they complement each other. While the exact procedures (especially SALOME-2 with its Local Lower Bound Method) are able to obtain sharp lower bounds, the tabu search procedures are very successful in finding good upper bounds even if they are applied for a short computation time only.

7.2.4 Comparative Evaluation of SALBP-E Procedures

In this section, we examine search methods for SALBP-E which are based on solution procedures for SALBP-1 and SALBP-2, respectively.

7.2.4.1 Data Set for SALBP-E

No data set is available for SALBP-E in the open literature. In Table 7.28, we define a data set with 256 instances which contains the same precedence graphs as our data sets for SALBP-1 and SALBP-2 (cf. Scholl (1993)). Each instance is characterized by the used precedence graph and a range $[m_{min}, m_{max}]$ of possible values for the number m of stations. Optimal solutions are described by the parameters idle*, m*, and c* denoting the minimal sum of idle times, the corresponding number of stations, and the cycle time, respectively. If the minimal idle time is not yet known, an interval is given. Currently, optimal solutions are available for 229 instances. If several optimal (m,c)-combinations exist, one is chosen arbitrarily.

Note that the maximal line efficiency E* is computed by $t_{sum}/(m^* \cdot c^*)$. We prefer specifying the minimal total idle time ($idle^* = m^* \cdot c^* - t_{sum}$) instead of E* due to its integral values.

Table 7.28. Data Set for SALBP-E

Name	m_{min}	m_{max}	idle*	m*	c*	m_{min}	m_{max}	idle*	m*	c*
Mertens	3	5	1	3	10	4	5	6	5	7
Bowman	3	5	9	3	28	4	5	10	5	17
Jaeschke	3	7	2	3	13	4	7	3	4	10
	5	7	8	5	9	6	7	11	6	8
Jackson	3	7	2	3	16	4	7	2	4	12
	5	7	4	5	10					
Mansoor	3	5	1	3	62	4	5	7	4	48
Mitchell	3	9	0	3	35	4	9	0	5	21
	6	9	3	6	18	7	9	7	8	14
Rosenberg	3	10	1	3	42	4	10	1	6	21
	7	10	3	8	16	9	10	15	10	14
Heskiaoff	3	10	0	4	256	5	10	1	5	205
	6	10	2	6	171	7	10	5	7	147
	8	10	8	8	129	9	10	20	9	116
Buxey	3	13	0	3	108	4	13	1	5	65
	6	13	4	8	41	6	7	5	7	47
	9	13	9	9	37	10	11	16	10	34
	10	13	12	12	28					

Table 7.28. Data Set for SALBP-E (continued)

Name	m_{min}	m_{max}	idle*	m*	c*	m_{min}	m_{max}	idle*	m*	c*
Sawyer	3	13	0	3	108	4	13	0	4	81
	5	13	1	5	65	6	13	4	8	41
	6	7	5	7	47	9	13	9	9	37
	10	11	16	10	34	10	13	12	12	28
Lutz 1	3	11	156	4	3574	5	11	220	5	2872
	6	11	236	6	2396	7	11	532	7	2096
	8	11	602	9	1638	10	11	1120	10	1526
Gunther	3	13	0	3	161	4	13	1	4	121
	5	13	2	5	97	6	13	3	9	54
	6	8	21	6	84	10	13	17	10	50
	11	13	45	11	48	12	13	45	12	44
Kilbrid	3	11	0	3	184	4	11	0	4	138
	5	11	0	6	92	7	11	0	8	69
	9	11	6	9	62	10	11	8	10	56
Hahn	3	4	335	3	4787	3	8	89	5	2823
	6	8	374	6	2400	7	8	1230	8	1907
Warnecke	3	30	0	3	516	4	30	0	4	387
	5	30	0	6	258	7	30	0	9	172
	10	30	2	10	155	11	13	12	13	120
	11	30	6	14	111	15	30	12	15	104
	16	17	16	17	92	19	20	32	20	79
	19	30	27-32	20	79	22	30	36	24	66
	25	30	52-72	27	60	27	30	72	27	60
	28	30	76	29	56					
Tonge	3	23	0	3	1170	4	23	0	5	702
	6	23	0	6	585	7	23	2	8	439
	10	23	4	14	251	11	12	10	11	320
	14	23	4	14	251	16	17	26	17	208
	16	23	18	18	196	19	23	24	19	186
	20	23	30	20	177	21	23	54	22	162
Wee-Mag	3	38	1	3	500	4	38	1	4	375
	5	38	1	5	300	6	38	1	6	250
	7	38	1	10	150	7	9	4	9	167
	11	38	1	12	125	13	14	9	13	116
	13	38	1	15	100	16	18	5	16	94
	16	19	5	16	94	16	38	5	16	94
	21	24	13	21	72	21	38	5-13	21	72
	22	24	19	22	69	26	29	191	26	65
	26	31	113	31	52	26	38	5-37	32	48
	27	29	229-256	27	65	32	38	5-37	32	48
	35	38	111	35	46	36	38	157	36	46
Arcus 1	3	21	1	3	25236	4	21	1	4	18927
	5	21	3	5	15142	6	21	13	6	12620
	7	21	75	7	10826	8	21	93	10	7580
	8	9	725	8	9554	11	12	1237	12	6412
	11	13	525	13	5864	11	21	467	14	5441
	15	17	853	15	5104	15	21	853	15	5104
	16	17	1065	17	4516	19	21	1585	19	4068

Table 7.28. Data Set for SALBP-E (continued)

Name	m_{min}	m_{max}	idle*	m*	c*	m_{min}	m_{max}	idle*	m*	c*
Lutz 2	3	49	1	3	162	4	49	1	6	81
	7	49	1	9	54	10	49	5	10	49
	11	13	7	12	41	11	49	5	14	35
	15	16	10	15	33	15	49	8	17	29
	18	28	9	19	26	18	49	8	29	17
	20	25	15	20	25	20	26	9	26	19
	21	24	19	21	24	21	25	15	25	20
	22	24	19	24	21	30	34	11	31	16
	30	49	11	31	16	32	34	25	34	15
	36	49	33	37	14	42	44	43	44	12
	42	49	43	44	12	46	48	67	46	12
Lutz 3	3	23	0	3	548	4	23	0	4	411
	5	23	1	5	329	6	23	6	10	165
	7	9	8	7	236	8	9	12	9	184
	11	23	6	15	110	12	14	8	14	118
	12	15	6	15	110	17	23	22	17	98
	18	21	30	18	93	18	23	28	22	76
	19	20	36-47	19	89	19	21	36	21	80
Arcus 2	3	27	0	3	50133	4	27	0	9	16711
	4	8	1	4	37600	5	8	1	5	30080
	6	7	3	6	25067	6	8	1	8	18800
	10	16	1-4	11	13673	10	27	1-4	11	13673
	11	16	4	11	13673	12	16	9	12	12534
	13	16	11	13	11570	14	16	59	14	10747
	15	16	126-141	15	10036	18	19	81-233	19	7928
	18	27	1-121	20	7526	21	24	59-499	22	6859
	21	27	59-499	22	6859	22	23	59-499	22	6859
	22	24	59-499	22	6859	26	27	1545-1857	26	5856
Bart-holdi 1	3	15	0	3	1878	4	15	0	6	939
	4	5	1	5	1127	7	15	0	9	626
	7	8	1	7	805	10	15	6	10	564
	11	15	6	12	470	13	15	8	13	434
	14	15	8	14	403					
Bart-holdi 2	3	28	1	5	847	3	52	0	29	146
	6	28	1	7	605	8	10	5	9	471
	8	28	1	11	385	12	28	2	12	353
	13	18	4	13	326	13	28	3	19	223
	14	18	6	16	265	20	25	6	20	212
	20	28	4	26	163	21	25	8	21	202
	22	25	12	22	193	23	25	14	24	177
	27	28	5	27	157	30	34	13	31	137
	30	52	1	35	121	32	33	22	32	133
	32	34	16	34	125	36	39	14	36	118
	36	52	6	40	106	37	39	17	39	109
	41	52	8	42	101	43	49	23	43	99
	43	52	16	50	85	44	48	34	44	97
	44	49	29	49	87	45	47	41	45	95
	45	48	38	48	89	51	52	50	51	84
Scholl	3	4	1	4	17414	3	51	0	5	13931
	6	51	1	8	8707	6	7	2	7	9951
	9	20	5	9	7740	9	51	2	21	3317
	10	20	5	10	6966	11	20	5	12	5805

Table 7.28. Data Set for SALBP-E (continued)

Name	m_{min}	m_{max}	idle*	m*	c*	m_{min}	m_{max}	idle*	m*	c*
Scholl	13	14	9	14	4976	13	20	5	15	4644
	16	17	9	16	435	16	20	5	18	3870
	19	20	5	20	3483	22	26	12	23	3029
	22	28	5	27	2580	22	30	3	29	2402
	22	51	2	31	2247	24	26	17	24	2903
	25	26	20	25	2787	32	35	8	33	2111
	32	40	5	36	1935	32	51	4-5	36	1935
	34	35	11	34	2049	37	40	16	37	1883
	38	39	37	38	1834	38	40	25	40	1742
	42	51	23-41	44	1584	44	45	41	44	1584
	44	48	35-41	44	1584	44	51	35-41	44	1584
	47	48	41	48	1452	50	51	45	50	1394

7.2.4.2 Priority Based Search Methods

We consider an Upper Bound Search for SALBP-E (maximize the line efficiency E) which applies priority based heuristics for SALBP-1 to each trial (m,c)-combination (cf. Section 4.3.2). In order to examine different rules and planning directions, we perform the same test with 4 forward rules, 4 backward rules, and 5 bidirectional rules as for SALBP-1 (cf. Table 7.15, p. 249).

The initial values LE and UE for the lower and upper bound on the line efficiency, respectively, are computed as described in Section 2.2.4.3. To be more specific, UE is determined by (2.70) with the simple lower bound LC_1 of SALBP-2 (cf. Remark 2.7). The initial LE is obtained by computing upper bounds UC(m) of SALBP-2 as described in Remark 2.5 (see pp. 60) for each possible value of m.

For each (m,c)-combination within the search process, the cycle time c is fixed and the respective heuristic is applied. A feasible solution, which may improve the current lower bound LE, is obtained when the requisite number of stations does not exceed m_{max}. If the solution requires no more than m stations, the procedure stops. A solution is proven to be optimal only if the procedure stops before the upper bound is decreased in step 2 of the search procedure (cf. Figure 4.25, p. 172). Otherwise, an optimum may be found but cannot be proven, and an interval [LE,UE] remains. Note that UE is the initial value of the upper bound, because the updating of UE in step 2 is only based on heuristic examinations of the respective SALBP-F instances. Therefore, the quality of solutions is measured by relative deviations

$$\text{rel.LE} = (E^* - LE) / E^* \cdot 100\,\%$$

of the lower bound LE from the optimal line efficiency E*.

Table 7.29 shows a summary of the results for the data set of Section 7.2.4.1 with 256 instances. It contains the following information:

- # f.opt: number of instances for which an optimal solution is found but not necessarily proven
- # best: number of instances for which the respective rule provides the best solution found by any of the rules applied in the test
- av.rel: average relative deviation of LE in % (average value of rel.LE)
- max.rel: maximal relative deviation of LE in %
- av.cpu: average computation time per instance (in seconds)
- max.cpu: maximal computation time per instance (in seconds)

The results show that all rules obtain acceptable solutions with average relative deviations between 1.53 % and 2.28 %. The best results are provided by the bidirectional rules Bi-MaxTimeSlack and Bi-MaxTSAvail as well as the forward rules F-MaxTimeL and F-MaxTimeSlack. The ranking of rules is similar to that of Table 7.15 (see p. 249) but the rules Bi-MaxF and Bi-MaxPW perform worse. This is presumably due to the different compositions of the data sets for SALBP-1 and SALBP-E with respect to the number of instances of each precedence graph.

Table 7.29. Results of different Priority Rules

Rule	Forward				Backward				Bidirectional				
	MaxF	Max PW	Max TimeL	Max Time Slack	MaxF	Max PW	Max TimeL	Max Time Slack	MaxF	Max PW	Max TimeL	Max Time Slack	Max TS Avail
# f.opt	20	19	36	36	21	17	33	32	21	19	34	37	34
# best	69	52	102	104	74	49	71	68	41	49	76	88	99
av.rel	1.85	1.91	1.60	1.61	1.83	1.92	1.97	2.01	2.22	2.28	1.75	1.53	1.58
max.rel	11.71	11.71	9.09	9.09	9.26	8.33	8.16	10.20	12.50	11.71	8.42	8.33	9.09
av.cpu	3.34	3.55	3.66	3.74	2.91	3.23	4.62	4.59	4.85	5.03	5.19	4.92	4.94
max.cpu	68.5	66.0	66.9	71.0	44.8	53.9	77.7	76.3	75.3	97.0	91.3	89.5	85.4

Even though some rules show the best solutions for about 100 instances, they do not really dominate other rules. This is underlined by the average (maximal) relative deviation of 0.96 % (7.41 %) obtained by a composite heuristic which applies all 13 rules for each trial cycle time. The composite procedure finds 57 optimal solutions whereas the best single rule Bi-MaxTimeSlack achieves 37 ones.

The computation times are similar for all unidirectional rules. However, the rules based on MaxPW and MaxF have slightly smaller time requirements. Furthermore, the obtained solution quality is reflected by the computation times, because the less (m,c)-combinations are considered in the search process, the better the solution obtained is. The bidirectional rules show only slightly increased computation times instead of being much larger than those of unidirectional rules as in case of

SALBP-1 (cf. Table 7.15). This indicates that most of the iterations consider (m,c)-combinations for which the respective priority based procedure immediately recognizes that it cannot find a feasible solution and, thus, stops. That means, the search process itself, including the sorting of (m,c)-combinations, becomes more relevant with respect to computation times.

The results show that task time related rules and bidirectional planning are useful for solving SALBP-E in the framework of an Upper Bound Search procedure. However, it is recommendable to use a composite heuristic with several rules in order to obtain really good solutions.

7.2.4.3 Search Methods with Exact Procedures

We examine the performance of the Upper Bound Search if exact solution procedures for SALBP-1 or SALBP-2 are applied to each (m,c)-combination, respectively. The branch and bound procedures FABLE, SALOME-1, and SALOME-2 are selected for the comparison and are applied in different versions (see pp. 173):

- FABLE: FABLE for SALBP-F instances
- SAL1-1: SALOME-1 for SALBP-F instances
- SAL1-2: SALOME-1 with fathoming according to condition (4.18)
- SAL2-1: SALOME-2 for SALBP-F instances
- SAL2-2: SALOME-2 with fathoming according to condition (4.18)
- SAL2-3: SALOME-2 in its original version until global lower bound is increased

All methods are applied in a basic version which iteratively solves SALBP-F instances (FABLE, SAL1-1, SAL2-1), i.e., all nodes of the enumeration trees which cannot lead to a solution for the corresponding (m,c)-combination are fathomed. Therefore, no feasible solution of SALBP-E can be found before an optimal (m,c)-combination has been considered. For SALOME-1 and SALOME-2, modified versions (SAL1-2, SAL2-2) which attempt to find feasible solutions by *flexible fathoming* are included in the test. These procedures apply the fathoming rule with condition (4.18). Furthermore, a third version of SALOME-2 (SAL2-3) which tries to find *enlarged increments* of the cycle time is used. Contrary to the procedures outlined previously, this version allows enlargements of the local bounds in all nodes. It terminates when the (global) lower bound of the root node has to be increased. The new global bound serves as modified cycle time c' in step 4 of the search procedure (cf. Figure 4.25, p. 172).

Each of the six procedures included in the test is applied to the data set of Section 7.2.4.1 with a time limit of 500 seconds for each instance. Since exact procedures are able to improve upper bounds, the solution quality is measured by rel.LB (see p. 263) and relative deviations rel.UE of the upper bound UE from the optimal line efficiency E* defined as follows:

$$\text{rel.UE} = (\text{UE} - \text{E}^*)/\text{E}^* \cdot 100\,\%$$

Table 7.30 shows a summary of the results including information about:

- # opt: number of instances for which an optimal solution is found and proven
- av.rel.LE: average relative deviation of LE from optimality in %
- max.rel.LE: maximal relative deviation of LE from optimality in %
- av.rel.UE: average relative deviation of UE from optimality in %
- av.cpu: average computation time per instance (in seconds)

The results show that all procedures but SAL2-2 and SAL2-3 show average relative lower bound deviations (av.rel.LE) which are worse than those of the simple priority based methods (see Table 7.29). This is due to the fact that for some instances no improvement of the initial lower bound, i.e., no feasible solution, is obtained by those procedures (cf. the large values of max.rel.LE), because they focus on lowering the upper bound UE. Therefore, very small upper bound deviations (av.rel.UE) are obtained by SAL1-1 and SAL2-1. Compared to these procedures, FABLE performs clearly worse. This confirms the results for SALBP-1 and SALBP-2 (cf. Sections 7.2.2.2 and 7.2.3.2).

Table 7.30. Results of Exact Approaches

	SALBP-1 Procedures			SALBP-2 Procedures		
	FABLE	SAL1-1	SAL1-2	SAL2-1	SAL2-2	SAL2-3
# opt	146	192	189	191	187	184
av.rel.LE	9.88	6.85	2.52	7.05	0.67	0.66
max.rel.LE	47.22	47.22	45.02	47.22	9.62	9.62
av.rel.UE	0.16	0.01	0.12	0.08	0.15	0.16
av. cpu	228.64	138.48	142.53	140.85	150.81	155.72

Even though SAL1-2 achieves smaller lower bound deviations than SAL1-1 and SAL2-1, the solutions are not satisfying relative to those of the priority based heuristics. However, SAL1-2 and the other exact procedures provide much more proven optimal solutions. (Note that the priority rule based procedures find up to 37 optimal solutions (cf. Table 7.29) but prove at most 15 solutions to be optimal).

Only the versions SAL2-2 and SAL2-3 of SALOME-2 perform well with respect to all measures. The included strategies for finding feasible solutions are very successful and result in a slight deterioration of the upper bound deviations only.

The superiority of the modified SALBP-2 procedures can be explained by considering only such instances of the data set which cannot be solved to optimality. Then the SALBP-F based procedures fail to improve the initial solution, and only the methods SAL1-2, SAL2-2, and SAL2-3 are able to find feasible solutions at all. The best (additional) feasible solution which can be found by SAL1-2, while it enumerates for a certain (m,c)-combination, is (m+1,c). In terms of the total idle time, this corresponds to an increase of c time units. The SALBP-2 procedures may find a feasible solution for the combination (m,c+1) which increases the total idle time by m. Since the ranges of the cycle time are usually much larger than those of the number of stations, the feasible solutions found in SALBP-2 procedures are remarkably better than those in corresponding SALBP-1 procedures.[1]

Summary of the Results. Exact solution procedures for SALBP-E which apply an upper bound oriented search have to deal with the problem of finding no feasible solution when the computation time is restricted. Our tests show that flexible fathoming strategies are able to remove this disadvantage. Especially, SALBP-2 based methods are successful.

Furthermore, priority based heuristics could be added in order to combine their good feasible solutions with the sharp upper bounds provided by the pure Upper Bound Search performed by SAL1-1. In this context, the application of tabu search procedures seems to be very promising. However, no results confirming this conjecture are currently available.

7.2.5 Evaluation of UALBP Procedures

In Section 4.4, we have proposed the procedure ULINO which can be applied directly to UALBP-1 and within different search methods to UALBP-2 and UALBP-E. In the following, we report on some computational experiments testing the performance of ULINO. As data sets, we can use those for SALBP-1, SALBP-2 and SALBP-E, respectively. However, besides evaluating the performance of ULINO our tests are intended to give answers to the question which potential of improving the efficiency is provided by U-shaped assembly lines compared to straight lines.

7.2.5.1 Results for UALBP-1

The tests are performed in the same manner as in Section 7.2.2.2. Average results are contained in Table 7.31 which contains the following information:

---

[1] In our data set, the average ranges contain 1267 values for c and 10 values for m, respectively. The average cycle time value is 2151, while 19 stations are to be considered on the average.

- # opt: number of instances for which an optimal solution is found and proven
- av.rel: average relative deviation from the simple lower bound LB_1 in %
- max.rel: maximum relative deviation from LB_1 in %
- av.cpu: average computation time in seconds

We compute deviations from LB_1 instead of deviations from the optimal number of stations, because optimal solutions are presently unknown for a considerable number of instances and LB_1 is a valid bound for both UALBP-1 and SALBP-1.

Table 7.31. Results of ULINO

Data set	Combined	Talbot et al.	Hoffmann	Scholl
# opt	233	60	32	150
av.rel	0.59	0.39	1.99	0.08
max.rel	10.00	7.14	10.00	5.26
av.cpu	82.09	32.31	195.43	73.01

The results show that very good solutions are found for the Scholl data set which is the most complex data set known for SALBP-1. In particular, it is really noteworthy that ULINO quickly solves most of the large instances with up to 297 tasks. The Talbot and even more the Hoffmann data set, which are both quite easy to solve in case of SALBP-1, seem to be more complicating though they contain smaller problem instances. This is probably due to the fact that these instances have been chosen such that the total idle time $LB_1 \cdot c - t_{sum}$ is low. With SALBP-1, the precedence constraints often forbid a solution with LB_1 stations and the optimal solution having more stations is easily found. With UALBP-1, more solutions possibly including one with LB_1 stations become feasible such that finding the optimal solution or disproving LB_1 to be a feasible number of stations gets harder.

Miltenburg and Wijngaard (1994) also use the Talbot data set (except for three instances) for testing a dynamic programming procedure (based on Held et al. (1963)) and a modified ranked positional weight technique (based on Helgeson and Birnie (1961)). The dynamic programming procedure exactly solves only the 21 smallest problem instances with up to 11 tasks. For the remaining instances the ranked positional weight technique is applied. They obtain an average relative deviation of 3.61 % and find 39 optimal solutions, while ULINO gives a deviation of 0.40 % and gets 57 optimal solutions for these 61 instances. Excluding the four instances not solved this takes about 1.1 cpu seconds on the average.

In Table 7.32, we give a comparison between the optimal SALBP-1 solutions (best lower bound if optimum unknown yet) and the (heuristic) solutions of UALBP-1 found by ULINO. Since different problem types are compared, the relative deviations are again measured with respect to LB_1 which is a common lower bound of both problems. Furthermore, the comparison is restricted to those instances of each data set for which the SALBP-1 solution requires more than LB_1 stations in order to exclude instances where no saving is possible at all.

Table 7.32. Average Relative Deviations from LB_1

Data set	Combined	Talbot et al.	Hoffmann	Scholl
# instances	154	22	40	102
SALBP-1	11.66	12.58	6.24	13.03
UALBP-1	7.90	8.12	4.16	8.88

It appears that using the U-line concept sometimes gives considerable reductions in the number of stations compared to traditional straight lines (up to four stations can be saved). Further reductions may be possible, because ULINO is not able to find optimal solutions to all instances. The number of stations being optimal to SALBP-1 is easily found by ULINO to all but 6 instances within seconds.

To conclude, it seems to be very promising to utilize the U-line concept whenever possible. In addition to its organizational and social advantages described in Section 1.3.5, it provides considerable reductions in the number of stations required even if only heuristic UALBP-1 solutions are available.

7.2.5.2 Results for UALBP-2

ULINO is applied within search methods (Lower Bound Search: LBS, Binary Search: BS, Fibonacci Binary Search: FBS) as described in Section 4.4.2 and tested in the same manner as the exact SALBP-2 procedures in Section 7.2.3.2.

The results are summarized in Table 7.33 which contains the same type of information as Table 7.31 with LB_1 being replaced by the simple UALBP-2 bound LC_1.

Table 7.33. Results of Search Methods with ULINO

Procedure	LBS	BS	FBS
# opt	258	247	254
av.rel	6.87	0.31	6.53
max.rel	89.45	9.09	89.45
av.cpu	89.46	114.59	93.42

It appears that BS gives very good results with small average and maximal relative deviations from the best known lower bound values. LBS finds the most proven optima, because it has to consider only a single trial cycle time whenever the optimal cycle time is equal to the initial lower bound LC_1. On the other hand, it fails to find feasible solutions at all when it runs out of time. Then only the initial upper bound with a worst-case deviation of 89.45% is available producing very large average deviations. FBS is faced with the same basic difficulty but finds improved upper bounds whenever it has time enough to finish the Fibonacci search phase. To summarize, only BS gives satisfactory results. These findings are similar to those made in Section 7.2.3.2 for SALBP-2 search methods.

In Table 7.34, we give a comparison between the optimal (or best known) SALBP-2 solutions and the (heuristic) solutions of UALBP-2 found by the search methods with ULINO. By analogy with SALBP-1, we measure the relative deviations with respect to the theoretical minimum cycle time LC_1 and restrict the comparison to those 139 instances for which the SALBP-2 solution requires a cycle time greater than LC_1. The average deviation is reduced by almost 50% compared to traditional straight lines if BS is chosen. So, we can conclude that also in case of minimizing the cycle time the U-line concept provides considerable improvement potentials.

Table 7.34. Average Relative Deviations from LC_1

	optimal SALBP-2 solutions	LBS	BS	FBS
av.rel	2.56	12.95	1.38	12.15

7.2.5.3 Results for UALBP-E

We examine the performance of the search procedure described in Section 4.4.3. It is applied to each of the 256 instances of the SALBP-E data set described in Section 7.2.4.1 with a time limit of 1000 seconds. As a basis of comparing solutions we consider the line capacity $T = m \cdot c$, which is to be minimized for maximizing the line efficiency.

As is the case with the Lower Bound Search for UALBP-2, our search procedure for UALBP-E fails to find a feasible solution whenever a time-out occurs before finding the optimal solution. Therefore, an initial upper bound is determined heuristically. For each combination (m,c) initially in the list, a priority rule based procedure is applied given the corresponding cycle time c. The minimal value of the line capacity obtained in this manner is taken as initial upper bound. The priority rule consists of sorting the tasks in order of non-increasing operation times (tie-breaker: largest number of preceding and following tasks). The stations are constructed in forward direction one after the other. The tasks being assignable to the current station are assigned in priority order until the station is filled maximally.

Applied to the 256 instances of the data set, the search procedure finds 223 proven optimal solutions. As measures of solution quality we compute the relative surplus capacity $100\% \cdot (T^* - t_{sum}) / t_{sum}$, where T^* is the best objective function value found by the procedure. The average and maximal surplus percentages (av.sp and max.sp) as well as the average computation times are given in Table 7.35 (second column).

Table 7.35. Results for UALBP-E

	search method with ULINO	optimal SALBP-E solutions	search method with ULINO	optimal SALBP-E solutions
# instances	256 (complete data set)		223 (optimally solved by ULINO)	
av.sp	3.41	1.86	0.98	1.80
max.sp	89.58	29.73	29.73	29.73
av.cpu	149.69	-	21.69	-

Obviously, in some time-out cases the initial feasible solutions are rather weak with large surplus capacities. Therefore, the average values are much worse than those of the optimal (or best known) SALBP-E solutions (third column). These bad solutions stem from instances where the range of possible station numbers is rather narrow. That is, finding feasible solutions is very hard for the initial priority based procedure as well as ULINO. Solutions found for a certain cycle time require often one or two stations more than maximally allowed. However, if we consider only the 223 instances solved to optimality by our search method (columns 4 and 5), the potential of improvement provided by U-lines becomes apparent as in case of UALBP-1 and UALBP-2.

Note that using a personal computer with a modern AMD K6-200 processor (which is approximately seven times faster than the 80486 machine) results only in five instances being additionally solved to optimality. This impressively shows the complexity of some hard instances.

7.3 Mixed-Model Sequencing Procedures

We report on computational experiments with different exact and heuristic procedures for the mixed-model sequencing problem MSP-W. This problem is formulated in Section 3.3.3. The tests include the branch and bound procedure described in Section 6.2.1.2 and various versions of the hybrid heuristic of Section 6.2.4.3. Furthermore, the tabu search procedure contained in the hybrid heuristic is applied to the original problem as an independent procedure.

7.3.1 Test Problems and Test Conditions

The procedures are tested on the basis of a data set with 100 problem instances. It contains different settings with respect to size and structure of the problem in order to cover at least a part of possible problem specifications.

Problem Generation. Problem instances are built by a generation procedure of Undt and Scholl (1995). We outline this procedure and specify parameter values used to construct the data set.

The size of a problem instance is characterized by the parameters M (number of models), K (number of stations), and I (length of the sequence). We consider instances with M = 5, 10, 20 and K = 5, 10, 20. For each used combination (M,K), different values of I ranging from 10 for the small problems (5,5) to 400 for the larger problems are defined.

The problem data consist of the following integer parameters: cycle time c, station extents l_k (k = 1,...,K), demands d_m (m = 1,...,M), and processing times τ_{mk}. In all instances, the cycle time is arbitrarily set to c = 90. With respect to the station lengths, three different settings are considered:

- all stations have a constant length $l_k = 110$ (in time units)
- all stations have a constant length $l_k = 150$ (in time units)
- the station extents are randomly chosen from the interval [88, 132] (in time units)

The demand values d_m are randomly chosen from the interval $[0.5 \cdot I/M, 1.5 \cdot I/M]$, i.e., they vary around the expected demand per model (I/M) by at most 50 %.

The processing times are determined as follows. For each model m, a value T_m is randomly selected from the interval $[0.75 \cdot c \cdot K, c \cdot K]$. T_m serves as a target figure for the sum of processing times of model m (given by $\Sigma_k \tau_{mk}$). The single processing times τ_{mk} are randomly chosen from the interval $[0.5 \cdot T_m/K, \min\{l_k, 1.5 \cdot T_m/K\}]$ such that the desired total processing time T_m is approximately reached for every model m.

Note that all non-fixed parameter values are obtained by means of integer random numbers which are uniformly distributed in the respective intervals. They are computed by the pseudo-random number generator of Borland's Pascal 7.0 (cf. Borland (1992, pp. 188)).

Data Set. Table 7.36 gives a summary of the instance specifications included in the data set. The instances get numbers j = 1,...,100 and are grouped according to the values of M, K, and l_k. The first column (# inst.) gives the number of instances in every group. The last column specifies the values used for the sequence length I. In parentheses, the numbers of respective instances are denoted.

Table 7.36. Randomly Generated Data Set for MSP-W

j	# inst.	M	K	l_k	different sequence lengths I
1-5	5	5	5	110	10 (2), 20 (2), 30 (1)
6-10	5	5	5	[88,132]	10 (2), 20 (2), 30 (1)
11-40	30	10	10	110	50, 80, 110, 140, 170, 200, 250, 300, 350, 400 (3 each)
41-50	10	20	10	110	140, 200, 250, 300, 350 (2 each)
51-60	10	10	20	110	140, 200, 250, 300, 350 (2 each)
61-70	10	20	20	110	140, 200, 250, 300, 350 (2 each)
71-80	10	10	10	[88,132]	140, 200, 250, 300, 350 (2 each)
81-90	10	10	10	150	140, 200, 250, 300, 350 (2 each)
91-100	10	10	10	110	140, 200, 250, 300, 350 (2 each) / basic model

The instances 91-100 differ from the other instances of the data set with respect to a particular model. The first model (m = 1) is defined as a *basic model* with its demand d_1 being randomly chosen from the interval $[0.3 \cdot I, 0.6 \cdot I]$. Furthermore, its processing times are minimal with respect to all stations, i.e., $\tau_{1k} \leq \min\{\tau_{2k},...,\tau_{Mk}\}$ for all k. The demands and processing times of the remaining models m=2,...,M are determined as described above. This type of problem reflects an industrial situation in which a basic model of a product (e.g., a basic version of a car) is produced in large quantities. The other models differ from the basic product only with respect to optional extras such as radios, air bags, etc., which lead to larger processing times at one or more stations. That is, these models require all operations necessary for the basic product and further processing in order to install the extras.

Measures for the Quality of Solutions. For each problem instance j = 1,...,100, a *lower bound* LW_j on the work overload is determined by the TPP-relaxation described in Section 6.2.1.2 (see pp. 209). So far, proven optimal solutions with *optimal values* WO_j^* of the work overload are known for 7 instances only. This is mainly due to the weakness of the lower bounds which prevent exact procedures from finding optimal solutions even for small problems. Therefore, we use the work overloads of the best known solutions instead of the weak lower bounds as estimates for WO_j^* when comparing procedures. The weakness of the lower bounds is shown for our data set with 100 instances by dividing the total of the lower bound values TotLW by the total of the best known objective function values TotWO*:

$$TotLW/TotWO^* = \sum_{j=1}^{100} LW_j \Big/ \sum_{j=1}^{100} WO_j^* = 0.22$$

A solution, obtained by some procedure for an instance j, is characterized by its objective function value which provides an *upper bound* $UW_j \geq WO_j^*$ on the optimal work overload. The relative deviation of UW_j from optimality can be estimated by its relative deviation from the best known objective function value WO^*_j:

$$rel.UW_j := (UW_j - WO^*_j) / WO^*_j \cdot 100\,\%$$

However, this measure is not defined for $WO^*_j = 0$, i.e., an optimal sequence for the problem instance j does not include any work overload. Even by adding a small number ϵ to WO^*_j, this difficulty is not removed in a proper manner, because $rel.UW_j$ gets very large values even for small absolute deviations. Therefore, we use the following aggregate measure which represents the relative deviation of the total of the upper bound values ($TotUW = \Sigma_j UW_j$) from TotWO*:

$$rel.UW := (TotUW - TotWO^*) / TotWO^* \cdot 100\,\%$$

This value is defined for any data set containing instances which require utility work in their optimal sequences.

In the tests described below, we impose a time limit of 600 seconds per instance for each procedure. However, this time limit cannot be held in all cases in order to obtain a reasonable solution. Besides the aggregate relative deviation (rel.UW), we use the following measures to compare the performance of the different procedures:

- # best: number of instances for which the best known or optimal solution is found
- av.cpu: average computation time per instance (in seconds)
- max.cpu: maximal computation time per instance (in seconds)
- # excess: number of instances for which the time limit is exceeded by at least 5 sec.

7.3.2 Branch and Bound versus Tabu Search

We compare the branch and bound procedure described in Section 6.2.1.2 with several tabu search procedures (cf. phase 3 of the hybrid heuristic in Section 6.2.4.3). The latter are applied to the original problem instead of being used for solving the pattern sequencing problem. This is equivalent to specifying a maximal pattern length of 1 in the first two phases of the hybrid heuristic.

The branch and bound procedure (BB for short) computes a global lower bound by solving the corresponding TPP-relaxation. Due to the weakness of this type of bound, local lower bounds are not computed. Nodes are built according to rule TB4.

The considered tabu search procedures differ with respect to the move evaluation strategy (see pp. 228):

- TS-TSP: Tabu search with TSP-evaluation
- TS-Pre: Tabu search with pre-selection
- TS-Com: Tabu search with complete move evaluation

The initial solutions are computed as described for the pattern sequencing problem (see p. 227) by defining a single pattern of length 1 for each requisite model unit. In all procedures, the basic length of the tabu list is initially set to 5. The computations are terminated when the time limit of 600 seconds is exceeded (time-out) or no move is available for the actual length of the tabu list though the basic length is already zero. However, a time-out is not always detected immediately, because the timer is checked only once per iteration.

The results are summarized in Table 7.37. Besides the measures defined above, it contains the average and minimal number of iterations performed by the tabu search procedures (av.iter and min.iter).

Table 7.37. Results of Branch and Bound and Tabu Search

	BB	TS-TSP	TS-Pre	TS-Com
# best	7	10	30	15
rel.UW	28.01	16.02	14.60	39.34
av.cpu	564.55	547.9	579.0	658.1
max.cpu	600	601	681	1069
# excess	0	0	43	83
av.iter	-	10,060	4,722	2,252
min.iter	-	1,158	20	2

The results of the branch and bound procedure (BB) are rather poor. Besides the large aggregate deviation of 28.01 %, BB finds the best known solutions only for 7 instances (6 of these instances are proven to be optimal). That is, the parts of the enumeration trees which can be examined by BB within the time limit are too small. This drawback is already true for small problems with 5 models and a sequence length $I = 20$.

The performance of TS-Com is even poorer with respect to solution quality. Furthermore, it exceeds the time limit for most of the instances (up to 469 seconds). That is, one iteration (evaluating all admissible moves) takes several hundreds of seconds for the instances with sequence length $I = 400$ due to the expensive evaluation method. Then only 2 iterations can be performed within the time limit (min.iter=2). However, some best known solutions are found for smaller problems.

The two other versions of the tabu search procedure show clearly better results. However, TS-Pre is partially subject to the same difficulties as TS-Com with respect to the small number of iterations for instances with long sequences and/or many stations. Only TS-TSP performs a considerable number of iterations for all instances. Nevertheless, it is slightly outperformed by TS-Pre, because it does not really select the best admissible move in many iterations.

7.3.3 The Hybrid Heuristic

We investigate the performance of several versions of the hybrid heuristic procedure for MSP-W described in Section 6.2.4.3.

Maximal Pattern Length. In a first test, we examine the influence of the maximal pattern length MaxPatt on the efficiency of the hybrid procedure. Five versions are defined by setting MaxPatt to the values 3,4,5,6,and 7, respectively. In each case, the procedure is specified by the following setting of the parameters.

Phase 1. The *average fit strategy* is used as pattern selection rule. After finding a first column with negative reduced costs, the column search is terminated whenever no improvement has been achieved during a time interval of 3 seconds. The whole column generation procedure is terminated when the time limit of 600 seconds is exceeded.

Phase 2. A fractional solution is transformed into an integer solution by the procedure *round-all*.

Phase 3. The pattern sequencing problem (PSP) is solved by the tabu search procedure with *TSP-evaluation* which is applied as described in Section 7.3.2. In order to achieve reasonable solutions, the tabu search procedure gets at least 60 seconds even if the phases 1 and 2 have already exhausted the available computation time.

Table 7.38 summarizes the results of the five procedures (Col3, Col4, Col5, Col6, Col7 for short) for the data set with 100 instances of Section 7.3.1.

Table 7.38. Different Maximal Pattern Lengths

	Col3	Col4	Col5	Col6	Col7
# best	18	35	29	24	17
rel.UW	2.24	1.48	1.19	2.39	3.59
av.cpu	480.2	477.8	508.5	513.7	531.4
max.cpu	600	662	672	667	671
# excess	0	16	20	20	22

All procedures definitely outperform the branch and bound procedure as well as the pure tabu search procedures (cf. Table 7.37). The best solutions are obtained by Col5 and Col4. This finding confirms the complexity trade-off between the column generation and the pattern sequencing part of the hybrid procedure (see pp. 229). While for MaxPatt = 3 the column generation is rather inexpensive, the pattern sequencing problem becomes more complex. Furthermore, the obtained patterns cannot sufficiently utilize compensation effects between the contained units. The opposite is true for MaxPatt = 7. The column generation procedure takes much time, because a sequencing problem with length 7 has to be solved in order to find the entering column in each iteration. Therefore, the time limit is exceeded for all 20 instances with M = 20 models and 2 further instances, and extra time is required for the tabu search procedure. For most of those instances, the column generation procedure already runs out of time in Col4. As a consequence, Col4 and even Col3 show better results than Col5 for the instances with M = 20, while the procedures Col6 and Col7 perform clearly worse in comparison with the whole data set. The same tendencies can be observed for the problems with K = 20 stations.

Modifications. We consider modifications of the above procedure Col5 with respect to the pattern selection rule, the rounding procedure as well as the tabu search procedure:

- Col5-F: first fit pattern selection instead of average fit
- Col5-B: best fit pattern selection instead of average fit
- Col5-R1: round-one instead of round-all
- Col5-Pre: tabu search with pre-selection instead of TSP-evaluation
- Col5-Com: tabu search with complete move evaluation instead of TSP-evaluation

For the procedure Col5-F, an additional cut-off is used to terminate the column generation procedure, because it tends to produce very small improvements of the objective function value per iteration and, hence, a large number of iterations. In order to avoid highly fractionated solutions and numerical problems as well, the column generation process is stopped when the decrease of the objective function value is less than $\epsilon = 0.00000001$ during the last 10 iterations.

Table 7.39. Modifications of Col5

	Col5-F	Col5-B	Col5-R1	Col5-Pre	Col5-Com
# best	28	28	29	30	30
rel.UW	2.50	3.66	3.38	1.22	1.35
av.cpu	453.2	553.2	520.8	534.33	536.35
max.cpu	660	663	673	670	690
# excess	1	44	29	24	40

The results are summarized in Table 7.39. They show that only the procedures modified with respect to the tabu search part (phase 3) are competitive to Col5. In case of Col5-F, the first fit strategy leads to a decrease of the average computation time but to worse solutions. This may result from small quantities (remarkably lower than 1) with which patterns enter the basis. Then only few patterns remain in the integer solution after the rounding process. Col5-B shows worse results relative to Col5 due to the large computation times required per iteration of the column generation procedure. As a consequence, Col5-B runs out of time before the best LP-solution has been found for 44 instances. The procedure Col5-R1 leads to more time-out situations than Col5, because it transforms the LP-solution to an integer one in several steps requiring the repeated application of the column generation procedure.

The procedures Col5, Col5-Pre, and Col5-Com differ only with respect to the move evaluation strategy within the tabu search part. In contrast to the pure tabu search procedures (cf. Table 7.37, p. 282), they obtain similar results. This is due to the fact that the pattern sequencing problems are considerably smaller than the original MSP-W instances. Furthermore, only short computation times remain after the column generation process. Even though all procedures perform well, Col5-Com is slightly outperformed by the others because of its larger time requirements.

Instances with Basic Models. In Table 7.40, we present a summary of the results for the ten instances (j = 91,...,100) with a basic model (M = 10, K = 10), because they differ from those for the other instances of the data set considerably. For all procedures considered above, the aggregate relative deviations (rel.UW) are given.

Table 7.40. Results for Instances with Basic Models

	BB	TS-TSP	TS-Pre	TS-Com	
rel.UW	231.45	211.09	488.73	932.14	
	Col3	Col4	Col5	Col6	Col7
rel.UW	24.96	4.43	0.43	8.24	17.07
	Col5-F	Col5-B	Col5-R1	Col5-Pre	Col5-Com
rel.UW	0.89	1.50	0.56	0.94	0.15

Table 7.40 shows that the results of the branch and bound procedure as well as the pure tabu search procedures are miserable compared to those of the hybrid procedure. This is due to the special structure of the considered instances. The basic model has the smallest operation times in all stations and covers a large portion of the total demand, while the other models have large operation times at one or more stations. Therefore, the sequence positions of the units of the basic model (basic units), which are always the best choice to equalize cycle time violations, are most critical.

However, the problem may be simple if $d_1 \geq 0.5 \cdot I$. Whenever the basic model is able to compensate the cycle time violations of any other model, optimal solutions are achieved by assigning units of the basic models to all even-numbered positions of the sequence and the remaining units to the odd-numbered positions in an arbitrary manner. Nevertheless, the branch and bound procedure is only able to solve two out of three instances with that specification contained in the data set, while all other procedures find optimal solutions without work overload. This is due to the way the descending nodes are ordered within the branching process (according to minimal additional work overload and the rule TB4). Since these rules tend to assign units of the basic model to neighbouring positions at the first levels of the enumeration tree, only models which are incompatible to each other remain for the last positions of the sequence. The rules TB1, TB2, and TB3 even reinforce this tendency (see p. 208). Due to the depth-first-search strategy, disadvantageous decisions at the first levels are not removed before large subtrees are completely enumerated.

This basic difficulty of the branch and bound procedure is strengthened for instances with $d_1 < 0.5 \cdot I$, where the positions of the basic units become even more relevant. A similar problem arises for the pure tabu search procedures, because they are started with an initial solution which is computed in a greedy manner, i.e., basic units are cumulated in the first positions. Then the tabu search procedures require many iterations to obtain improved solutions. Since these procedures (especially TS-Com) perform only few iterations within the available computation time, the solutions are far away from the best ones. Furthermore, the chosen definition of tabu moves (a position i is forbidden for all units of a model) complicates the search process due to the large number of basic units.

The hybrid procedure performs well for a maximal pattern length of 5. This is due to the fact that the demand of the basic model is roughly $2/5 \cdot I$ in most of the examined instances. That means, good patterns tend to consist of two basic units and three units of other models, the cycle time violations of which are equalized by the basic units. As a consequence, Col5 and its versions find patterns with small or without work overload, while the procedures with other values of MaxPatt perform worse. Contrary to the whole data set, the best solutions for the instances with basic models are obtained by Col5-Com, because only few patterns contain work overload. Therefore, the choice of really improving moves is difficult for the other evaluation strategies (Col5 with TSP-evaluation and Col5-Pre with pre-selection).

Summary of the Results. The tests show that the hybrid procedure achieves clearly better results than a simple branch and bound procedure and pure tabu search approaches. This is especially true for problem instances with long sequences

(I ≥ 200), because the length of the sequence does not influence the column generation process (for given values of MaxPatt, M, and K), whereas it leads to exponentially increasing time requirements for the other procedures. As a consequence, the hybrid procedure is particularly suited to solving master sequencing problems with a limited number of model families but large demands (cf. the hierarchical planning approach in Section 3.4).

The number M of models is a restricting factor for all procedures, because more models come into question for each sequence position. Since only restricted sequences (patterns) are considered in the hybrid procedure, its relative success (with respect to the quality of solutions) becomes, however, larger for increasing M compared to the other procedures. Due to the increased complexity for determining patterns, Col4 outperforms Col5 for the instances with M = 20, while Col5 is better for M = 5 and M = 10.

The same statements with respect to the behaviour of the algorithms are true for the number K of stations. For the instances with K = 20 stations, Col4 outperforms Col5 and all other procedures, because the calculation of sequences requires more time than for instances with less stations.

Beyond its good performance for the whole data set, the hybrid procedure is especially suited to solve problem instances with a basic model as shown above. For such instances, the parameter MaxPatt may be chosen depending on the proportion with which the basic model is required.

8 Summary and Conclusions

The book on hand deals with decision problems which arise when flow-line production systems are installed and operated (Part I). Corresponding solution methods are developed and investigated by computational experiments (Part II).

Flow-line production is a basic principle in industrial manufacturing. Such production systems are utilized to manufacture a large variety of products. In particular, standardized commodities are assembled in high quantities on assembly lines. In managing such production systems, different decision problems have to be considered. Among these, assembly line balancing problems are important tasks in medium-term production planning. This type of problem consists of partitioning the total work necessary to assemble one or several products among different stations of an assembly line. In the short-term, sequencing problems have to be solved in order to operate the production system efficiently.

Chapter 1 introduces basic terms of flow-line production systems and discusses relationships of balancing and sequencing problems to other decision problems arising in such systems (Sections 1.1 and 1.2). Furthermore, a comprehensive classification of balancing problems with respect to different industrial conditions is given in Section 1.3. A main characteristic is the number of products to be assembled on one or several assembly lines. Correspondingly, single-model, mixed-model, and multi-model balancing problems have to be distinguished. A mixed-model production system assembles several models (versions) of a basic product in an intermixed order on the same line. In multi-model systems, different products are manufactured in batches, because set-up activities are necessary when changing from one to another product.

In **Chapter 2**, single-model assembly line balancing problems are described in detail based on notations and definitions given in Section 2.1. The main part of the chapter (Section 2.2) is devoted to the simple assembly line balancing problem (SALBP) which is extensively discussed in the literature. It is to assign operations to work stations such that precedence constraints between the operations are satisfied and the efficiency of the production line system is maximized. The line efficiency depends on the combination of the production rate and the number of stations. The production rate determines the cycle time, i.e., the time available in each station for performing operations on a single unit of the product. Several versions of this fundamental combinatorial optimization problem are distinguished and are discussed with respect to their specific properties. Lower and upper bounds on the respective objective function values are introduced, including known and new calculating methods. Furthermore, mathematical model formulations are given.

Section 2.3 describes single-model problems which are generalized with respect to stochastic operation times, the line layout (U-shaped and parallel lines) and pacing aspects. These generalized problems are shown to be strongly related to SALBP. Hence, it is concluded that efficient solution procedures are required for SALBP in order to be able to compute acceptable solutions for such generalized problems.

Chapter 3 is concerned with mixed-model assembly line production. Based on a discussion of the relationships between medium-term balancing and short-term sequencing problems in Section 3.1, these fundamental problem types are comprehensively represented.

In Section 3.2, the mixed-model assembly line balancing problem (MALBP) is shown to be related to the corresponding single-model problem SALBP. Therefore, the presentation concentrates on additional objectives for minimizing the variations of work loads. These variations are important with respect to the short-term sequencing problem, because large variations may cause work overloads and idle times when operating the line for a certain short-term production program.

For a given line balance, the mixed-model sequencing problem (MSP) is to find a sequence of units which satisfies the demands for all models contained in a short-term production program. In Section 3.3, the fundamental characteristics of MSP and possible objectives are discussed. A version of this problem (MSP-W) is to minimize the sum of times during which utility workers are required to compensate work overloads. For MSP-W, a mathematical model formulation is given.

Reflecting the strong relationship of several problems with different planning horizons, a hierarchical planning approach is proposed in Section 3.4. This approach attempts to integrate short-term aspects into medium-term decisions by defining and utilizing a so-called master sequencing problem.

Part II of the book is devoted to solution methods for assembly line balancing and sequencing problems. **Chapter 4** provides a comprehensive survey and description of exact procedures for different versions of SALBP.

For the well-known problem SALBP-1, which is to minimize the number of work stations for a given cycle time, a large number of exact solution procedures are proposed in the literature. These procedures mainly utilize the branch and bound technique, a common solution methodology in combinatorial optimization. Different approaches with respect to branching and bounding strategies are identified and comparatively discussed in Section 4.1. By carefully analyzing specific strengths and shortcomings of two efficient procedures (FABLE and EUREKA), a new procedure, which is called SALOME-1, is developed. It contains a new enumeration technique (Local Lower Bound Method) which systematically constructs the enumeration tree by a depth-first-search strategy according to non-decreasing lower bounds. SALOME-1 is complemented by new components which utilize the structure of the problem. In particular, a bidirectional branching strategy is well suited to derive advantage from the structure of precedence relations of tasks.

In Section 4.2, procedures for the problem version SALBP-2, which is to minimize the cycle time for a given number of stations, are presented. Due to its strong relationship to SALBP-1, almost all solution procedures presently known consist of repeatedly solving particular instances of SALBP-1 in the framework of a general search method. After an explanation of such search procedures, two new branch and bound procedures which directly solve SALBP-2 are introduced. One method (TBB-2) is a very flexible approach which can be specified by defining different optional components. Such options concern the generation of initial solutions, the branching process, dominance rules, and reduction tests. The other method (SALOME-2) is a non-trivial modification of the procedure SALOME-1. In particular, the Local Lower Bound Method has to be adapted to SALBP-2.

Section 4.3 proposes search methods for solving SALBP-E, the most general version of SALBP. This problem consists of maximizing the line efficiency (the proportion of the total available time which is utilized for production). Since the sum of operation times is fixed, the line efficiency can be maximized by minimizing the product of the cycle time and the number of stations. Therefore, SALBP-E can be optimized by repeatedly solving instances of SALBP-1 or SALBP-2, respectively. Corresponding search methods are described, especially with respect to the application of the procedures SALOME-1 and SALOME-2.

In Section 4.4, the exact solution procedures for the different versions of SALBP are transferred to corresponding versions of the U-line assembly line balancing problem (UALBP).

In **Chapter 5**, heuristic procedures for the versions of SALBP are considered. Most approaches presently on hand are based on priority rules which order operations and assign them to stations according to priority values. A survey of used priority rules is given in Section 5.1. In particular, two different planning strategies are identified and compared. Furthermore, new concepts with respect to dynamic rules and bidirectional planning are introduced.

Section 5.2 is devoted to improvement procedures which iteratively perform small changes (moves) on current solutions of a balancing problem in order to achieve improved balances. In the literature, steepest descent procedures are proposed for SALBP-2. In each iteration, they choose a most improving move among all possible ones. Such approaches show the fundamental difficulty of getting stuck in a local optimum soon. This problem is overcome by tabu search, a recent heuristic meta-strategy. A corresponding improvement procedure is described for SALBP-2. Furthermore, this approach is applied within a search method for solving the related problem SALBP-1.

Chapter 6 deals with solution methods for mixed-model problems. In Section 6.1, it is shown that different versions of the mixed-model balancing problem (MALBP) can be solved by modified procedures for related single-model problems.

Procedures for mixed-model sequencing problems (MSP) are described in Section 6.2. While only a few exact algorithms are available in the literature, a large number of heuristic procedures is proposed. For either class of methods a survey of existing approaches is given. Furthermore, a branch and bound procedure as well as a new hybrid heuristic procedure is developed for the problem version MSP-W. The branch and bound procedure contains several branching rules and a lower bound derived from a relaxation of MSP-W to the linear transportation problem. The hybrid procedure contains three phases and utilizes a reformulation of the problem based on so-called patterns (subsequences). In phase 1, a column generation procedure is used to obtain a fractional solution for the reformulated problem. This solution is transformed to an integral (feasible) solution in phase 2 by some rounding procedures. The final phase builds a sequence of the generated patterns in order to obtain a solution for MSP-W. This is achieved by a tabu search procedure which can also be applied to MSP-W directly.

Chapter 7 reports on a large number of computational experiments performed in order to evaluate the solution methods developed in this work. All tested procedures are implemented by the author in a unique hardware and software environment which is described in Section 7.1.

In Section 7.2, comprehensive numerical investigations are described which show the effectiveness of the new exact and heuristic methods for SALBP-1, SALBP-2, and SALBP-E proposed in Chapter 4 and 5, respectively. The tests are based on benchmark data sets provided in the open literature. Further data sets which contain harder test problems with up to 297 tasks have been collected by the author.

For SALBP-1, the new branch and bound procedure SALOME-1 is compared to other procedures and shown to be most effective. An analysis of the results indicates that its good performance is mainly due to the enumeration technique, the use of various dominance rules, and the bidirectional branching strategy. Applying SALOME-1 in a modified manner to the U-line balancing problem UALBP-1, which is a modification of SALBP-1, also gives very promising results.

A comparison of priority rule based heuristic procedures shows that task-oriented approaches are outperformed by stationoriented ones. Furthermore, task time related priority rules as well as dynamic bidirectional rules turn out to be effective. Nevertheless, the results of priority based methods are not always satisfactory. Better solutions are obtained by applying a tabu search procedure for SALBP-2 to SALBP-1 based on a dual strategy. When good initial solutions are provided, this approach is even competitive to exact procedures.

In tests for SALBP-2 (and UALBP-2), the efficiency of SALOME-1 is underlined by applying it within the framework of a search procedure. Its derivative SALOME-2 which directly solves SALBP-2 achieves similar results. Both approaches clearly outperform other indirect and direct exact solution procedures, namely FABLE, EUREKA, and TBB-2. Again, the enumeration technique as well as the bidirectional branching strategy appear to be responsible for the superiority of the SALOME procedures.

Further tests concern the application of priority based heuristic procedures within different general search methods. The results confirm the findings observed for SALBP-1 with respect to the efficiency of dynamic bidirectional rules. However, precedence related priority rules turn out to be better than task time related ones for SALBP-2.

Very promising results are obtained for different versions of the tabu search procedure for SALBP-2. In particular, strategies for dynamically modifying the size of the tabu list and for managing conflicts are effective. By combining the tabu search procedure with the exact procedures SALOME-2 and TBB-2, further considerable improvements are achieved.

In computational experiments for SALBP-E (and UALBP-E), different procedures for SALBP-1 and SALBP-2 are applied within a common search method. The results show that priority based procedures for SALBP-1 provide acceptable solutions in many cases. On the contrary, the use of exact procedures is difficult due to a trade-off between finding good lower bounds and good upper bounds on the line efficiency, respectively. That is, strictly aiming at improved upper bounds prevents from finding good feasible solutions, and vice versa. With respect to this difficulty, SALOME-2 appears to be superior to SALOME-1.

Section 7.3 reports on computational experiments with exact and heuristic procedures for the mixed-model sequencing problem MSP-W. They are based on randomly generated problem instances which cover different structures of the problem data. The hybrid procedure clearly outperforms the branch and bound procedure as well as different tabu search procedures. This is mainly due to the complexity of calculating the objective function value for a given sequence. In each iteration of the tabu search procedures, a large number of neighbouring solutions has to be evaluated with respect to their objective function value. Therefore, only a relatively small number of iterations can be performed within a given amount of computation time. The branch and bound procedure additionally suffers from the weakness and the little number of known bound arguments. These computational problems are avoided by the hybrid procedure, because it concentrates on finding efficient patterns (short subsequences) in its first two phases such that the remaining (pattern) sequencing problem is sufficiently small in the third phase. The application of tabu search procedures to this aggregate problem is much more promising than a direct application to the original problem.

The results make clear that considerable improvements with respect to the efficiency of solution procedures can be achieved by new approaches even though a lot of research work has already been performed on simple assembly line balancing problems. This is particularly important, because generalized problems such as U-line and mixed-model problems can usually be solved by (modified) SALBP procedures. Therefore, it is necessary to have very efficient procedures on hand although SALBP is often too restrictive to represent industrial problems completely. Besides the possibility to apply such approaches to generalized problems directly, they provide insights into the basic structure of the problem which can be utilized for developing specialized procedures. For example, the application of tabu search procedures to MALBP seems to be very promising. Additionally, tabu search procedures are well suited to incorporate additional aspects like assignment restrictions.

By the results of the mixed-model sequencing procedures, it becomes evident that a large potential of optimization is contained in those problems. The application of simple priority rules or of a truncated branch and bound procedure produces solutions for MSP-W instances which may deviate from the best known ones by several hundred per cent. The hybrid heuristic procedure achieves very promising results and is particularly suited to problems with relatively few models and large demands. Such problems arise in the context of the hierarchical planning approach as a master sequencing problem. In the opinion of the author, such a planning approach is very useful with respect to capacity considerations. It allows incorporating short-term capacity aspects when medium-term decisions are made. In this context, adequate objective functions, which tend to equalize work load variations in mixed-model balancing solutions with respect to possible consequences on the short-term sequencing, are very important.

References

Aarts, E. and J. Korst (1989): Simulated Annealing and Boltzmann Machines. Wiley, Chichester.

Adamczyk, J. (1969): Der Fließzusammenbau elektrotechnischer Bauteile und Erzeugnisse. Beuth, Berlin.

Agnetis, A.; A. Ciancimino, M. Lucertini and M. Pizzichella (1995): Balancing Flexible Lines for Car Components Assembly. International Journal of Production Research 33, 333 - 350.

Agrawal, P.K. (1985): The Related Activity Concept in Assembly Line Balancing. International Journal of Production Research 23, 403 - 421.

Ahmadi, R.H.; S. Dasu and C.S. Tang (1992): The Dynamic Line Allocation Problem. Management Science 38, 1341 - 1353.

Akagi, F.; H. Osaki and S. Kikuchi (1983): A Method for Assembly Line Balancing with more than one Worker in each Station. International Journal of Production Research 21, 755 - 770.

Anderson, D.R. and C.L. Moodie (1969): Optimal Buffer Storage Capacity in Production Line Systems. International Journal of Production Research 7, 233 - 240.

Aneke, N.A.G. and A.S. Carrie (1984): A Comprehensive Flowline Classification Scheme. International Journal of Production Research 22, 281 - 297.

Aneke, N.A.G. and A.S. Carrie (1986): A Design Technique for the Layout of Multi-Product Flowlines. International Journal of Production Research 24, 471 - 481.

Arcus, A.L. (1963 a): COMSOAL: A Computer Method of Sequencing Operations for Assembly Lines. In: Buffa, E.S. (ed.): Readings in Production and Operations Management, Wiley, New York, 336 - 360.

Arcus, A.L. (1963 b): An Analysis of a Computer Method of Sequencing Assembly Line Operations. Ph.D. Dissertation, University of California, Berkely.

Arcus, A.L. (1966): COMSOAL: A Computer Method of Sequencing Operations for Assembly Lines. International Journal of Production Research 4, 259 - 277.

Auwärter, A. and H.-P. Lentes (1977): EDV-unterstützte Leistungsabstimmung von Montagelinien. wt - Zeitschrift für industrielle Fertigung 67, 43 - 48.

Baker, K.R. (1974): Introduction to Sequencing and Scheduling. Wiley, New York.

Baker, K.R.; S.G. Powell and D.F. Pyke (1990): Buffered and Unbuffered Assembly Systems with Variable Processing Times. Journal of Manufacturing and Operations Management 3, 200 - 223.

Baker, K.R. and L.E. Schrage (1978): Finding an Optimal Sequence by Dynamic Programming: An Extension to Precedence-Related Tasks. Operations Research 26, 111 - 120.

Balas, E. (1965): An Additive Algorithm for Solving Linear Programs with Zero-One Variables. Operations Research 13, 517 - 546.

Balas, E. and M.W. Padberg (1979): Set Partitioning - A Survey. In: Christofides, N.; A. Mingozzi, P. Toth and C. Sandi (eds.): Combinatorial Optimization, Wiley, Chichester, 151 - 210.

Bard, J.F. (1989): Assembly Line Balancing with Parallel Workstations and Dead Time. International Journal of Production Research 27, 1005 - 1018.

Bard, J.F.; E. Dar-El and A. Shtub (1992): An Analytic Framework for Sequencing Mixed Model Assembly Lines. International Journal of Production Research 30, 35 - 48.

Bard, J.F.; A. Shtub and S.B. Joshi (1994): Sequencing Mixed-Model Assembly Lines to Level Parts Usage and Minimize Line Length. International Journal of Production Research 32, 2431 - 2454.

Barnes, R.M. (1980): Motion and Time Study: Design and Measurement of Work. 7th ed., Wiley, New York.

Bartholdi, J.J. (1993): Balancing Two-Sided Assembly Lines: A Case Study. International Journal of Production Research 31, 2447 - 2461.

Basu, R.N. (1977): The Interstage Buffer Storage Capacity of Non-Powered Assembly Lines - A Simple Mathematical Approach. International Journal of Production Research 15, 365 - 382.

Bauer, G. and G. Scholz (1984): Iterative Entwicklung von Heuristiken zur Montagesteuerung bei Mehrprodukt-Fließfertigung. In: Steckhan, H.; W. Bühler, K.E. Jäger, Ch. Schneeweiß and J. Schwarze (eds.), OR Proceedings 1983, Springer, Berlin, 37 - 45.

Baybars, I. (1985): On "Currently Practiced Formulations of the Assembly Line Balance Problem". Journal of Operations Management 5, 449 - 453.

Baybars, I. (1986 a): A Survey of Exact Algorithms for the Simple Assembly Line Balancing Problem. Management Science 32, 909 - 932.

Baybars, I. (1986 b): An Efficient Heuristic Method for the Simple Assembly Line Balancing Problem. International Journal of Production Research 24, 149 - 166.

Bedworth, D.D. and J.E. Bailey (1982): Integrated Production Control Systems: Management, Analysis, Design. Wiley, New York.

Bellman, R. (1957): Dynamic Programming. Princeton University Press, Princeton.

Bellman, R. and S.E. Dreyfus (1962): Applied Dynamic Programming. Princeton University Press, Princeton.

Bennett, G.B. and J. Byrd (1976): A Trainable Heuristic Procedure for the Assembly Line Balancing Problem. AIIE Transactions 8, 195 - 201.

Berger, I.; J.-M. Bourjolly and G. Laporte (1992): Branch-and-Bound Algorithms for the Multi-Product Assembly Line Balancing Problem. European Journal of Operational Research 58, 215 - 222.

Betts, J. and K.I. Mahmoud (1989): A Method for Assembly Line Balancing. Engineering Costs and Production Economics 18, 55 - 64.

Bhattacharjee, T.K. and S. Sahu (1987): A Critique of Some Current Assembly Line Balancing Techniques. International Journal of Operations and Production Management 7/6, 32 - 43.

Bhattacharjee, T.K. and S. Sahu (1988): A Heuristic Approach to General Assembly Line Balancing. International Journal of Operations & Production Management 8, 67 - 77.

Bhattacharjee, T.K. and S. Sahu (1990): Complexity of Single Model Assembly Line Balancing Problems. Engineering Costs and Production Economics 18, 203 - 214.

Blumenfeld, D.E. (1989): Performance Comparison of Assembly Systems with Fixed and Flexible Cycle Times. International Journal of Flexible Manufacturing Systems 1, 207 - 222.

Boctor, F.F. (1995): A Multiple-rule Heuristic for Assembly Line Balancing. Journal of the Operational Research Society 46, 62 - 69.

Boguschewski, A.; H. Gehring and K. Köstermann (1990): Montagelinien in der Pkw-Montage abstimmen. Arbeitsvorbereitung 27, 130 - 133.

Bolat, A. (1994): Sequencing Jobs on an Automobile Assembly Line: Objective and Procedure. International Journal of Production Research 32, 1219 - 1236.

Bolat, A.; M. Savsar and M.A. Al-Fawzan (1994): Algorithms for Real-Time Scheduling of Jobs on Mixed Model Assembly Lines. Computers & Operations Research 21, 487 - 498.

Bonney, M.C.; N.A. Schofield and A. Green (1976): Assembly Line Balancing and Mixed Model Line Sequencing. In: Project Implementation and Management, Bridging the Gap: Proceedings of the Fifth INTERNET World Congress, Birmingham, 112 - 121.

Borland (1992): Borland Pascal mit Objekten 7.0 - Referenzhandbuch. Borland GmbH, Langen.

Boucher, T.O. (1987): Choice of Assembly Line Design under Task Learning. International Journal of Production Research 25, 513 - 524.

Bowman, E.H. (1960): Assembly-Line Balancing by Linear Programming. Operations Research 8, 385 - 389.

Brennecke, D. (1968): Two Parameter Assembly Line Balancing Model. In: Hottenstein, M.P. (ed.): Models and Analysis for Production Management, International Textbook Company, Scranton.

Buffa, E.S. (1983): Modern Production/Operations Management. Wiley, New York.

Burns, L.D. and C.F. Daganzo (1987): Assembly Line Job Sequencing Principles. International Journal of Production Research 25, 71 - 99.

Bussmann, K.F.; H.-S. Beissel, U. Brachvogel, J. Hänisch, R. Koxholt, P. Kraus and P. Mertens (1968): Ein Vergleich von Fließbandabstimmungsverfahren. In: Bussmann, K.F. and P. Mertens (eds.): Operations Research und Datenverarbeitung bei der Produktionsplanung, Poeschel, Stuttgart, 313 - 356.

Buxey, G.M. (1974): Assembly Line Balancing with Multiple Stations. Management Science 20, 1010 - 1021.

Buxey, G. (1979): The Nature of Manual, Moving Belt Flowlines with Overlapping Stations. International Journal of Production Research 17, 143 - 154.

Buxey, G.M. and D. Sadjadi (1976): Simulation Studies of Conveyor-Paced Assembly Lines with Buffer Capacity. International Journal of Production Research 14, 607 - 624.

Buxey, G.M.; N.D. Slack and R. Wild (1973): Production Flow Line System Design - A Review. AIIE Transactions 5, 37 - 48.

Buzacott, J.A. (1967): Automatic Transfer Lines with Buffer Stocks. International Journal of Production Research 5, 183 - 200.

Buzacott, J.A. (1968): Prediction of the Efficiency of Production Systems without Internal Storage. International Journal of Production Research 6, 173 - 188.

Buzacott, J.A. (1971): The Role of Inventory Banks in Flow-Line Production Systems. International Journal of Production Research 9, 425 - 436.

Buzacott, J.A. (1990): Abandoning the Moving Assembly Line: Models of Human Operators and Job Sequencing. International Journal of Production Research 28, 821 - 839.

Buzacott, J.A. and L.E. Hanifin (1978): Models of Automatic Transfer Lines with Inventory Banks: A Review and Comparison. AIIE Transactions 10, 197 - 207.

Buzacott, J.A. and J.G. Shanthikumar (1993): Stochastic Models of Manufacturing Systems. Prentice Hall, Englewood Cliffs.

Carlier, J. (1982): The One-Machine Sequencing Problem. European Journal of Operational Research 11, 42 - 47.

Carraway, R.L. (1989): A Dynamic Programming Approach to Stochastic Assembly Line Balancing. Management Science 35, 459 - 471.

Carter, J.C. and F.N. Silverman (1984): A Cost-Effective Approach to Stochastic Line Balancing with Off-Line Repairs. Journal of Operations Management 4, 145 - 157.

Caruso, F.R. (1965): Assembly Line Balancing for Improved Profits. Automation 12, 48 - 52.

Cauley, J.M. (1968): A Review of Assembly Line Balancing Algorithms. Proceedings of the 19th Annual Institute Conference and Convention of the American Institute of Industrial Engineers, Tampa, 23 - 29.

Chakravarty, A.K. (1988): Line Balancing with Task Learning Effects. IIE Transactions 20, 186 - 193.

Chakravarty, A.K. and A. Shtub (1985): Balancing Mixed Model Lines with In-Process Inventories. Management Science 31, 1161 - 1174.

Chakravarty, A.K. and A. Shtub (1986 a): Dynamic Manning of Long Cycle Assembly Lines with Learning Effect. IIE Transactions 18, 392 - 397.

Chakravarty, A.K. and A. Shtub (1986 b): A Cost Minimization Procedure for Mixed Model Production Lines with Normally Distributed Task Times. European Journal of Operational Research 23, 25 - 36.

Charlton, J.M. and C.C. Death (1969): A General Method for Machine Scheduling. International Journal of Production Research 7, 207 - 217.

Chow, W.-M. (1987): Buffer Capacity Analysis for Sequential Production Lines with Variable Process Times. International Journal of Production Research 25, 1183 - 1196.

Chow, W.-M. (1990): Assembly Line Design: Methodology and Applications. Dekker, New York.

Coffman, E.G.; M.R. Garey and D.S. Johnson (1978): An Application of Bin-Packing to Multiprocessor Scheduling. SIAM Journal on Computing 7, 1 - 17.

Coffman, E.G.; M.R. Garey and D.S. Johnson (1984): Approximation Algorithms for Bin-Packing - An Updated Survey. In: Ausiello, G.; M. Lucertini and P. Serafini (eds.): Algorithm Design for Computer System Design, Springer, Berlin, 49 - 106.

Conway, R.; W. Maxwell, J.O. McClain and L.J. Thomas (1988): The Role of Work-In-Process Inventory in Serial Production Lines. Operations Research 36, 229 - 241.

Cook, S.A. (1971): The Complexity of Theorem-Proving Procedures. Proceedings of the 3rd Annual ACM Symposium on Theory of Computing, New York, 151 - 158.

Cooper, L. and M.W. Cooper (1981): Introduction to Dynamic Programming. Pergamon Press, Oxford.

Daganzo, C.F. and D.E. Blumenfeld (1994): Assembly System Design Principles and Tradeoffs. International Journal of Production Research 32, 669 - 681.

Dammeyer, F. and S. Voß (1993): Dynamic Tabu List Management Using the Reverse Elimination Method. Annals of Operations Research 41, 31 - 46.

Dar-El, E.M. (1973): MALB - A Heuristic Technique for Balancing Large Single-Model Assembly Lines. AIIE Transactions 5, 343 - 356.

Dar-El, E.M. (1975): Solving Large Single-Model Assembly Line Balancing Problems - A Comparative Study. AIIE Transactions 7, 302 - 310.

Dar-El, E.M. (1978): Mixed-Model Assembly Line Sequencing Problems. OMEGA 6, 313 - 323.

Dar-El, E.M. and R.F. Cother (1975): Assembly Line Sequencing for Model Mix. International Journal of Production Research 13, 463 - 477.

Dar-El, E.M. and S. Cucuy (1977): Optimal Mixed-Model Sequencing for Balanced Assembly Lines. OMEGA 5, 333 - 342.

Dar-El, E.M. and A. Nadivi (1981): A Mixed-Model Sequencing Application. International Journal of Production Research 19, 69 - 84.

Dar-El, E.M. and M. Rabinovitch (1988): Optimal Planning and Scheduling of Assembly Lines. International Journal of Production Research 26, 1433 - 1450.

Dar-El, E.M. and Y. Rubinovitch (1979): MUST - A Multiple Solutions Technique for Balancing Single Model Assembly Lines. Management Science 25, 1105 - 1114.

Dar-El, E.M. and J. Rubinovitz (1991): Using Learning Theory in Assembly Lines for New Products. International Journal of Production Economics 25, 103 - 109.

Davis, L.E. (1966): Pacing Effects on Manned Assembly Lines. International Journal of Production Research 4, 171 - 184.

Decker, M. (1993 a): Variantenfließfertigung. Physica, Heidelberg.

Decker, M. (1993 b): Capacity Smoothing and Sequencing for Mixed-Model Lines. International Journal of Production Economics 30-31, 31 - 42.

Deckro, R.F (1989): Balancing Cycle Time and Workstations. IIE Transactions 21, 106 - 111.

Deckro, R.F. and S. Rangachari (1990): A Goal Approach to Assembly Line Balancing. Computers & Operations Research 17, 509 - 521.

Dempster, M.A.H.; M.L. Fisher, L. Jansen, B.J. Lageweg, J.K. Lenstra and A.H.G. Rinnooy Kan (1981): Analytical Evaluation of Hierarchical Planning Systems. Operations Research 29, 707 - 716.

Denardo, E.V. (1982): Dynamic Programming: Models and Applications. Prentice Hall, Englewood Cliffs.

Denardo, E.V. and B.L. Fox (1979): Shortest-Route Methods: 1. Reaching, Pruning, and Buckets. Operations Research 27, 161 - 186.

Denardo, E.V. and L.G. Mitten (1967): Elements of Sequential Decision Processes. Journal of Industrial Engineering 18, 106 - 112.

Deutsch, D.F. (1971): A Branch and Bound Technique for Mixed-Product Assembly Line Balancing. Ph.D. Dissertation, Arizona State University.

de Werra, D. and A. Hertz (1989): Tabu Search Techniques: A Tutorial and an Application to Neural Networks. OR Spektrum 11, 131 - 141.

Dick, R. (1966): Zur Prozeßplanung der Fließfertigung. Dissertation, Universität zu Köln.

Ding, F.-Y. and L. Cheng (1993): A Simple Sequencing Algorithm for Mixed-Model Assembly Lines in Just-in-Time Production Systems. Operations Research Letters 13, 27 - 36.

Dobson, G. and C.A. Yano (1994): Cyclic Scheduling to Minimize Inventory in a Batch Flow Line. European Journal of Operational Research 75, 441 - 461.

Domschke, W. (1995): Logistik: Transport. 4th ed., Oldenbourg, München.

Domschke, W. (1997): Logistik: Rundreisen und Touren. 4th ed., Oldenbourg, München.

Domschke, W. and A. Drexl (1991): Kapazitätsplanung in Netzwerken: Eine Übersicht über neuere Modelle und Verfahren. OR Spektrum 13, 63 - 76.

Domschke, W. and A. Drexl (1998): Einführung in Operations Research. 4th ed., Springer, Berlin.

Domschke, W.; P. Forst and S. Voß (1992): Tabu Search Techniques for the Quadratic Semi-Assignment Problem. In: Fandel, G.; T. Gulledge and A. Jones (eds.): New Directions for Operations Research in Manufacturing, Springer, Berlin, 389 - 405.

Domschke, W.; R. Klein and A. Scholl (1996 a): Antizipative Leistungsabstimmung bei moderner Variantenfließfertigung. Zeitschrift für Betriebswirtschaft 66, 1465 - 1490.

Domschke, W.; R. Klein and A. Scholl (1996 b): Tabu Search - Durch Verbote zum schnellen Erfolg. c't - Magazin für Computer Technik, Issue 12, 326 - 332.

Domschke, W.; R. Klein and A. Scholl (1996 c): Tabu Search - Eine intelligente Lösungsstrategie für komplexe Optimierungsprobleme. WiSt - Wirtschaftswissenschaftliches Studium 25, 606 - 610.

Domschke, W.; A. Scholl and S. Voß (1994 a): Verfahren zur Abstimmung von Einprodukt-Fließbändern. WiSt - Wirtschaftswissenschaftliches Studium 23, 106 - 111.

Domschke, W.; A. Scholl and S. Voß (1994 b): FABLE: Ein effizientes Verfahren zur Abstimmung von Einprodukt-Fließbändern. WiSt - Wirtschaftswissenschaftliches Studium 23, 147 - 150.

Domschke, W.; A. Scholl and S. Voß (1997): Produktionsplanung - Ablauforganisatorische Aspekte. 2nd ed., Springer, Berlin.

Downey, B.S. and M.S. Leonard (1992): Assembly Line with Flexible Work-Force. International Journal of Production Research 30, 469 - 483.

Drexl, A. (1990): Fließbandaustaktung, Maschinenbelegung und Kapazitätsplanung in Netzwerken - Ein integrierender Ansatz. Zeitschrift für Betriebswirtschaft 60, 53 - 70.

Drexl, A.; B. Fleischmann, H.-O. Günther, H. Stadtler and H. Tempelmeier (1994): Konzeptionelle Grundlagen kapazitätsorientierter PPS-Systeme. Zeitschrift für betriebswirtschaftliche Forschung 46, 1022 - 1045.

Drexl, A. and C. Jordan (1995): Materialflußorientierte Produktionssteuerung bei Variantenfließfertigung. Zeitschrift für betriebswirtschaftliche Forschung 47, 1073 - 1087.

Driscoll, J. and A.A.A. Abdel-Shafi (1985): A Simulation Approach to Evaluating Assembly Line Balancing Solutions. International Journal of Production Research 23, 975 - 985.

Easton, F.F. (1990): A Dynamic Program with Fathoming and Dynamic Upper Bounds for the Assembly Line Balancing Problem. Computers & Operations Research 17, 163 - 175.

Easton, F.; B. Faaland, T.D. Klastorin and T. Schmitt (1989): Improved Network Based Algorithms for the Assembly Line Balancing Problem. International Journal of Production Research 27, 1901 - 1915.

Eiselt, H.A.; G. Pederzoli and C.-L. Sandblom (1987): Continuous Optimization Models. de Gruyter, Berlin.

Ellegard, K.; D. Jonsson, T. Engström, M.I. Johansson, L. Medbo and B. Johansson (1992): Reflective Production in the Final Assembly of Motor Vehicles - An Emerging Swedish Challenge. International Journal of Operations & Production Management 12, 117 - 133.

Elmaghraby, S.E. and W.S. Herroelen (1980): On the Measurement of Complexity in Activity Networks. European Journal of Operational Research 5, 223 - 234.

El-Rayah, T.E. (1979 a): The Efficiency of Balanced and Unbalanced Production Lines. International Journal of Production Research 17, 61 - 75.

El-Rayah, T.E. (1979 b): The Effect of Inequality of Interstage Buffer Capacities and Operation Time Variability on the Efficiency of Production Line Systems. International Journal of Production Research 17, 77 - 89.

Engström, T. and L. Medbo (1992): Preconditions for Long Cycle Time Assembly and its Management - Some Findings. International Journal of Operations & Production Management 12, 134 - 146.

Felbecker, O. (1980): Ein Beitrag zur Reihenfolgeplanung bei Mehrprodukt-Linienfertigung. Dissertation, RWTH Aachen.

Foulds, L.R. (1992): Graph Theory Applications. Springer, New York.

Freeman, D.R. (1968): A General Line Balancing Model. Proceedings of the 19th Annual Institute Conference and Convention of the American Institute of Industrial Engineers, Tampa, 230 - 235.

Freeman, D.R. and J.V. Jucker (1967): The Line Balancing Problem. Journal of Industrial Engineering 18, 361 - 364.

Freeman, D.S. and R.W. Swain (1986): A Heuristic Technique for Balancing Automated Assembly Lines. Research Report No. 86-6, Industrial & Systems Engineering Department, University of Florida, Gainesville.

Frieze, A. and I. Baybars (1986): Expected Behaviour of Line-Balancing Heuristics. IMA Journal of Mathematics in Management 1, 1 - 11.

Gagnon, R.J. and S. Ghosh (1991): Assembly Line Research: Historical Roots, Research Life Cycles and Future Directions. OMEGA 19, 381 - 399.

Gagnon, R.J. and S. Ghosh (1992): A Comprehensive Decision Framework for Assembly Systems. Industrial Engineering (May), 40 - 45.

Garey, M.R. and D.S. Johnson (1979): Computers and Intractability: A Guide to the Theory of NP-Completeness. Freemann, New York.

Gaul, W. and M. Schader (eds.) (1986): Classification as a Tool of Research. North-Holland, Amsterdam.

Gehring, H. and G. Schütz (1994): Zwei genetische Algorithmen zur Lösung des Bandabgleichproblems. In: Werners, B. and R. Gabriel (eds.): Operations Research - Reflexionen aus Theorie und Praxis, Springer, Berlin.

Gehrlein, W.V. and J.H. Patterson (1975): Sequencing for Assembly Lines with Integer Task Times. Management Science 21, 1064 - 1070.

Gehrlein, W.V. and J.H. Patterson (1978): Balancing Single Model Assembly Lines: Comments on a Paper by E.M. Dar-El (Mansoor). AIIE Transactions 10, 109 - 112.

Geoffrion, A.M. and G.W. Graves (1976): Scheduling Parallel Production Lines with Changeover Costs: Practical Application of a Quadratic Assignment/LP Approach. Operations Research 24, 595 - 610.

Geoffrion, A.M. and R.E. Marsten (1972): Integer Programming Algorithms: A Framework and State-of-the-Art Survey. Management Science 18, 465 - 491.

Ghosh, S. and R.J. Gagnon (1989): A Comprehensive Literature Review and Analysis of the Design, Balancing and Scheduling of Assembly Systems. International Journal of Production Research 27, 637 - 670.

Gilmore, P.C. and R.E. Gomory (1961): A Linear Programming Approach to the Cutting-Stock Problem. Operations Research 9, 849 - 859.

Gilmore, P.C. and R.E. Gomory (1963): A Linear Programming Approach to the Cutting-Stock Problem - Part II. Operations Research 11, 863 - 888.

Globerson, S. and A. Tamir (1980): The Relationship between Job Design, Human Behavior and System Response. International Journal of Production Research 18, 391 - 400.

Glover, F. (1989): Tabu Search - Part I. ORSA Journal on Computing 1, 190 - 206.

Glover, F. (1990): Tabu Search - Part II. ORSA Journal on Computing 2, 4 - 32.

Glover, F. and M. Laguna (1997): Tabu Search. Kluwer Academic Publishers, Boston.

Görke, M. (1978): Rechnerunterstütztes Verfahren zur Leistungsabstimmung von Mehrmodell-Montagesystemen. Krausskopf, Mainz.

Görke, M. and H.-P. Lentes (1976): Leistungsabstimmung von Montagelinien. Arbeitsvorbereitung 13, 71 - 77, 106 - 113, and 147 - 153.

Görke, M. and H.-P. Lentes (1981): Modellfolgebestimmung bei gemischter Produktfertigung. wt - Zeitschrift für industrielle Fertigung 71, 153 - 160.

Goldberg, D.E. (1989): Genetic Algorithms in Search, Optimization and Machine Learning. Addison Wesley, Reading.

Golenko-Ginzburg, D. and A. Goldfeld (1991): Feeding Parts to an Assembly Line by Several Production Units. International Journal of Production Economics 24, 77 - 90.

Graves, S.C. and B.W. Lamar (1983): An Integer Programming Procedure for Assembly System Design Problems. Operations Research 31, 522 - 545.

Graves, S.C. and C.H. Redfield (1988): Equipment Selection and Task Assignment for Multiproduct Assembly System Design. International Journal of Flexible Manufacturing Systems 1, 31 - 50.

Gunter, S.I. and L.A. Swanson (1985): Inspector Location in Convergent Production Lines. International Journal of Production Research 23, 1153 - 1169.

Gunther, R.E.; G.D. Johnson and R.S. Peterson (1983): Currently Practiced Formulations for the Assembly Line Balance Problem. Journal of Operations Management 3, 209 - 221.

Gupta, J.N.D. (1988): Two-Stage, Hybrid Flowshop Scheduling Problem. Journal of the Operational Research Society 39, 359 - 364.

Gupta, J.N.D. and E.A. Tunc (1991): Schedules for a Two Stage Hybrid Flowshop with Parallel Machines at the Second Stage. International Journal of Production Research 29, 1489 - 1502.

Gupta, J.N.D. and E.A. Tunc (1994): Scheduling a Two-Stage Hybrid Flowshop with Separable Setup and Removal Times. European Journal of Operational Research 77, 415 - 428.

Gutjahr, A.L. and G.L. Nemhauser (1964): An Algorithm for the Line Balancing Problem. Management Science 11, 308 - 315.

Hackman, S.T.; M.J. Magazine and T.S. Wee (1989): Fast, Effective Algorithms for Simple Assembly Line Balancing Problems. Operations Research 37, 916 - 924.

Hackstein, R. and O. Felbecker (1981): Bedeutung der Reihenfolgeplanung bei Varianten-Linienfertigung. Arbeitsvorbereitung 18, 76 - 82 and 149 - 153.

Hahn, R. (1972): Produktionsplanung bei Linienfertigung. de Gruyter, Berlin.

Hahn, R.; L. Lutz and K. Roschmann (1968): Die Bandabgleichung - ein Problem bei Fließfertigung. Industrielle Organisation 37/2, 85 - 101.

Haider, S.W. and C.L. Moodie (1978): An Investigation of the Use of an Interactive Computer Mode for Balancing Paced Assembly Lines. Computers & Industrial Engineering 2, 83 - 89.

Hall, R.W. (1983): Zero Inventories. Dow Jones - Irwin, Homewood.

Hardeck, W. (1974): Rechnerunterstützte Austaktung von Fließbandlinien. Zeitschrift für Operations Research 18, B237 - B249.

Hautsch, K.; H. John and H. Schürgers (1972): Taktabstimmung bei Fließarbeit mit dem Positionswert-Verfahren. REFA-Nachrichten 25, 451 - 464.

Hax, A.C. and D. Candea (1984): Production and Inventory Management. Prentice Hall, Englewood Cliffs.

Hax, A.C. and H.C. Meal (1975): Hierarchical Integration of Production Planning and Scheduling. In: Geisler, M. (ed.): TIMS Studies in Management Science Vol. 1, 53 - 69.

Heinrici, A. (1994): A Comparison between Simulated Annealing and Tabu Search with an Example from the Production Planning. In: Dyckhoff, H.; U. Derigs, M. Salomon and H.C. Tijms (eds.): Operations Research Proceedings 1993, Springer, Berlin, 498 - 503.

Held, M.; R.M. Karp and R. Shareshian (1963): Assembly Line Balancing - Dynamic Programming with Precedence Constraints. Operations Research 11, 442 - 459.

Helgeson, W.B. and D.P. Birnie (1961): Assembly Line Balancing Using the Ranked Positional Weight Technique. Journal of Industrial Engineering 12, 394 - 398.

Henig, M.I. (1986): Extensions of the Dynamic Programming Method in the Deterministic and Stochastic Assembly-Line Balancing Problems. Computers & Operations Research 13, 443 - 449.

Hentze, J. and A. Kammel (1994): Implementing Lean Production Systems at German Automobile Manufacturers: An Exploratory Study. Zeitschrift für Planung 5, 39 - 48.

Heskiaoff, H. (1968): A Heuristic Method for Balancing Assembly Lines. The Western Electric Engineer 12/3, 9 - 13.

Hillier, F.S. and R.W. Boling (1966): The Effect of Some Design Factors on the Efficiency of Production Lines with Variable Operation Times. Journal of Industrial Engineering 12, 651 - 658.

Hillier, F.S. and R.W. Boling (1979): On the Optimal Allocation of Work in Symmetrically Unbalanced Production Line Systems with Variable Operation Times. Management Science 25, 721 - 728.

Hillier, F.S. and G.J. Lieberman (1988): Operations Research: Einführung. 4th ed., Oldenbourg, München.

Hillier, F.S. and K.C. So (1991): The Effect of Machine Breakdowns and Interstage Storage on the Performance of Production Line Systems. International Journal of Production Research 29, 2043 - 2055.

Hillier, F.S. and K.C. So (1993): Some Data for Applying the Bowl Phenomenon to Large Production Line Systems. International Journal of Production Research 31, 811 - 822.

Hillier, F.S.; K.C. So and R.W. Boling (1993): Toward Characterizing the Optimal Allocation of Storage Space in Production Line Systems with Variable Processing Times. Management Science 39, 126 - 133.

Hiraki, S.; K. Aoki and R. Muramatsu (1977): A Study on Designing Buffer Capacity in the Mixed-Model Assembly Line. Proceedings of the 4th International Conference on Production Research, Tokyo, 431 - 450.

Hochbaum, D.S. and D.B. Shmoys (1987): Using Dual Approximation Algorithms for Scheduling Problems: Theoretical and Practical Results. Journal of the ACM 34, 144 - 162.

Hoffmann, T.R. (1963): Assembly Line Balancing with a Precedence Matrix. Management Science 9, 551 - 562.

Hoffmann, T.R. (1990): Assembly Line Balancing: A Set of Challenging Problems. International Journal of Production Research 28, 1807 - 1815.

Hoffmann, T.R. (1992): EUREKA: A Hybrid System for Assembly Line Balancing. Management Science 38, 39 - 47.

Hoffmann, T.R. (1993): Response to Note on Microcomputer Performance of "FABLE" on Hoffmann's Data Sets. Management Science 39, 1192 - 1193.

Hopfield, J.J. and D. Tank (1985): Neural Computation of Decisions in Optimization Problems. Biological Cybernetics 52, 141 - 152.

Hopp, W.J.; N. Pati and P.C. Jones (1989): Optimal Inventory Control in a Production Flow System with Failures. International Journal of Production Research 27, 1367 - 1384.

Hu, T.C. (1961): Parallel Sequencing and Assembly Line Problems. Operations Research 9, 841 - 848.

Hübscher, R. and F. Glover (1994): Applying Tabu Search with Influential Diversification to Multiprocessor Scheduling. Computers & Operations Research 21, 877 - 884.

Ignall, E.J. (1965): A Review of Assembly Line Balancing. Journal of Industrial Engineering 16, 244 - 254.

Inman, R.R. and R.L. Bulfin (1991): Sequencing JIT Mixed-Model Assembly Lines. Management Science 37, 901 - 904.

Inman, R.R. and R.L. Bulfin (1992): Quick and Dirty Sequencing for Mixed-Model Multi-Level JIT Systems. International Journal of Production Research 30, 2011 - 2018.

Inman, R.R. and M. Leon (1994): Scheduling Duplicate Serial Stations in Transfer Lines. International Journal of Production Research 32, 2631 - 2644.

Iskander, W.H. and J. Chou (1990): Unbalanced Production Line Scheduling with Partial Job Specialization. Naval Research Logistics 37, 789 - 805.

Jackson, J.R. (1956): A Computing Procedure for a Line Balancing Problem. Management Science 2, 261 - 271.

Jacobson, R.E. and R.P. Sadowski (1977): A Scheduling Algorithm for the Unbalanced Production Line. AIIE Transactions 9, 36 - 43.

Jaeschke, G. (1964): Branching and Bounding: Eine allgemeine Methode zur Lösung kombinatorischer Probleme. Ablauf- und Planungsforschung 5, 133 - 155.

Jafari, M.A. and J.G. Shanthikumar (1989): Determination of Optimal Buffer Storage Capacities and Optimal Allocation in Multistage Automatic Transfer Lines. IIE Transactions 21, 130 - 135.

Johnson, R.V. (1981): Assembly Line Balancing Algorithms: Computation Comparisons. International Journal of Production Research 19, 277 - 287.

Johnson, R.V. (1983): A Branch and Bound Algorithm for Assembly Line Balancing Problems with Formulation Irregularities. Management Science 29, 1309 - 1324.

Johnson, R.V. (1988): Optimally Balancing Large Assembly Lines with "FABLE". Management Science 34, 240 - 253.

Johnson, R.V. (1991): Balancing Assembly Lines for Teams and Work Groups. International Journal of Production Research 29, 1205 - 1214.

Johnson, R.V. (1993): Note: Microcomputer Performance of "FABLE" on Hoffmann's Data Sets. Management Science 39, 1190 - 1193.

Kälberer, G. (1976): Praktischer Wert heuristischer Auswahlregeln zur Fließbandabstimmung. Zeitschrift für Operations Research 20, B195 - B205.

Kao, E.P.C. (1976): A Preference Order Dynamic Program for Stochastic Assembly Line Balancing. Management Science 22, 1097 - 1104.

Kao, E.P.C. (1979): Computational Experience with a Stochastic Assembly Line Balancing Algorithm. Computers & Operations Research 6, 79 - 86.

Kao, E.P.C. and M. Queyranne (1982): On Dynamic Programming Methods for Assembly Line Balancing. Operations Research 30, 375 - 390.

Karp, R.M. (1972): Reducibility among Combinatorial Problems. In: Miller, R.E. and J.W. Thatcher (eds.): Complexity of Computer Computation, Plenum Press, New York, 85 - 103.

Kilbridge, M.D. and L. Wester (1961 a): The Balance Delay Problem. Management Science 8, 69 - 84.

Kilbridge, M.D. and L. Wester (1961 b): A Heuristic Method for Assembly Line Balancing. Journal of Industrial Engineering 12, 292 - 298.

Kilbridge, M.D. and L. Wester (1962): A Review of Analytical Systems of Line Balancing. Operations Research 10, 626 - 638.

Klein, M. (1963): On Assembly Line Balancing. Operations Research 11, 274 - 281.

Klein, R. (1998): Bidirectional Planning: Improving Priority Rule Based Heuristics for Scheduling Resource-Constrained Projects. Schriften zur Quantitativen Betriebswirtschaftslehre 10/98, TU Darmstadt.

Klein, R. and A. Scholl (1996): Maximizing the Production Rate in Simple Assembly Line Balancing - A Branch and Bound Procedure. European Journal of Operational Research 91, 367 - 385.

Klein, R. and A. Scholl (1998 a): PROGRESS: Optimally Solving the Generalized Resource-Constrained Project Scheduling Problem. Schriften zur Quantitativen Betriebswirtschaftslehre 5/98, TU Darmstadt.

Klein, R. and A. Scholl (1998 b): Scattered Branch and Bound: An Adaptive Search Strategy Applied to Resource-Constrained Project Scheduling. Schriften zur Quantitativen Betriebswirtschaftslehre 6/98, TU Darmstadt.

Klein, R. and A. Scholl (1999): Computing Lower Bounds by Destructive Improvement - An Application to Resource-Constrained Project Scheduling. European Journal of Operational Research 112, 322 - 346.

Klenke, H. (1977): Ablaufplanung bei Fließfertigung. Gabler, Wiesbaden.

Koenigsberg, E. (1959): Production Lines and Internal Storage - A Review. Management Science 5, 410 - 433.

Koether, R. (1985): Improving Productivity in Model-Mix Assembly. In: Bullinger, H.-J. and H.J. Warnecke (eds.): Toward the Factory of the Future: Proceedings of the Eight International Conference on Production Research, New York, 761 - 766.

Koether, R. (1986): Verfahren zur Verringerung von Modell-Mix-Verlusten in Fließmontagen. Springer, Berlin.

Koether, R. and K. Thaler (1986): Wirtschaftliche Online-Modellfolgebestimmung in Karossenlagern. VDI-Zeitschrift 128, 905 - 909.

Kottas, J.F. and H.-S. Lau (1973): A Cost-Oriented Approach to Stochastic Line Balancing. AIIE Transactions 5, 164 - 171.

Kottas, J.F. and H.-S. Lau (1976): A Total Operating Cost Model for Paced Lines with Stochastic Task Times. AIIE Transactions 8, 234 - 240.

Kottas, J.F. and H.-S. Lau (1981): A Stochastic Line Balancing Procedure. International Journal of Production Research 19, 177 - 193.

Kovach, J. (1969): CALB: Systematic Assembly Management. Assembly Engineering 12, 20 - 25.

Krafcik, J.F. (1988): Triumph of the Lean Production System. Sloan Management Review 30 (Fall), 41 - 52.

Kubiak, W. (1993): Minimizing Variation of Production Rates in Just-in-Time Systems: A Survey. European Journal of Operational Research 66, 259 - 271.

Kubiak, W. and S. Sethi (1991): A Note on "Level Schedules for Mixed-Model Assembly Lines in Just-in-Time Production Systems". Management Science 37, 121 - 122.

Küpper, H.-U. and S. Helber (1995): Ablauforganisation in Produktion und Logistik. 2nd ed., Schäffer-Poeschel, Stuttgart.

Kusiak, A. (1987): The Production Equipment Requirements Problem. International Journal of Production Research 25, 319 -325.

Labbé, M.; G. Laporte and H. Mercure (1991): Capacitated Vehicle Routing on Trees. Operations Research 39, 616 - 622.

Laguna, M. and F. Glover (1993): Bandwidth Packing: A Tabu Search Approach. Management Science 39, 492 - 500.

Lau, H.-S. and A. Shtub (1987): An Exploratory Study on Stopping a Paced Line when Incompletions Occur. IIE Transactions 19, 463 - 467.

Lawler, E.L. (1979): Efficient Implementation of Dynamic Programming Algorithms for Sequencing Problems. Report BW 106/79, Stichting Mathematisch Centrum, Amsterdam.

Lawler, E.L. (1991): Computing Shortest Paths in Networks Derived from Recurrence Relations. Annals of Operations Research 33, 363 - 377.

Lawler, E.L.; J.K. Lenstra, A.H.G. Rinnooy Kan and D.B. Shmoys (1985): The Traveling Salesman Problem. Wiley, Chichester.

Lehman, M. (1969): On Criteria for Assigning Models to Assembly Lines. International Journal of Production Research 7, 269 - 285.

Levary, R.R. and M.D. Renfro (1991): Application of Assembly Line Balancing Techniques to Installment Lending Operations of Commercial Banks. Computers & Industrial Engineering 20, 105 - 109.

Liao, C.-J. (1994): A New Node Selection Strategy in the Branch-and-Bound Procedure. Computers & Operations Research 21, 1095 - 1101.

Link, C. (1992): Bandabgleich bei variantenreicher Fließfertigung. Diploma Thesis, Universität Mannheim, Lehrstuhl für Allgemeine Betriebswirtschaftslehre und Unternehmensforschung.

Lorenz, J.D. (1982): The ALPACA Line Balancing System. Computers & Industrial Engineering 6, 115 - 123.

Lutz, C.M.; K.R. Davis and C.F. Turner (1994): Development of Operator Assignment Schedules: A DSS Approach. OMEGA 22, 57 - 67.

Lutz, L. (1974): Abtakten von Montagelinien. Krausskopf, Mainz.

Macaskill, J.L.C. (1969): Application of Computers to the Analysis of Mixed-Product Assembly Lines. Proceedings of the Fourth Australian Computer Conference, Adelaide, South Australia, 1969, 61 - 67.

Macaskill, J.L.C. (1972): Production-Line Balances for Mixed-Model Lines. Management Science 19, 423 - 434.

Macaskill, J.L.C. (1973): Computer Simulation for Mixed-Model Production Lines. Management Science 20, 341 - 348.

Magad, E.L. (1971): Computerized Line Balancing. Proceedings of the 34th Annual IMS National Industrial Engineering and Management Clinic, Chicago, 1970, Industrial Management Society, 137 - 141.

Magazine, M.J. and G.L. Silver (1978): Heuristics for Determining Output and Work Allocations in Series Flow Lines. International Journal of Production Research 16, 169 - 181.

Malakooti, B. (1991): A Multiple Criteria Decision Making Approach for the Assembly Line Balancing Problem. International Journal of Production Research 29, 1979 - 2001.

Malakooti, B. (1994): Assembly Line Balancing with Buffers by Multiple Criteria Optimization. International Journal of Production Research 32, 2159 - 2178.

Mansoor, E.M. (1964): Assembly Line Balancing - An Improvement on the Ranked Positional Weight Technique. Journal of Industrial Engineering 15, 73 - 77 and 322 - 323.

Mansoor, E.M. (1967): Improvement on Gutjahr and Nemhauser's Algorithm for Line Balancing Problems. Management Science 14, 250 - 254.

Mansoor, E.M. and S. Ben-Tuvia (1966): Optimizing Balanced Assembly Lines. Journal of Industrial Engineering 17, 126 - 132.

Mansoor, E.M. and M. Yadin (1971): On the Problem of Assembly Line Balancing. In: Avi-Itzhak, B. (ed.): Developments in Operations Research, Gordon & Breach, New York, 361 - 375.

Martello, S. and P. Toth (1990 a): Knapsack Problems - Algorithms and Computer Implementations. Wiley, New York.

Martello, S. and P. Toth (1990 b): Lower Bounds and Reduction Procedures for the Bin Packing Problem. Discrete Applied Mathematics 28, 59 - 70.

Martin, G.E. (1994): Optimal Design of Production Lines. International Journal of Production Research 32, 989 - 1000.

Masson, E. and Y.-J. Wang (1990): Introduction to Computation and Learning in Artificial Neural Networks. European Journal of Operational Research 47, 1 - 28.

Mastor, A.A. (1970): An Experimental Investigation and Comparative Evaluation of Production Line Balancing Techniques. Management Science 16, 728 - 746.

McCormick, S.T.; M.L. Pinedo, S. Shenker and B. Wolf (1989): Sequencing in an Assembly Line with Blocking to Minimize Cycle Time. Operations Research 37, 925 - 935.

McNaughton, R. (1959): Scheduling with Deadlines and Loss Functions. Management Science 6, 1 - 12.

Mertens, P. (1967): Fließbandabstimmung mit dem Verfahren der begrenzten Enumeration nach Müller-Merbach. Ablauf- und Planungsforschung 8, 429 - 433.

Michalewicz, Z. (1992): Genetic Algorithms + Data Structures = Evolution Programs. Springer, Berlin.

Milas, G.H. (1990): Assembly Line Balancing... Let's Remove the Mystery. Industrial Engineering (May), 31 - 36.

Miltenburg, J. (1989): Level Schedules for Mixed-Model Assembly Lines in Just-in-Time Production Systems. Management Science 35, 192 - 207.

Miltenburg, G.J. and T. Goldstein (1991): Developing Production Schedules which Balance Part Usage and Smooth Production Loads for Just-in-Time Production Systems. Naval Research Logistics 38, 893 - 910.

Miltenburg, J. and G. Sinnamon (1989): Scheduling Mixed-Model Multi-Level Just-in-Time Production Systems. International Journal of Production Research 27, 1487 - 1509.

Miltenburg, J. and G. Sinnamon (1992): Algorithms for Scheduling Multi-Level Just-in-Time Production Systems. IIE Transactions 24, 121 - 130.

Miltenburg, J. and J. Wijngaard (1994): The U-line Line Balancing Problem. Management Science 40, 1378 - 1388.

Minoux, M. (1986): Mathematical Programming - Theory and Algorithms. Wiley, Chichester.

Mitchell, J. (1957): A Computational Procedure for Balancing Zoned Assembly Lines. Research Report 6-94801-1-R3, Westinghouse Research Laboratories, Pittsburgh.

Mitsumori, S. and K. Takada (1972): Optimum Schedule Control of Conveyor Line in Mixed Production. Electrical Engineering in Japan 92, 121 - 129.

Mitten, L.G. (1970): Branch-and-Bound Methods: General Formulation and Properties. Operations Research 18, 24 - 34.

Miyazaki, S. and H. Ohta (1990): Line Balancing Problem with Capital Investments in the Flexible Assembly Line. Transactions of the Institute of Systems, Control and Information Engineers (Japan) 3/6, 167 - 175.

Monden, Y. (1998): Toyota Production System - An Integrated Approach to Just-in-time. 3rd ed., Kluwer, Dordrecht.

Moodie, C.L. and H.H. Young (1965): A Heuristic Method of Assembly Line Balancing for Assumptions of Constant or Variable Work Element Times. Journal of Industrial Engineering 16, 23 - 29.

Müller-Merbach, H. (1981): Heuristics and their Design: A Survey. European Journal of Operational Research 8, 1 - 23.

Mukherjee, S.K. and S.K. Basu (1964): An Application of Heuristic Method of Assembly Line Balancing in an Indian Industry. Proc. Instn. Mech. Engrs. 178/1/11, 277 - 292.

Muth, E.J. (1973): The Production Rate of a Series of Work Stations with Variable Service Times. International Journal of Production Research 11, 155 - 169.

Muth, E.J. and A. Alkaff (1987): The Bowl Phenomenon Revisited. International Journal of Production Research 25, 161 - 173.

Nahmias, S. (1989): Production and Operations Analysis. Irwin, Homewood.

Nanda, R. and J.M. Scher (1975): Assembly Lines with Overlapping Work Stations. AIIE Transactions 7, 311 - 318.

Nanda, R. and J.M. Scher (1976): Nonparallelability Constraints in Assembly Lines with Overlapping Work Stations. AIIE Transactions 8, 343 - 349.

Nawara, G.M. and H.A. Elezaby (1980): Assembly Line Balancing - A Modified Algorithm with Comparative Applications. Journal of Engineering Sciences 6, 63 - 69.

Nevins, A.J. (1972): Assembly Line Balancing Using Best Bud Search. Management Science 18, 529 - 539.

Ng, W.C. and K.L. Mak (1994): A Branch and Bound Algorithm for Scheduling Just-in-Time Mixed-Model Assembly Lines. International Journal of Production Economics 33, 169 - 183.

Nourie, F.J. and E.R. Venta (1987): The Packing Heuristic for Assembly Line Balancing Problems. Journal of Applied Business Research 3, 122 - 128.

Nourie, F.J. and E.R. Venta (1991): Finding Optimal Line Balances with OptPack. Operations Research Letters 10, 165 - 171.

Nourie, F.J. and E.R. Venta (1996): Microcomputer Performance of OptPack on Hoffmann's data sets: Comparison with Eureka and FABLE. Management Science 42, 304 - 306.

Okamura, K. and H. Yamashina (1979): A Heuristic Algorithm for the Assembly Line Model-Mix Sequencing Problem to Minimize the Risk of Stopping the Conveyor. International Journal of Production Research 17, 233 - 247.

Papadimitriou, C.H. and K. Steiglitz (1982): Combinatorial Optimization: Algorithms and Complexity. Prentice Hall, Englewood Cliffs.

Papadopoulos, H.T. and M.E.J. O'Kelly (1993): Exact Analysis of Production Lines with no Intermediate Buffers. European Journal of Operational Research 65, 118 - 137.

Parrello, B.D.; W.C. Kabat and L. Wos (1986): Job-Shop Scheduling Using Automated Reasoning: a Case Study of the Car-Sequencing Problem. Journal of Automated Reasoning 2, 1 - 42.

Patterson, J.H. and J.J. Albracht (1975): Assembly-Line Balancing: Zero-One Programming with Fibonacci Search. Operations Research 23, 166 - 172.

Pearl, J. (1984): Heuristics - Intelligent Search Strategies for Computer Problem Solving. Addison-Wesley, Reading.

Pesch, E. (1994): Learning in Automated Manufacturing - A Local Search Approach. Physica, Heidelberg.

Pierce, J.F. (1963): Some Large-Scale Production Scheduling Problems in the Paper Industry. Prentice Hall, Englewood Cliffs.

Pike, R. and G.E. Martin (1994): The Bowl Phenomenon in Unpaced Lines. International Journal of Production Research 32, 483 - 499.

Pinto, P.A.; D.G. Dannenbring and B.M. Khumawala (1975): A Branch and Bound Algorithm for Assembly Line Balancing with Paralleling. International Journal of Production Research 13, 183 - 196.

Pinto, P.A.; D.G. Dannenbring and B.M. Khumawala (1978): A Heuristic Network Procedure for the Assembly Line Balancing Problem. Naval Research Logistics Quarterly 25, 299 - 307.

Pinto, P.A.; D.G. Dannenbring and B.M. Khumawala (1981): Branch and Bound and Heuristic Procedures for Assembly Line Balancing with Paralleling of Stations. International Journal of Production Research 19, 565 - 576.

Pinto, P.A.; D.G. Dannenbring and B.M. Khumawala (1983): Assembly Line Balancing with Processing Alternatives: An Application. Management Science 29, 817 - 830.

Pourbabai, B. (1989): Optimal Selection of the Local Storage Sizes of an Assembly System. International Journal of Production Research 27, 1625 - 1636.

Powell, S.G. (1994): Buffer Allocation in Unbalanced Three-Station Serial Lines. International Journal of Production Research 32, 2201 - 2217.

Prenting, T.O. and R.M. Battaglin (1964): The Precedence Diagram: A Tool for Analysis in Assembly Line Balancing. Journal of Industrial Engineering 15, 208 - 213.

Pulleyblank, W.R. (1989): Polyhedral Combinatorics. In: Nemhauser, G.L.; A.H.G. Rinnooy Kan and M.J. Todd (eds.): Optimization. Handbooks in Operations Research and Management Science, Vol. 1, North-Holland, Amsterdam, 371 - 446.

Queyranne, M. (1985): Bounds for Assembly Line Balancing Heuristics. Operations Research 33, 1353 - 1359.

Rachamadugu, R. and J.G. Shanthikumar (1991): Layout Considerations in Assembly Line Design. International Journal of Production Research 29, 755 - 768.

Rachamadugu, R. and B. Talbot (1991): Improving the Equality of Workload Assignments in Assembly Lines. International Journal of Production Research 29, 619 - 633.

Rachamadugu, R.M.V. and C.A. Yano (1994): Analytical Tool for Assembly Line Design and Sequencing. IIE Transactions 26, 2 - 11.

Ramsing, K. and D. Downing (1970): Assembly Line Balancing with Variable Element Times. Industrial Engineering (January), 41 - 43.

Rao, D.M. (1971): Single and Mixed-Model Assembly Line Balancing Methods for Both Deterministic and Normally Distributed Work Element Times. M.S. Thesis, Industrial Engineering Department, Oregon State University.

Raouf, A. and C. Tsui (1982): A New Method for Assembly Line Balancing Having Stochastic Work Elements. Computers & Industrial Engineering 6, 131 - 148.

Raouf, A.; C.L. Tsui and E.A. El-Sayed (1980): A New Heuristic Approach to Assembly Line Balancing. Computers & Industrial Engineering 4, 223 - 234.

Reeve, N.R. and W.H. Thomas (1973): Balancing Stochastic Assembly Lines. AIIE Transactions 5, 223 - 229.

REFA (1975): Methodenlehre des Arbeitsstudiums. Teil 2: Datenermittlung. 4th ed., Hanser, München.

REFA (1992): Methodenlehre der Betriebsorganisation - Ablauforganisation im Bürobereich. Hanser, München.

Roberts, S.D. and C.D. Villa (1970): On a Multiproduct Assembly Line-Balancing Problem. AIIE Transactions 2, 361 - 364.

Robinson, L.W.; J.O. McClain and L.J. Thomas (1990): The Good, the Bad and the Ugly: Quality on an Assembly Line. International Journal of Production Research 28, 963 - 980.

Roodman, G.M. (1986): Near-Optimal Solutions to One-Dimensional Cutting Stock Problems. Computers & Operations Research 13, 713 - 719.

Rosenberg, O. and H. Ziegler (1992): A Comparison of Heuristic Algorithms for Cost-Oriented Assembly Line Balancing. Zeitschrift für Operations Research 36, 477 - 495.

Rosenblatt, M.J. and R.C. Carlson (1985): Designing a Production Line to Maximize Profit. IIE Transactions 17, 117 - 121.

Sahgal, V.K. (1970): A Mixed Model Sequencing Algorithm. M.S. Thesis, Mississippi State University, Mississippi.

Saltzman, M.J. and I. Baybars (1987): A Two-Process Implicit Enumeration Algorithm for the Simple Assembly Line Balancing Problem. European Journal of Operational Research 32, 118 - 129.

Salveson, M.E. (1955): The Assembly Line Balancing Problem. Journal of Industrial Engineering 6, 18 - 25.

Sarin, S.C. and S.K. Das (1987): Determination of Optimal Part Delivery Dates in a Stochastic Assembly Line. International Journal of Production Research 25, 1013 - 1028.

Sarin, S.C. and E. Erel (1990): Development of Cost Model for the Single-Model Stochastic Assembly Line Balancing Problem. International Journal of Production Research 28, 1305 - 1316.

Sarker, B.R. and J.G. Shanthikumar (1983): A Generalized Approach for Serial or Parallel Line Balancing. International Journal of Production Research 21, 109 - 133.

Sawyer, J.H.F. (1970): Line Balancing. The Machinery Publishing, Brighton.

Schad, G. (1986): Entwicklung und Einsatz eines interaktiven Verfahrens zur Leistungsabstimmung von Montagesystemen. Springer, Berlin.

Scheer, A.-W. (1994): CIM - Computer Integrated Manufacturing, Towards the Factory of the Future. 3rd ed., Springer, Berlin.

Schildt, B. (1994): Strategische Produktions- und Distributionsplanung: Betriebliche Standortoptimierung bei degressiv verlaufenden Produktionskosten. Deutscher Universitäts-Verlag, Wiesbaden.

Schneeweiß, Ch. (1974): Dynamisches Programmieren. Physica, Würzburg.

Schneeweiß, Ch. (1992): Planung 2. Springer, Berlin.

Schneeweiß, Ch. (1994): Elemente einer Theorie hierarchischer Planung. OR Spektrum 16, 161 - 168.

Schneeweiß, Ch. and V. Söhner (1991): Kapazitätsplanung bei moderner Fließfertigung. Physica, Heidelberg.

Schofield, N.A. (1979): Assembly Line Balancing and the Application of Computer Techniques. Computers & Industrial Engineering 3, 53 - 69.

Scholl, A. (1993): Data of Assembly Line Balancing Problems. Schriften zur Quantitativen Betriebswirtschaftslehre 16/93, TH Darmstadt.

Scholl, A. (1994): Ein B&B-Verfahren zur Abstimmung von Fließbändern bei gegebener Stationsanzahl. In: Dyckhoff, H.; U. Derigs, M. Salomon and H.C. Tijms (eds.): Operations Research Proceedings 1993, Springer, Berlin, 175 - 181.

Scholl, A. and R. Klein (1997 a): SALOME: A Bidirectional Branch and Bound Procedure for Assembly Line Balancing. INFORMS Journal on Computing 9, 319 - 334.

Scholl, A. and R. Klein (1997 b): ULINO: Optimally Balancing U-shaped JIT Assembly Lines. to appear in: International Journal of Production Research.

Scholl, A. and R. Klein (1999): Balancing Assembly Lines Effectively - A Computational Comparison. European Journal of Operational Research 114, 50 - 58.

Scholl, A.; R. Klein and W. Domschke (1995): Pattern Based Vocabulary Building for Effectively Sequencing Mixed Model Assembly Lines. to appear in: Journal of Heuristics.

Scholl, A.; R. Klein and C. Jürgens (1997 a): BISON: A Fast Hybrid Procedure for Exactly Solving the One-Dimensional Bin Packing Problem. Computers & Operations Research 24, 627 - 645.

Scholl, A.; G. Krispin, R. Klein and W. Domschke (1997 b): Branch and Bound - Optimieren auf Bäumen: je beschränkter, desto besser. c't - Magazin für Computer Technik, Issue 10, 336 - 345.

Scholl, A. and S. Voß (1994): A Note on Fast, Effective Heuristics for Simple Assembly Line Balancing. Schriften zur Quantitativen Betriebswirtschaftslehre 13/94, TH Darmstadt.

Scholl, A. and S. Voß (1995): Kapazitätsorientierte Leistungsabstimmung in der Fließfertigung. Zeitschrift für betriebswirtschaftliche Forschung 47, 919 - 935.

Scholl, A. and S. Voß (1996): Simple Assembly Line Balancing - Heuristic Approaches. Journal of Heuristics 2, 217 - 244.

Schonberger, R. (1982): Japanese Manufacturing Techniques - Nine Hidden Lessons in Simplicity. Free Press, New York.

Schrage, L. and K.R. Baker (1978): Dynamic Programming Solution of Sequencing Problems with Precedence Constraints. Operations Research 26, 444 - 449.

Sculli, D. (1979): Dynamic Aspects of Line Balancing. OMEGA 7, 557 - 561.

Seeland, B. (1967): Optimale Anzahl von Umstellungen bei Montagewechselfließreihen. Fertigungstechnik und Betrieb 17, 415 - 417.

Shin, D. and H. Min (1991 a): Uniform Assembly Line Balancing with Stochastic Task Times in Just-in-Time Manufacturing. International Journal of Operations & Production Management 11/8, 23 - 34.

Shin, D. and H. Min (1991 b): Flexible Line Balancing Practices in a Just-in-Time Environment. Production and Inventory Management Journal 32/4, 38 - 41.

Shtub, A. (1984 a): The Effect of Incompletion Cost on Line Balancing with Multiple Manning of Work Stations. International Journal of Production Research 22, 235 - 245.

Shtub, A. (1984 b): On the Natural Cycle Time of Assembly Lines. Journal of Operations Management 5, 19 - 25.

Shtub, A. and E.M. Dar-El (1989): A Methodology for the Selection of Assembly Systems. International Journal of Production Research 27, 175 - 186.

Shtub, A. and E.M. Dar-El (1990): An Assembly Chart Oriented Assembly Line Balancing Approach. International Journal of Production Research 28, 1137 - 1151.

Silver, E.A. and R. Peterson (1985): Decision Systems for Inventory Management and Production Planning. 2nd ed., Wiley, New York.

Silverman, F.N. and J.C. Carter (1986): A Cost-Based Methodology for Stochastic Line Balancing with Intermittent Line Stoppages. Management Science 32, 455 - 463.

Skorin-Kapov, J. and A.J. Vakharia (1993): Scheduling a Flow-Line Manufacturing Cell: A Tabu Search Approach. International Journal of Production Research 31, 1721 - 1734.

Smunt, T.L. and W.C. Perkins (1985): Stochastic Unpaced Line Design: Review and Further Experimental Results. Journal of Operations Management 5, 351 - 373.

Sniedovich, M. (1981): Analysis of a Preference Order Assembly Line Problem. Management Science 27, 1067 - 1080.

So, K.C. (1989): On the Efficiency of Unbalancing Production Lines. International Journal of Production Research 27, 717 - 729.

Sphicas, G.P. and F.N. Silverman (1976): Deterministic Equivalents for Stochastic Assembly Line Balancing. AIIE Transactions 8, 280 - 282.

Stadtler, H. (1988): Hierarchische Produktionsplanung bei losweiser Fertigung. Physica, Heidelberg.

Stadtler, H. (1990): A One-Dimensional Cutting Stock Problem in the Aluminium Industry and its Solution. European Journal of Operational Research 44, 209 - 223.

Stadtler, H. (1996): Hierarchische Produktionsplanung. In: Kern, W.; H.-H. Schröder and J. Weber (eds.): Handwörterbuch der Produktionswirtschaft, 2nd ed., Schäffer Poeschel, Stuttgart, 631-641.

Steffen, R. (1977): Produktionsplanung bei Fließbandfertigung. Gabler, Wiesbaden.

Steven, M. (1994): Hierarchische Produktionsplanung. 2nd ed., Physica, Heidelberg.

Suhail, A. (1983): Reliability and Optimization Considerations in a Conveyor-Paced Assembly Line System. International Journal of Production Research 21, 627 - 640.

Suliman, S.M.A.; A.M. Al-Tamimi and G.M. Nawara (1985): Computational Methods for the Mixed-Model Assembly Line Problem: A Review. Journal of Engineering Sciences 11, 241 - 271.

Sumichrast, R.T. and R.S. Russell (1990): Evaluating Mixed-Model Assembly-Line Sequencing Heuristics for Just-in-Time Production Systems. Journal of Operations Management 9, 371 - 390.

Sumichrast, R.T.; R.S. Russell and B.W. Taylor (1992): A Comparative Analysis of Sequencing Procedures for Mixed-Model Assembly Lines in a Just-in-Time Production System. International Journal of Production Research 30, 199 - 214.

Suresh, G. and S. Sahu (1994): Stochastic Assembly Line Balancing Using Simulated Annealing. International Journal of Production Research 32, 1801 - 1810.

Szendrovits, A.Z.; G.J. Miltenburg and Z. Goldstein (1990): Optimizing Total Safety Inventory for Large Transfer Lines. OMEGA 18, 299 - 313.

Talbot, F.B. and J.H. Patterson (1984): An Integer Programming Algorithm with Network Cuts for Solving the Assembly Line Balancing Problem. Management Science 30, 85 - 99.

Talbot, F.B; J.H. Patterson and W.V. Gehrlein (1986): A Comparative Evaluation of Heuristic Line Balancing Techniques. Management Science 32, 430 - 454.

Taylor, B.W. and R.S. Russell (1982): A Simulation Approach for Adapting a Production Line Balancing Procedure to a Probabilistic Environment. International Journal of Production Research 20, 787 - 801.

Tempelmeier, H. and H. Kuhn (1993): Flexible Manufacturing Systems - Decision Support for Design and Operation. Wiley, New York.

Terno, J.; R. Lindemann and G. Scheithauer (1987): Zuschnittprobleme und ihre praktische Lösung: Mathematische Modelle von Layout - Problemen. Deutsch, Thun.

Thangavelu, S.R. and C.M. Shetty (1971): Assembly Line Balancing by Zero-One Integer Programming. AIIE Transactions 3, 61 - 68.

Thomopoulos, N.T. (1967): Line Balancing - Sequencing for Mixed-Model Assembly. Management Science 14, B59 - B75.

Thomopoulos, N.T. (1968): Some Analytical Approaches to Assembly Line Problems. The Production Engineer (July), 345 - 351.

Thomopoulos, N.T. (1970): Mixed Model Line Balancing with Smoothed Station Assignments. Management Science 16, 593 - 603.

Thomopoulos, N.T. and M. Lehman (1969): The Mixed Model Learning Curve. AIIE Transactions 1, 127 - 132.

Tonge, F.M. (1960): Summary of a Heuristic Line Balancing Procedure. Management Science 7, 21 - 39.

Tonge, F.M (1961): A Heuristic Program for Assembly Line Balancing. Prentice Hall, Englewood Cliffs.

Tonge, F.M. (1965): Assembly Line Balancing Using Probabilistic Combinations of Heuristics. Management Science 11, 727 - 735.

Undt, H.-J. and A. Scholl (1995): Erzeugung von Testdaten für Reihenfolgeprobleme bei Variantenfließfertigung. Schriften zur Quantitativen Betriebswirtschaftslehre 2/95, TH Darmstadt.

Van Assche, F. and W.S. Herroelen (1979): An Optimal Procedure for the Single-Model Deterministic Assembly Line Balancing Problem. European Journal of Operational Research 3, 142 - 149.

Voß, S. (1990): Steiner - Probleme in Graphen. Hain, Frankfurt/Main.

Voß, S. (1993 a): The Two-Stage Hybrid-Flowshop Scheduling Problem with Sequence-Dependent Setup Times. In: Fandel, G.; Th. Gulledge and A. Jones (eds.): Operations Research in Production Planning and Control. Springer, Berlin, 336 - 352.

Voß, S. (1993 b): Intelligent Search. Manuscript, TH Darmstadt.

Voß, S. (1994): Tabu Search in Manufacturing. In: Dyckhoff, H.; U. Derigs, M. Salomon and H.C. Tijms (eds.): Operations Research Proceedings 1993, Springer, Berlin, 183 - 194.

Vrat, P. and A. Virani (1976): A Cost Model for Optimal Mix of Balanced Stochastic Assembly Line and the Modular Assembly System for a Customer Oriented Production System. International Journal of Production Research 14, 445 - 463.

Wäscher, G. and T. Gau (1996): Heuristics for the Integer One-Dimensional Cutting Stock Problem: A computational Study. OR Spektrum 18, 131 - 144.

Wang, F. and R.C. Wilson (1986): Comparative Analyses of Fixed and Removable Item Mixed Model Assembly Lines. IIE Transactions 18, 313 - 317.

Warnecke, H.J. (1971): Anwendung mathematischer Methoden bei der Leistungsabstimmung von Montagelinien. CIRP-Annals 20, 99 - 100, and Fertigung 2/5, 173 - 177.

Warnecke, H.J.; H.-J. Bullinger and H.-P. Lentes (1978): The Design of Manual Flow-Lines with the Use of Simulation. CIRP-Annals 27, 429 - 434.

Wedekind, H. (1963): Ein linearer Programmansatz für das Fließbandproblem. Ablauf- und Planungsforschung 4, 245 - 248.

Wee, T.S. and M.J. Magazine (1981 a): An Efficient Branch and Bound Algorithm for Assembly Line Balancing - Part I: Minimize the Number of Work Stations. Working Paper No. 150, University of Waterloo, Waterloo.

Wee, T.S. and M.J. Magazine (1981 b): An Efficient Branch and Bound Algorithm for an Assembly Line Balancing Problem - Part II: Maximize the Production Rate. Working Paper No. 151, University of Waterloo, Waterloo.

Wee, T.S. and M.J. Magazine (1982): Assembly Line Balancing as Generalized Bin Packing. Operations Research Letters 1/2, 56 - 58.

Wester, L. and M.D. Kilbridge (1962): Heuristic Line Balancing: A Case. Journal of Industrial Engineering 13, 139 - 149.

Wester, L. and M. Kilbridge (1964): The Assembly Line Model-Mix Sequencing Problem. Proceedings of the Third International Conference on Operations Research, Oslo 1963, Dunod, Paris, 247 - 260.

White, W.W. (1961): Comments on a Paper by Bowman. Operations Research 9, 274 - 276.

Wild, R. (1972): Mass-Production Management: The Design and Operation of Production Flow-Line Systems. Wiley, London.

Wilhelm, B. (1979): Beitrag zur optimalen Steuerung von Montagelinien mit Modell-Mix in der Automobilindustrie. Dissertation, TU Braunschweig.

Wilson, J.M. (1986): Formulation of a Problem Involving Assembly Lines with Multiple Manning of Work Stations. International Journal of Production Research 24, 59 - 63.

Yamashina, H. and K. Okamura (1983): Analysis of In-Process Buffers for Multi-Stage Transfer Line Systems. International Journal of Production Research 21, 183 - 195.

Yano, C.A. and A. Bolat (1989): Survey, Development, and Application of Algorithms for Sequencing Paced Assembly Lines. Journal of Manufacturing and Operations Management 2, 172 - 198.

Yano, C.A. and R. Rachamadugu (1991): Sequencing to Minimize Work Overload in Assembly Lines with Product Options. Management Science 37, 572 - 586.

Zäpfel, G. (1975): Ausgewählte fertigungswirtschaftliche Optimierungsprobleme von Fließfertigungssystemen. Beuth, Berlin.

Zäpfel, G. (1989): Taktisches Produktions-Management. de Gruyter, Berlin.

Zanakis, S.H.; J.R. Evans and A.A. Vazacopoulos (1989): Heuristic Methods and Applications: A Categorized Survey. European Journal of Operational Research 43, 88 - 110.

Ziegler, H. (1990): Produktionssteuerung bei Mehrproduktfließlinien. Manuscript, Universität-GH-Paderborn, Fachbereich Wirtschaftswissenschaften.

Index

Made in the USA
Monee, IL
13 September 2026

62973753R00188